Into the Lion's Mouth

The Story of the *Wildrake* Diving Accident

Michael Smart

Lion's Mouth Publishing, LLC.
Medford, Oregon

This is a work of non-fiction. The events and people described in this story are real. All sources of information can be found at the back of the book in the sections labeled Endnotes, and Bibliography and Works Cited.

Published in the United States by Lion's Mouth Publishing, LLC.

Book design by Michael Smart using Adobe® InDesign®
Front cover photo courtesy of Marcus Taylor, Ltd.
Back cover photo of Thistle Alpha courtesy of Marcus Taylor, Ltd; photo of author courtesy of Jon Sears.
Marine Technology logo courtesy of Santa Barbara City College
Unless otherwise stipulated, illustrations in this book were created with Adobe® Illustrator® by Michael Smart.

First Edition
Library of Congress Control Number: 2011915008

Publisher's Cataloging-in-Publication data

Smart, Michael, 1950 –
Into the lion's mouth : the story of the Wildrake diving accident / Michael Smart.
p. cm.
ISBN 978-0-615-52838-0
Includes bibliographical references and index.
1. Deep diving –Accidents --North Sea. 2. Deep diving --North Sea. 3. Marine accidents –North Sea. 4. Offshore oil industry --North Sea. 5. Petroleum industry and trade--Great Britain. 6. Petroleum --Prospecting --North Sea. I. Title.

VM981 .S63 2011
797.2/3 –dc22 2011915008

To purchase additional copies visit the web at:
Lionsmouthpublishing.com
Inquiries: contact info@lionsmouthpublishing.com
In Europe visit the web at: buyhistoryprints.com

Printed in the United States of America

— What divers are saying —

It's an outrageous story! Finally, it comes to light.

— Skip Gordon, Technip, North Sea

A devastating and heartbreaking tragedy that needed to be told; it will shock and stir your every emotion—make you angry—and even make you weep.

— Kevin Casey, diver, North Sea

It was the most emotionally difficult book that I have ever read...Michael Smart has done us all a great service in uncovering and bringing to light all of the reasons for the deaths of Richard and Skip. I owe him a debt of gratitude for his years of work in pursuing such a difficult task and illuminating for me the circumstances of a friend's horrible death.

— Louie Meader, former Infabco diver

Because this accident was so preventable it punctuates the importance that this story must be read and passed on to every commercial dive school student, tender, and journeyman diver who either wishes to make a career, or continue working, in the diving industry. The unfortunate truth is that a disregard in following safe diving practices, and the perceived threat of losing one's livelihood, is what killed my two good friends on the *Wildrake*. If anyone wishes to learn from other people's mistakes, you'll need to pay close attention while reading this book.

— Tom Belcher, President, Underwater Resources, Inc.

Into the Lion's Mouth is a very well researched, well written and factual story of the tragic events leading up to and causing the deaths of Richard and Skip. Even though the story is over 30 years in the telling, it's a story long overdue.

— Steve Nye, former Sub Sea International diver
BW Offshore, Houston

This is truly an amazing story about the deaths of two good friends of mine that should be read by everyone in the diving industry. I was working offshore in the middle of all this and had no idea of the back-room deals that were taking place. I thought it was supposed to be all about our performance offshore and how fast we got the bell in the water—boy was I wrong! Thanks for your great work in getting the truth out.

— Gary Kane, aka Snork
former Infabco diving supervisor
President, The Kane Kompany, Inc.

— What divers are saying —

From page one you are catapulted into the life of the offshore commercial diver as it was in the 1970s. Be prepared to go through a range of emotions from apprehension, incredulity, anger, disbelief, disgust and deep sadness as you live through the events that lead to the unnecessary deaths of two experienced professionals. The cover-ups that followed and the despicable manner in which the families were treated will leave you wanting to pound on doors and demand justice.

The book has been meticulously researched, and is difficult to put down from the turn of the first page until the bitter bitter end.

It is a story that demands to be told!

— Mark Longstreath, owner, Longstreath.com

Years ago, when I entered Santa Barbara City College, I'd heard the rumors about the *Wildrake* accident. Now, so many years later, Michael Smart takes us back to the scene of the accident and elicits long suppressed feelings of dread and joy in those of us who spent our lives working on the bottom of the sea.

This is a story of incompetence, greed and deceit. It's also a story about the bonds of courage and cooperation that connect divers to ensure each other's safety and survival. It details the heavy price paid for the energy we need, and the advancement of diving knowledge and technology.

The *Wildrake* tragedy did not have to happen; but it would be an even greater tragedy if this story were not told and we did not learn from the sacrifice of Richard Walker and Victor Guiel, and all those who perished in our industry. This book should be required reading for managers, supervisors, divers and students alike.

— Geoff Thielst
Director, Marine Diving Technologies
Santa Barbara City College

For Richard and Skip and their families...

Contents

Contents

The Past Comes Calling

That accident with Dick and Skip was one that would give you nightmares and make you drunk thinking about it. No doubt it was a horrible, horrible thing for them to go through because they had to be aware of what was happening, that they were running out of time.

—— one of the *Wildrake* divers during interview

Over the many years it took me to write this book, I have often been asked why I wanted to write about an incident that happened so long ago. To answer that question one has to look back in time.

In August 1979 I was on board a construction barge called the *Balder* when I first heard about the *Wildrake* accident. Having just arrived in the North Sea three months earlier, I was new to the industry. In fact I had no working dives in my log book and therefore no way to evaluate the ordeal that Richard Walker and Skip Guiel had gone through. Bell diving to 500 feet was still a mystery to me, and when the sad announcement came, it came with no details, only that the bell had been lost and when it was finally recovered, the two divers inside were found dead.

In the months and years that followed, I did not read any newspaper articles or reports about the accident. Nor did I discuss it with any of the people involved. What happened to the *Wildrake* divers had to be pushed to one side in order for me to get on with the demands of becoming a diver in the North Sea.

The next time I came across the word *Wildrake* was in 1983. I was in a boarding house in Aberdeen when I read a small article in the local newspaper about Richard's widow receiving an award in damages for the death of her husband. The amount left me stunned. It didn't seem like much money for a life and I wondered how this mother of a young daughter had coped with the tragedy. It was a fleeting thought. Once again I pushed the accident to the back of my mind and went on with my career.

In 1986 I left the North Sea and moved back to the United States. I settled down, bought a small farm, and eventually opened a computer training business which required me to take on help. I hired a man named Jack Johnston and one night in 1991, he led me to the obsession that resulted in this book. I didn't know it at the time, but Jack had a Master's degree in psychology and had taken an interest in my behavior. He wanted to know why I paid so much attention to detail in just about everything I did.

At first, I deflected his question, but under his persistent inquisition I looked into my more recent past and offered that I had once been a diver which required a certain diligence to perform one's job well.

Jack reacted in the way I'd seen others when I mentioned the word diver–thinking in terms of Jacques Cousteau, Lloyd Bridges, Sea Hunt—that I must have had an exotic job swimming among colored fish in clear water. It took a while, but I managed to disabuse him of that picture, replacing it with one closer to reality—pitch black freezing water, a gloomy moonscape seabed, huge wolf eels lurking about, claustrophobic diving bells, critical gas tolerances, week-long decompressions, and the omnipresence of danger.

The more I unveiled the commercial diving world, the more Jack kept firing questions at me. I told him about the overwhelming awe of working alone at 600 feet, the physiological and psychological impact, the stress, the money, the constant dependency upon other members of the team, the use of fatalism as an antidote to fear, and the technique of saturation diving in which men breathe helium-oxygen mixtures and never see the light of day for as much as 40 days unless it's through the porthole of the bell. That fact alone really struck home with Jack. That after a dive you didn't get to walk around on deck and fill your lungs with fresh air. Instead, you remained cocooned inside a chamber complex in the bowels of the ship until the next dive.

Who would live like that? was the expression on his face. This was all new to him, but he was beginning to have an inkling as to why I was the way I was. We talked for more than an hour while I drew pictures on the classroom board to quicken his introduction into the world I had left behind, painting a picture of a life that very few people know about when they pull up to a gas station to fill up their tank. The central theme of what I was telling him was that, in the commercial diving industry, one always had to know—*not assume*—what he was doing every minute of the working day. The consequences for not knowing were too horrific to contemplate.

It was at this point that I began talking about a fatal "lost bell" diving accident that occurred the year I arrived in the North Sea. It was known throughout the industry as the *Wildrake* (*Wil-drake*) accident because it was named after the ship upon which it occurred. I related to Jack what I could from fragments of memory. The bell had gone into the water with Richard Walker and Skip Guiel (*Geel*) inside and during the dive something went wrong which prevented the crew from recovering it. The bell was ultimately set on the seabed in more than 500 feet of water with the two divers trapped inside. To make matters worse, the life-support umbilical had been destroyed—how I didn't know—but without hot water to keep them warm, it would be a matter of hours before they froze to death. What followed was a race against time to save the two men.

But the attempted rescue took too long, and by the time they pulled the bell out of the water, Richard and Skip had died of hypothermia. This was all I could relate to Jack that night because I never talked to any of the divers involved; didn't know the deceased, and after 12 years I couldn't remember the name of the diving company. Nor did I know why the rescue had failed. I merely knew it was a seminal moment in the history of North Sea diving and that it had a profound impact upon the way divers viewed safety.

When I finished talking, Jack was impressed by the story and suggested I try to publish it.

That was 1991.

As the months slipped by I thought occasionally about Richard and Skip. And then, for reasons I still don't fully understand, these thoughts snowballed into nightly visions of them trapped inside the bell on the bottom of the sea, slowly freezing to death. And every time I saw them the overriding question I had was—*What are they saying to each other?*

In 1979 when I first learned about the accident I had no bell diving experiences to draw from to fully appreciate the kind of nightmare these men had gone through. The only impact the accident had upon me was it hammered into my head the reality that I had just entered a world where people were actually getting killed on the job, something I'd never consciously considered while going through dive school.

But in 1991 I had thousands of hours in saturation and hundreds of dives behind me. I knew about the crippling cold and the risks and the faith one had to have in his supervisors to continue climbing into the bell for the next dive. I knew more than enough to know what Richard and Skip must have gone through. This is why my conversation with Jack had a kind of auspicious quality to it and perhaps why my mind began to dwell on their ordeal.

Months turned into years and still the images that came unsolicited refused to go away. By the end of 1993 I was haunted and knew I had to do something to put an end to these nightly visitations. In a way I suppose I used my own fantasies to carry Jack's idea forward. Sport diving was something familiar to the general public, but the type of diving done in the North Sea was virtually unknown. Perhaps the subject of deep-sea diving was, in and of itself, compelling enough to merit description.

But I had a serious problem. To me, writing had always been an exercise in masochism. It forced me to organize my thoughts in an unnatural, disciplined way, a viewpoint I'd cultivated in grade school after suffering through several humiliating experiences. But far beyond my childhood phobia I could see other staggering problems. How would I start such a project? How would I find the *Wildrake* divers? Would they speak to me? Where was the ship? Was it available for inspection? How would I document the story? Where was the dive log? These were just a few of the hundreds of questions floating around in my head, all of which were shouting at me that writing a book about this particular accident was preposterous, especially for someone with no writing background. I might just as well have decided to climb Mount Everest with no experience, no guide, and no equipment.

For a long time thereafter I considered abandoning the idea because of my terrible feelings of inadequacy and self-doubt. I suppose if I had any sense at all, I would have thrown off the project. But images of Richard and Skip were unrelenting, and I began to think more seriously about what Jack had suggested.

By April 1995 I was wrung out from dwelling on the death of two people whom I'd never known. I finally sat down and wrote the first of many awkward letters to the Sheriff Clerk's Office in Aberdeen, Scotland, requesting any information on the *Wildrake* accident. They referred me to the local prosecutor's office, so I wrote to them. My letter eventually found its way onto the desk of Mr. Graham Napier, an assistant prosecutor, who not only sent me a reply, he took it upon himself to contact a Mr. Frank Murray at the Health and Safety Executive, the branch of government that now oversees diving safety in the North Sea. Coincidentally, Murray had been a diver and remembered the *Wildrake* accident, too. And on the strength of Napier's phone call, Murray traveled across town on a mission to satisfy my request, because he knew exactly what I needed and where to get it.

In mid-May I was driving to the local college when I stopped by my mailbox on the side of the road. It was a beautiful spring day. The heavy scent of lilac was in the air and orange-breasted robins were dancing on my front lawn pulling up submerged earthworms.

I retrieved a series of letters and among them was a large brown envelope. I recognized immediately that it had come from abroad. In the upper left hand corner was the logo of the Health and Safety Executive. I threw the packet onto the seat next to me and prepared for the thirty-minute drive to the campus, then gave into my curiosity. I tore the packet open and pulled a handful of papers out. The cover page was from Murray, the Principal Operations Inspector, and he explained that after being contacted by Napier, he had managed to obtain a copy of a judge's findings into the death of the two *Wildrake* divers.

I pulled off the cover sheet and glanced at a set of stapled papers. The top line read, "Fatal Accident Inquiry—Victor Francis Guiel and Richard Arthur Walker." It was a lengthy document with an introduction and a series of titled sections. Excited, I stuffed the papers back into the envelope and drove quickly to the college where I took the findings to the second floor of the library, sat down and inspected them. The document was not written on letterhead; it gave no hint that it was "official," nor was there a date to indicate when it had been produced. Had it not been for Murray's cover letter telling me what I was reading, I might have questioned its authenticity.

The 22-page "Determinations" were battered and torn and here and there the single-spaced text had been underlined with pencil. On the last page was a scribbled signature which I couldn't make out, but later learned was that of Douglas James Risk, the judge who had presided over the Fatal Accident Inquiry. These were his findings.

In detached, almost clinical terms, Risk described the "Contract for Hire" between the diving company and the oil company, the personnel involved, the equipment and working practices, what triggered the accident, and the attempted rescue. The last three pages were devoted to his summary and conclusions. As I read along, red flags began popping into my mind. Within the first few pages, I was given the impression that there had been an attempt to circumvent the law with respect to Richard and Skip's salary arrangements. There was also mention of a criminal trial associated with the accident, something I was unaware of. There were problems with leadership, modifications to the diving system, abandonment of critical safety gear, and questions about the security of the bell when it went into the water. The further I read, the more disturbing the report became. Equipment that should have been installed was not, and the diving vessel was brand-new, another red flag in my mind.

At one point Risk concluded that the diving contractor had been "more concerned with speed than with safety." This jibed with what I knew about the industry in general, that pressure to get the job done as fast as possible was pervasive. And then Risk explained what triggered the accident, something so simple I shook my head in disbelief, and yet it set into motion a series of wrong-headed decisions that led to one of the most awful scenarios I could have ever imagined.

The Determinations were of course only a summary of what appeared to be a very complex sequence of events; and understandably, there were gaps in the story. Motivations for critical acts were not explained, and I could not discern what the time sequences were. This missing information only generated more questions in my mind. What I needed were the dive log and the transcript of the court proceedings, if they still existed.

When I put the papers down I knew I'd stumbled upon something more than just a simple accident report. Risk's narrative was both compelling and horrifying. Reading between the lines, I sensed there was a kind of immortal human-interest story here that went far beyond the boundaries of an industrial accident. In retrospect, I think my reading of this document was a defining moment for me because I realized I couldn't walk away from knowing what had happened—my nightmares would have only intensified—and so I made a conscious decision to find out what went wrong and why.

Over the next eight years, as I searched for every available record and hunted down witnesses to the accident, the picture that Risk had described began to flesh out and change shape. Some of the assumptions I started with dissolved under the weight of evidence. Some remained fixed. Rumors that couldn't be ignored were investigated and discarded if they couldn't be confirmed, while others proved to be true. Gradually as a clearer understanding of the accident emerged, I began to feel as though I had entered into a cave. The further away from the entrance I got, the darker the story; and occasionally I found myself thinking, *This can't be true.* And yet the evidence kept stacking up to say that it was.

Eventually, after years of investigation, I managed to uncover all the failures that had contributed to the dual fatality, except one. I still couldn't understand why Richard and Skip had not been saved. I'd made dozens of seabed recoveries myself and knew what was involved. How could a fairly straightforward 1-2 hour recovery operation turn into a 17-hour-long disaster? That was the one unanswered question that remained illusive. As a result, the latter part of my investigation became a detective's quest to unravel a mystery long hidden from public view.

Along the way, the vast majority of those I approached were extremely helpful, and to my surprise I found that many of them, even those with no emotional attachment to Richard and Skip, had saved letters, personal diaries, photographs, and documents relating to the accident. Considering the passage of time and the nature of the tragedy, that seemed odd to me, but no less so than the way in which this story had burrowed into my skin. There were, however, a few exceptions to the outpouring of aid I was receiving. One of the men responsible for the accident told me in an angry voice to leave the story alone, to "let it die."

But there are some stories that never die and must be told. This is one of them.

Michael Smart
August 2003

Into the Lion's Mouth

As a diving mecca, the North Sea is one of the most inhospitable and unforgiving places in the world to send a man to the bottom of the sea. The water is cold and deep and the weather so unpredictable that it can change from blazing sun to snowfall within a few hours. Watching the changes is almost like watching time-lapse photography. Fog appears out of nowhere and cloud formations look supernatural, as though a thousand volcanoes are spewing their dark ashen breath into the cosmos. Storm waves towering forty feet high sweep down out of the north without warning to slam into massive rigs hunkered down against hurricane-force winds that promise to blow a man overboard like a feather. These are fearsome waters.

But to a diver in the 1970s, the North Sea was an exciting place to be. It was where the action was, where the diving industry was expanding, and where the new frontiers were. The deep water, the harsh environment, and the money brought the best and the brightest from all over the world. And though they came from different social and economic backgrounds, they shared the same pioneering spirit, seeking adventure and a chance to test their courage against this formidable sea.

So they suppressed their fears and flew to the barges and rigs with legendary names like Thor, Hercules, and Haakon Magnus and braved the dangers and the loneliness and the cold, because they knew they couldn't let the challenge pass them by.

They were in the right place at the right time; they were there to make a name for themselves, and these were the glory days.

But, oh, they were the dark days too...

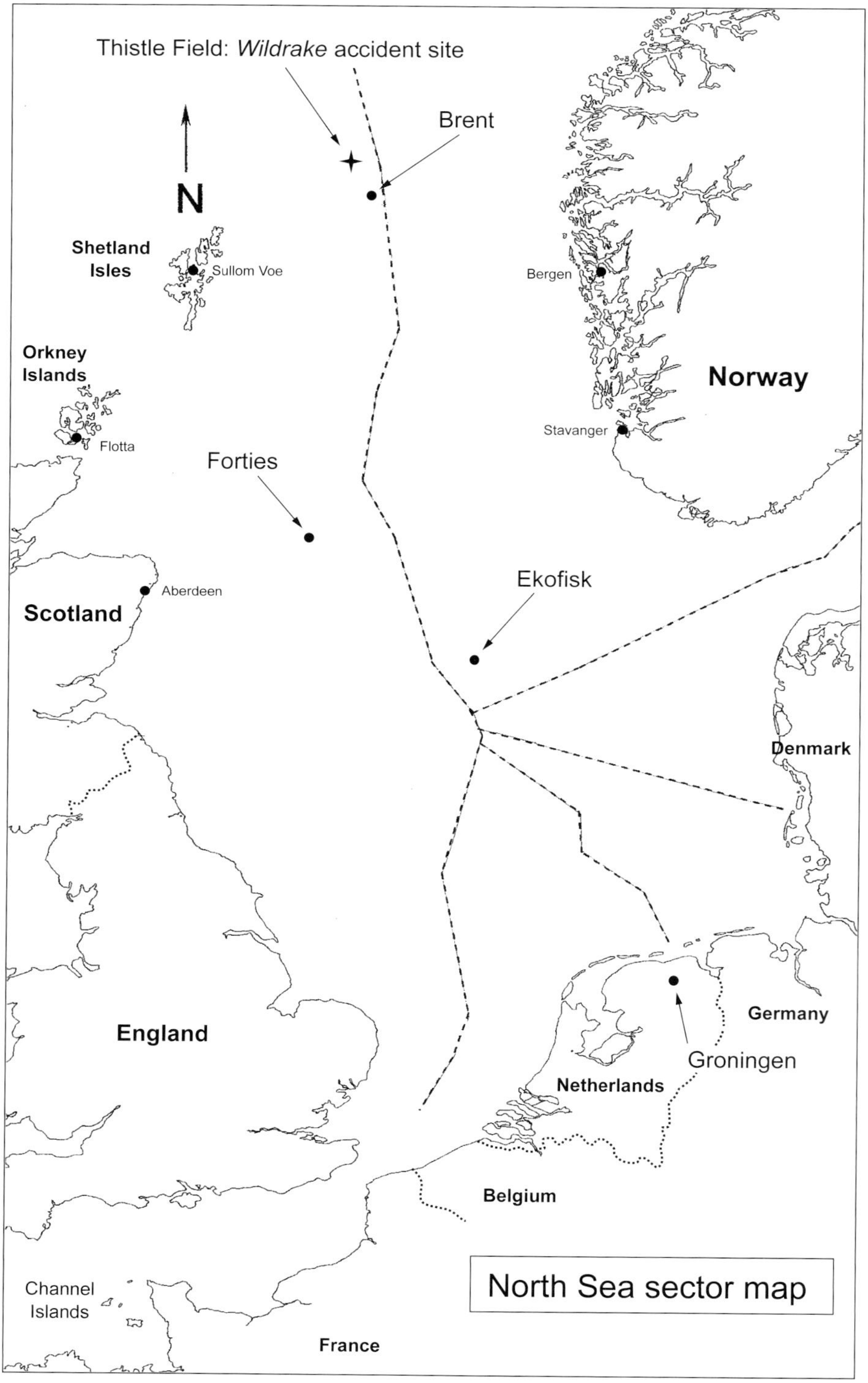

Thistle Field: *Wildrake* accident site
Brent
N
Shetland Isles
Sullom Voe
Orkney Islands
Flotta
Forties
Aberdeen
Scotland
Bergen
Norway
Stavanger
Ekofisk
Denmark
England
Germany
Groningen
Netherlands
Belgium
Channel Islands
France
North Sea sector map

Chapter One

Richard Walker

He chose this work because he had fallen in love with the ocean. He was a guy who often expressed joy, even in small everyday ways. Most of the time he walked around singing or whistling to himself. Humor was his best friend, beauty was his desire.

—— Mary Ellen Fletcher[1]

That first day at dive school was like the resurrection of a lost memory—it was kindergarten and innocence all over again—the hallways full of strange people, strange smells, strange everything. You had no idea what you were getting yourself into; deep-sea diving was at best a vague concept. But there you were standing in line with a bunch of other doe-eyed guys who were just as clueless as you were. The colored brochures, the advertisements, and the stories from other divers—it was all about adventure, travel, high pay. But it wasn't the full story, and no matter how much training they fed you, none of it would really prepare you for the challenges you were about to face. Only later would stark reality confront imagination; only then would you realize how naive you were, that the distance you'd traveled was no small step but a quantum leap, and that deep-sea diving was not working for Jacques Cousteau.

But the unknown and the money and the anticipation of adventure were part of the allure of becoming a diver, and every guy standing in line that first day had a different reason for turning professional. Some were dreamers, some had a special bond with the sea, and some were naturally dazzled by the stories of big money—*$500 a day! Well, my God, I could do that!* So, how do you join this smallest and least understood brotherhood?

You pay at the cashier's window.

Sometime in the fall of 1969, at the age of twenty-three, Richard Arthur Walker stepped across the threshold that separates the sport diver from deep-sea diver when he signed up for a Marine Technology program at Santa Barbara City College. Fresh out of the army with an honorable discharge, the tall, blonde-haired veteran had recently moved to southern California to be closer to his grandfather who lived in the nearby community of Montecito. He quickly landed a job at Kilovac Corporation in South Pasadena as a glass lathe operator, and when the company relocated to the Santa Barbara area in the fall, Richard followed, and so did that nagging question we all face in our twenties: What am I going to do with myself?

Born Richard Arthur Clinchy in Upland, California, he was the son of Art Clinchy and Frances Anne Fulmor. Very little is known about Richard's early childhood, even close friends struggle to recall what they know of it, but after Art and Frances divorced, she took her son back to Livonia, Michigan, where she later married Ron Walker, a personal injury attorney who adopted Richard at the age of eight and gave him his name.

At 14 he was given a series of aptitude tests and scored high in Verbal, Spatial, Mechanical, Spelling, and Sentences, but low in Clerical and Numerical skills which perhaps explains why his grades in high school were poor in typing and math. He had an IQ of 125, could express himself well in writing, and yet he barely applied himself scholastically in the core subjects. Fearful that her son might not graduate, Richard's mother forced him to take English and US History in summer school. But Art was where he excelled. Sculpting, painting, applied technique, his grades were high. He also excelled in sports as a member of the varsity track team, and sang in the school's award-winning Concert Choir. His high school senior picture displays a young man resembling one of the Righteous Brothers poised to sing "Unchained Melody."

In June 1964, Richard relieved his mother's anxieties by graduating George Bentley High School, and for the next sixteen months worked as an art apprentice in a large commercial studio in Detroit. In early 1966 he was taking art classes at Schoolcraft Community College when he came to the attention of his local draft board. By May, the army had claimed him and eventually stationed him in West Germany as an armored crewman, driving and gunning tanks. Twenty-three months later, in April 1968, he was discharged from the service.

At Kilovac, Richard liked making high voltage vacuum relay switches. It was technical and intricate work, but the standard 40-hour-work week didn't define him. It wasn't outside, it wasn't hard physical work, and it didn't leave him much time to pursue his other interests. That Richard was destined to become a deep-sea diver was not so much a matter of divine guidance, but that he happened to live in one of the strongest diving communities in America. The ocean, the available research facilities, and a burgeoning offshore oil industry made Santa Barbara an ideal location for training commercial divers, and City College was arguably the best training facility in the world. Graduates were so well trained, so sought after, and so numerous on some jobs in the North Sea that the British oftentimes referred to them as "the Santa Barbara Mafia."

When Richard discovered the Marine Technology program, the course subjects sounded so interesting: diving bells, decompression chambers, underwater welding, technical report writing, seamanship, boat handling, bounce diving, saturation diving. The intermittent nature of the work and the long periods of intensity and downtime appealed to him. The more he heard, the more it struck a chord within him. The industry was brand new; it was a ground-floor situation, and because he loved the ocean, he thought diving would be a great way to make a living.

In the fall of 1970 he started his prerequisite courses, and for the next several years, lived a life in sharp contrast to the two years he'd spent driving tanks around on a cold German landscape. Now he was living in paradise and attending school at the top of Cliff Drive overlooking one of the world's largest unconquered frontiers, the Pacific Ocean. It was blue and open; the college-town atmosphere relaxed and friendly, the air, warm and salty. There were beautiful women and good surfing beaches up and down the coast; and every year, when the city celebrated Fiesta, State Street blossomed with scarlet bougainvillea, parading partygoers, and showering snowstorms of confetti. This was his "land of the lotus," he said, a place he fell deeply in love with and described as having an "almost kinetic, divine, osmotic energy" to it.[2] The product of a broken home, Richard embraced Santa Barbara as fiercely as a man having discovered his star-crossed lover.

In the middle of his second semester he dropped out of the diving program for a year to earn money, and then rejoined the graduating class of 1973. The 43 other guys in his class were an interesting mix of people. They weren't kids just out of high school with heads full of idealism. Most had either served in Vietnam, had been to college, or were out working. They were centered, they had goals, there was a commitment to school and grades, and they were there because marine technology was what they wanted to do with their lives.

Where other training schools presented a more concentrated fast-track approach, the Marine Technology program was more even-paced, lasting two years. There was more time for Richard to ask questions and put theory into practice and learn the different techniques of commercial diving. Most importantly, because there was time, the instructors took deliberate steps to instill teamwork and responsibility in the students. These instructors knew that one day Richard and the rest of his class would be relying upon each other for their lives, and that a sense of duty and fellowship was a far more effective force against the inherent dangers of their profession than any poorly written regulation issued by a government body hundreds of miles away.

And there was something else the instructors advised. They suggested that each student keep a diary—not just to keep track of their pay, but to serve as a repository of daily events and personal experiences to draw from later—perhaps describing how a dive turned out and how it could have gone better or whether a new rigging procedure should replace an old one.

And Richard did that. He kept a diary.

Courtesy of Jeanne Walker

Richard (facing the camera) at SBCC.

As the months passed, his knowledge about commercial diving mushroomed into areas he never knew existed. He had an inquisitive mind and became fascinated by the science and technology of sending a man into deep water and getting him back alive. He learned the terminology, safe procedures to follow and bad practices to avoid. He especially liked the camaraderie that developed from working with people who had the same interests as he did. He was a natural problem solver and unafraid of trying new things. Besides being mechanically gifted with an inventor's heart, he had a refined side that expressed itself in his writing, his music, and his art. When he noticed that the Marine Technology program lacked an emblem, he designed one for the school that is still in use today. In one sense, whatever Richard did, he did it in a way to add beauty and joy to life.

For two years, he prepared himself for his new career and formed strong social and professional bonds with several members of his class, in particular, Burt Davis, Tom Belcher, and Craig Roberts. As they surfed the beaches of southern California and played softball at Stow Park in Goleta, the bond between these men steadily grew. Even at this early stage, they sensed they were special and had what it would take to meet the enormous challenges and responsibilities that lay before them.

Courtesy of SBCC

Richard's design.

And then one day, life on the edge of paradise came to an end.

On June 7, 1973, Richard, Tom, Burt, and Craig graduated, and within two months they were gone. Richard was recruited by Ocean Systems International and told to report to its Aberdeen office. Burt landed a job in South America, and Tom and Craig were off to the oil fields in the Gulf of Mexico. Like salmon fry spawned at the headwaters of a powerful ancestral river, they were venturing far out to sea into an uncertain future, unaware that during the next six years their paths would consistently cross in ways they never would have imagined.

By August, Richard had flown to Aberdeen, received his work assignment, and had gone offshore. Britain's offshore oil industry at that time was still very much in its infancy. Although significant gas discoveries had been developed in the shallower southern basin, only ten commercial oil deposits had been found since the discovery of oil in 1969. In the northern basin off Scotland, there was no archipelago of production platforms breaching the horizon, no vast network of subsea pipelines delivering crude to shore-based refineries, no flare stacks billowing black smoke across the pale northern sky—just a handful of nomadic drilling rigs wandering a gray liquid landscape in search of Britain's destiny far beyond her three-mile territorial limit.

In 1608, a Dutch scholar by the name of Hugo Grotius wrote a legal brief on behalf of the Dutch East India Company after a Dutch ship was captured in the Strait of Malacca by the Portuguese who claimed that body of water as their possession. In his treatise, Grotius argued that the ocean, like the air we breathe, could not be possessed but was itself the possessor of the earth and "the home of the human race."[3] His writings were so persuasive that they would eventually help to form the principle in international law that the High Seas should always remain free and common property to all.

As time passed, nations began to accept three nautical miles—the maximum distance over which a cannon ball could be hurled from shore—as the extent of their territorial jurisdiction. Everything outside that jurisdictional belt was considered to be the High Seas. But as countries grew more powerful and technological, particularly after the Second World War, many nations began to extend their sovereignty unilaterally beyond the accepted limit. Some nations desired a twelve-mile jurisdiction, others 200.[4] Very quickly crucial navigation lanes became threatened as issues erupted over such things as fishing rights, acts of piracy, right of innocent passage, and the exploitation of a country's Continental Shelf.

In 1956 the United Nations International Law Commission addressed these and other issues when it drew up a series of Articles codifying the law of the High Seas. But in order for the member nations to accept them and put them into practice, the Commission wanted the measures put to a vote by the international community. Two years later, the Conference on the Law of the Sea was convened in Geneva, Switzerland to do just that.

While the dispute over what constituted the breadth of territorial sea (three-miles or otherwise) was never resolved, the concept of a Continental Shelf and a nation's right to exploit its natural resources was newly adopted into international law.[5] Cleverly defined, the Continental Shelf was construed to be the seabed and subsoil extending into the High Seas to a depth of 200 meters, "or beyond that limit, to a depth allowing for the exploitation of natural resources."[6] Outside the territorial limit the seas were still free to all, but the resources that lay beneath them were viewed as the property of the coastal state.

Under Article 5, "installations and other devices" could be built, maintained, and operated offshore with safety zones established around them, and their protection provided for by whatever measures the coastal state deemed necessary. In the coming years, ships with names like *HMS Jura* and *Reward* would be dispatched to protect Britain's offshore assets.

A year after the Geneva Conference, Holland discovered a major gas field in the northern province of Groningen, which bordered the North Sea. The find was so massive—the second largest in the world—that it triggered a sudden interest in the North Sea basin which geologists believed was really a submerged part of the European continent. Seismic surveys were quickly conducted and the results looked favorable. Was it possible that the North Sea held untapped riches? Perhaps, but the oil companies were not about to embark upon an expensive High Seas drilling program without a legal framework to protect their property rights.[7]

In 1964, Britain's Parliament provided such a framework by passing the Continental Shelf Act which extended the Crown's right to grant licenses to "such persons as they think fit" to search for and extract oil from the Continental Shelf.[8] The only remaining question was: how should this potential economic pie be divided up by the five principal bordering countries?[a]

Article 6 of the 1958 Convention provided the necessary guidance. It stipulated that the boundaries should be determined by agreement between the coastal states. To Norway, dividing the pie down the middle seemed logical and fair. The average depth of the North Sea was about 94 meters and sloped towards deeper water in the northeastern direction. But when British ministers inspected a sounding map they noticed that a deep trench—much deeper than 200 meters—lay just off Norway's coast. In a somewhat laughable move to gain more than its fair share, they tried to suggest that Norway's claim extended only as far as the edge of the trench.[9] The Norwegians, however, were not buying that argument, and the matter was eventually settled by employing another provision of Article 6, namely, that in the event of a dispute, the territorial boundary should be determined by the "median point" equidistant between the countries.

Although the final agreement did not arrive until 1965, Britain had been selling exploration licenses for nearly a year. Once established, however, the boundaries gave rise to national sectors, with Britain and Norway capturing the lion's share.[b]

---

a Britain, Norway, Denmark, West Germany, and Netherlands

b Britain 46.7%, Norway 25.1%, Netherlands 10.7%, Denmark 9.2%, West Germany 6.8%. *Source*: D.I. MacKay and G.A. Mackay, *The Political Economy of the North Sea*, p. 21.

And it was in the Ekofisk Field in the Norwegian sector in the fall of 1973, on a pipelay barge called the *Choctaw*, that Richard experienced for the first time what life was going to be like as a diver in the North Sea.

Courtesy of Skip Gordon

Ekofisk Field, 1975.

At City College, everything he'd seen and touched was on a scale that made it easy for him to learn. The shackles, the wire rope, the cranes, the barges, the bolts, the pipe—it was all small and manageable. This was the imprint that Richard carried with him to Scotland. And like all those who came before him and all those who would follow, when Richard flew out to the barge that August day, the utter shock and surprise of what he was witnessing must have come tumbling out of his mouth: "Oh my God, look at this thing!"

Bigger than a football field, the *Choctaw* was an offshore factory giving birth to a new pipeline. Activity abounded everywhere: pipe davits were grappling concrete coated pipe, welders were welding, cranes were slewing, hydraulic motors were whining, sparks were flying, and gas lines were venting like gray whales spouting on the surface of the sea. The noise, at times, was ear-splitting. Men with white hard-hats were running around shouting orders in southern drawl at swarthy workers speaking a foreign tongue. In the midst of this organized chaos, more than 200 men were turning out about a mile of pipeline a day. In assembly-line fashion, forty-foot sections of precoated pipe were placed on a ramp, the ends beveled, butted end-to-end, welded together, X-rayed, then sealed with a protective coating. As each new "field joint" was completed, the barge winched itself forward with the pipeline ultimately trailing off the stern and down an articulated ramp called "the stinger," to the seabed nearly 300 feet below.

To Richard, this new world seemed strangely familiar but somehow distorted and magnified ten fold. It was as though he was Alice in Wonderland and had nibbled a piece of the mushroom and shrunk. The barges, the ships, the waves, the wind, the number of people—everything around him was at the extreme end of the scale and skewed. And in that brief instant, as he stood there, with the wind beating his face, he must have truly felt the magnificence, the majesty and the awe-inspiring power of the North Sea.

In the days ahead, he dressed in scuba gear and made periodic inspections of the stinger to check the rollers and load sensors to make sure the pipeline had the proper depth profile and that it wasn't going to buckle on its way to the bottom.[10] The temperature of the water was so cold it took his breath away, but it was absolutely crystal clear with visibility well over 100 feet. Far out at sea, with no land in sight, with no coral reefs or pretty fish to gawk at, the Santa Barbara diver swam in a blue liquid universe cut by silver curtains of light that ultimately cascaded down into a dark spooky abyss below.

In 1973, there were no safety regulations covering diving operations in the British sector of the North Sea. If a diver went offshore, he did so without any protection from the law. The closest set of regulations was for onshore operations thoroughly inadequate for the scope of diving currently taking place offshore. In the broadest of terms, they stipulated that the diver: a) should have previous knowledge and experience in diving (unless he is undergoing training), b) should be fit to dive, c) should be provided with properly maintained equipment, and d) should be supplied with enough air to make it back alive.[11] The regulations did not speak to such issues as: the use of breathing mixtures other than air, the responsibilities of the diving supervisor, the use of a diving bell, or the size of a decompression chamber. Nor did they place restrictions on the depth or duration of the dive.

As the search for oil moved northward into the deeper waters off the coast of Scotland, dives in excess of 200 feet were routinely conducted using surface-supplied diving gear that allowed the diver to not only breathe from a full-faced mask or helmet connected to a gas supply hose, but also to speak to his supervisor with hardwire communications. Appropriately branding the diver with the nickname "hose head," surface gear supported longer and deeper dives, and more work.

But the deeper and longer the dive, the greater the risk. Where decompression dives are frowned upon in sport diving, they are common in commercial diving. Beyond certain time constraints, the diver builds up a debt of inert gas in his blood and body tissues that has to be paid back through the slow process of decompression to surface pressure according to a prescribed schedule: the object being to get the inert gas to pass back out through the lungs the same way it went in. If the schedule is not followed precisely, the diver will get "the bends" (decompression sickness), sometimes with fatal results. There were three recorded diving fatalities in 1971, and in all three cases the divers were using surface gear at depths between 200 and 300 feet and died from complications due to inadequate decompression.[12] In one of the accidents, a highly trained former US Navy Lieutenant Commander unexpectedly rocketed to the surface (called a "blow up") while trying to free his fouled supply hose.[13]

Added to this setting were the harsh and unpredictable weather conditions which made long, in-water decompressions a dangerous, if not worrisome proposition. For these and other reasons, it became common practice to use a diving bell for transporting divers on deep dives to and from the job site, and hyperbaric chambers on board the vessel to decompress divers away from the hostilities of open water.

Like cars, there are as many types of deep diving systems as there are manufacturers that design them. Their configuration and layout vary from company to company, but they all function alike and have the same basic components. In a typical system, a spherical diving bell mates to a chamber called the entrance Lock, which in turn is connected to a series of chambers that provide living quarters for the divers when they are not actually diving. The whole complex, which resides on the vessel, resembles the International Space Station with the diving bell serving as the "space shuttle."

Courtesy of Skip Gordon

OSI system off the Ivory Coast, 1975.

To get the bell in and out of the water, Ocean Systems employed a large winch in conjunction with a davit, a kind of fixed boom mounted on a rotating pedestal. Divers were lowered in teams of two, and while one man remained inside (the bell man or standby diver), the other exited or "locked out" on the end of a life-support umbilical to do his work. From the control van on board the vessel, the supervisor spoke to the diver and directed his actions while life-support technicians regulated the environmental conditions within the chamber complex. Depending upon the operation, there could be as many as 35 divers on the job divided into two 12-hour shifts. At the top of the food chain was the superintendent, a god-like figure who laid down the law on his job. Below him were his two lieutenants, the day shift and night shift supervisors who conducted the dives and gave instructions to the rest of the crew.

Throughout the month of August and into the first half of September, Richard continued to perform stinger checks and helped launch the bell on deeper dives. Then, on September 14, he was transferred to a nearby jet barge called the *Cherokee* where he made his first bell dive.[14]

While the *Choctaw* laid pipe, the *Cherokee* buried it with a huge contraption called a "jet sled" in order to protect the pipe from anchors and bottom trawl fishing techniques. Overall the garage-sized sled was a tubular framework built on a pair of pontoons that supported "the claw," a pair of vertical pipes studded with water jets whose main business was to cut a trench deep enough for the pipeline to lie in. The sled straddled the pipeline and was attached to the stern of the barge by a towing bridle. As the barge winched itself forward using its anchoring system, the claw carved the trench. Periodically, divers were sent down to check the sled for alignment and to sample the seabed for hardness, and on September 25 Richard was selected, along with a small British diver by the name of Dave Abbott, to inspect the sled which was already in operation on the sea floor.[15]

In terms of planning, getting Richard and Dave to the bottom of the sea was the easy part; it was getting them back alive that took skill. To do that, the supervisor had to consider three critical elements: the working depth, the "bottom time," and the breathing mixture. The supervisor knew the sled was operating in 270 feet of water, which was fairly deep water, but not extreme; and he knew that the dive was going to be short, under an hour. Those two factors alone gave him the option to use a technique called "bounce diving," so called because, under stopwatch conditions, Richard was expected to pressurize the bell to the working depth, lock out, complete his task, and get back inside to begin his decompression schedule before the clock ran out. The whole process was very frenzied and tense; decompression back to surface pressure was accelerated, full of stops at different depths, and fairly lengthy compared to the actual time spent working. For example, in one recorded bounce dive to 500 feet for a mere 20 minutes, the bell team had to decompress for seven hours and ten minutes. Thus, bounce diving made sense for deep, short-term dives, but for jobs lasting days or even months it was impractical. In that instance, the supervisor would have chosen a different mode of diving.

The depth and duration of Richard's dive also determined his breathing gas, which was going to be a mixture of helium and oxygen, not air, because, at 270 feet, air mutates into a toxic gas, with the oxygen concentration snowballing far beyond tolerable limits, and the nitrogen content presenting dangerous problems as well. Beginning at roughly 100 feet, it produces a narcotic effect upon the human brain (nitrogen narcosis); at 270 feet, it would have left Richard completely dumbstruck and unable to concentrate on his work. Thus, breathing air at surface pressure is not toxic, but feeding it to a diver at 300 feet is extremely dangerous. So in order to avoid toxicity problems on deep dives, a synthetic gas mixture is supplied where the oxygen concentration is reduced, and helium, not nitrogen, is used as the diluent to carry the oxygen because it causes no narcotic side effect. Thus, the deeper the dive, the lower the oxygen percentage in the dive mix.

Compared to all the different modes of diving, this "Beat the Clock" method is the most problematic to the diver's health. The rapid changes in pressure, coupled with frequent errors in mismanaging the decompression, causes enormous damage to the diver's body and produces the most numerous and serious cases of the bends. The more bounce dives a diver makes during his career, the greater the risk he incurs in terms of lifelong injuries to joints, muscles, and nerves. Nevertheless, Richard was willing to endure these yo-yo dives because, like all beginning divers, he wanted to advance to The Promised Land of "saturation diving" where divers remain stabilized under pressure for much longer periods of time and ultimately make more money.

That he was a rookie and had been chosen to dive was not lost on Richard. He knew that supervisors were not inclined to put people down at 300 feet in dark water on a jet sled that they could not trust. The moment that he'd arrived offshore, his surface dives and deck work had been heavily scrutinized by everyone on the crew. In their eyes, even though he had been to City College, he was still a rookie. If he did well, he'd be able to stay. On the other hand, if he screwed up, he might be on the next chopper back to the beach. That's how it worked.

But Richard had been earning the respect of his crew mates which is why his supervisor was putting him in the water. In effect his boss was saying that when Richard was standing alone on the seabed in the dark and experiencing the cold, the lack of visibility, and the intensity of the job, he could be trusted to keep his cool and make all the right decisions and stay out of trouble, because every diver inevitably faces the moment when the very next decision he makes saves his own life.

Now, dressed in their dive suits, Dave was already in the bell speaking to the supervisor over the radio and running through internal bell checks when Richard came on deck and stood beside the bell. Like a giant white gumball machine, the four-ton spherical bell sat on top of a five-foot-tall tubular support frame called a stage. There were 15 tiny view ports and two black bumper guards, one running around the top of the stage, the other around the equator of the bell. To keep the divers alive and happy, the capsule was connected to the control van by a 1,200-foot long life-support umbilical which supplied, among other things, breathing gas, communications, electrical power, and hot water. Ocean Systems called it the ADS-IV.[a] By coincidence, Richard was not unfamiliar with it. He'd been trained in one at Santa Barbara City College.

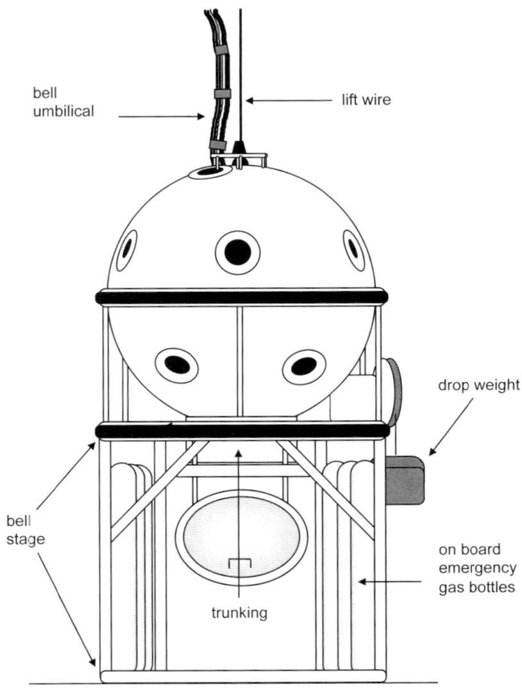

Ocean Systems bell.

After bell checks, Richard ducked through an opening in the stage and climbed inside through a short, cylindrical hatchway called the "trunking" which had two

a Advanced Diving Systems-IV.

heavy steel doors, a bottom door that opened outward, and an inside door which opened inward. The passageway was tight, only two feet in diameter, and once inside, Richard turned two large levers to "dog" the bottom door closed, but left the inside door open.

This bell, like all working bells, was crammed with gear. Two bulky dive umbilicals were coiled against the wall—one for Richard, the lockout diver, the other for Dave, the standby diver. Here and there a gaggle of valves, gauges, copper tubing, and electrical cables ran this way and that. Various "penetrators" on the hull allowed the service lines from the main bell umbilical to link up with the support facilities within. Next to the gas panel, a scrubber sat whirling and whining as it cleansed the atmosphere of carbon dioxide. There was also a small O_2 analyzer that indicated whether the gas in the bell had enough oxygen in it. Mounted at the top of the bell was a small "horn speaker." All Richard and Dave had to do to communicate with their supervisor was speak loudly into it. And stuffed in various nooks and crannies were such things as the bell emergency tool bag, water bottle, port cover, extra scrubber canister. Normally the bell came with two seats, but divers routinely ditched one of them to conserve space which there wasn't much of in this bell. In fact, throughout the entire North Sea, no bell was more claustrophobic than this one. It was only five feet, five inches in diameter.[16] Richard was six-foot-three and weighed 190 pounds.

Suddenly a voice over the horn speaker called out, "Okay, here we go. Coming up on the bell," and the gumball machine suddenly rocked off the deck and began to gently sway like a plumb bob. Through one of the ports Richard could see the deck crew operating the davit. Several minutes later the two men were rotated over the side of the barge and lowered into the gurgling sea. Fifteen feet under the surface, the bell was momentarily stopped in order for the duo to check the seal on the bottom door. Outside water pressure and a thick o-ring recessed into the hatch were what kept the water out. With the seal confirmed, topside continued their descent.

Up against the hull, and mounted next to the gas panel, were two depth gauges—one internal, the other external.

As the bell descended, the needle on the internal gauge still read zero, but the needle on the external gauge was slowly rotating clockwise—30, 40, 50 feet—then past 100 and beyond 150. The deeper they went, the darker it got outside.

On the way down Dave called out the depth every ten feet while Richard donned a small scuba tank. This "bailout bottle" was connected to his mask and was for emergency purposes only. If he lost his main gas supply (severed umbilical or topside mistakenly closes his supply), then all Richard had to do was reach up with his right hand and open a small valve on the side of his mask to make it back to the bell.

After roughly ten minutes, topside stopped the bell with a sudden jolt. It had reached its destination depth about 30 feet off the seabed where it would remain suspended throughout the dive. By this time Richard had soaped his mask to keep it from fogging up, clipped his umbilical into his harness, and put his gloves on. The two divers were still breathing air at surface pressure, but that was about to change.

Richard then lowered himself into a sitting position on the lip of the inside hatch. With his feet hanging in the well of the trunking, he undogged the door by kicking the hatch levers open. The door still remained closed because the outside pressure at 240 feet was more than eight times the internal pressure of the bell. Checking with topside, Richard and Dave let their supervisor know that they were ready to start the blow down.

When the supervisor gave the okay, Dave shouted into the speaker, "Starting blow down now," and then he opened the Blow Down Valve. Mixed gas immediately rushed into the bell with a deafening roar. The bell also began to heat up. Compressing gas alone will cause the temperature rise, but helium has enormous thermal conductivity properties, and the change in temperature for divers in a confined space can be dramatic—even fatally so. The helium had also changed Richard and Dave's speech pattern, so the supervisor had to use a special radio with a built in "unscrambler" to understand them.

While the needle on the internal gauge was racing to join its mate, both men wiggled their ears and jaws. In the bell van the supervisor started his stopwatch to track the bottom time of the dive. Richard had about 40 minutes to complete his inspection. The faster he blew down the more time he had for his lock out. And if he couldn't finish the job (which was a cultural taboo), he still had to scamper back to the bell to begin his decompression schedule.

After a number of minutes both gauges displayed identical depths, and magically the heavy steel door beneath Richard's feet dropped like a lid on a footlocker, turning the trunking into a small pool of quivering water. It was the gateway to another world.

It was also the doorway to the unknown.

In spite of everything you've done and know, it's difficult not to think what can go wrong at 300 feet in the dark on a jet sled whose business is to carve up the seabed. You have the knowledge and the training, but, still, you know you lack the experience. And so the fear and the anxiety and the apprehension begin to slip past the barricade. A sour feeling in your stomach suddenly appears. You close your eyes and mutter, "*No, no, not now,*" but the bile begins to flow anyway.

Part of the fear on this first dive is the depth and the darkness which tends to conjure up images of hideous demons and monsters from childhood. But mostly, what you're feeling is the fear of screwing up. You want to be a diver, and you want to get the job done, so you try to pump yourself up with unconvincing drivel: *Hey...come on, man, this is just like a night dive!* But from the moment you saw the barge lurching in the white-capped sea, and the serious looks on the faces of the dive crew, you knew that the fun and games of sport diving were over, that they weren't sending you down there to sing *Cumbaya* and to look for lobster.

So you try to be nonchalant and mask your feelings, but the kettledrum in your chest continues to pound in your ears. *Boom-Boom-Boom* it goes, and you can't stop it. You fiddle with your gear and take a few deep breaths, and then you sneak a look at your bell partner.

Oh for God sakes, he's reading a book!

The last thing Richard did was open his gas supply and pull on his mask. Shaped like a fiberglass catcher's mask, it was banded to a rubber hood that zipped down the back of his head. A thick neoprene facial guard circumnavigated his head just in front of his ears and under his chin. This rubber barrier, along with positive pressure inside the mask, kept the water out. Richard's mouth and nose were crammed inside a small oral-nasal mask connected to a demand regulator mounted on the outside. Surface gas was supplied when Richard inhaled. He could also free flow the mask if he chose to, but most divers avoided this because of the noise it created. Two ear speakers and a small microphone allowed him to speak with topside.

Slipping his fins on, he gave a quick comms check. "One, two, three, four, five, how do you read me topside?" he said in a voice pinched by helium.

"Loud and clear," came the response through his ear speakers.

From a wall-mounted rack, Dave peeled off a few coils of Richard's umbilical. Richard took these and shoved them down through the trunking, and then, like an Arctic seal diving through a hole in the ice, he suddenly vanished into the world that Neptune ruled.

In the water, bubbles exploded from Richard's exhaust manifold and reverberated against his ears. He sounded like a patient on a life-support machine. The noise was so loud he had to hold his breath to hear topside speaking to him. Despite what most people think, the sea is not a silent world at all. Every gurgle, every murmur and whisper from Mother Nature penetrated Richard's ears in full-volume stereo.

In the distance he heard the muffled roar of the jet sled.

"Diver in the water," he said.

"Roger," a voice came back.

Because the barge was anchored, the bell wasn't heaving up and down very much. Richard swung his body from under the trunking and took a quick look around to get his bearings. It was dark, but not pitch black. The bell lights reflected off a primordial soup like headlights on a foggy mountain road. A rope coming from above passed through a shackle on the side of the bell and down to the seabed. This was the "down-line" and it was attached to the sled.

Gathering the items he needed, Richard wasted no time in telling topside that he was leaving the bell for the job site. "Slack the diver as he goes," he said.

"Roger," topside answered.

He took hold of the down-line and began pulling himself along the rope through a suspension of tiny glassine life. Behind him, Dave was feeding his umbilical out through the trunking.

At every milestone during the dive—entering the water, leaving the bell, completing a specific task—Richard reported to his supervisor in short, terse statements. The supervisor in turn recorded the time and Richard's movements in a logbook. Occasionally Richard could hear his commands being echoed to Dave in the bell, but neither man could speak to each other directly. For safety reasons, all communications were relayed through the supervisor and they ran in one direction only at any given time. Unlike talking on a telephone, once the supervisor held his thumb down on the talk button, Richard was cut off. He could be trapped on the seabed with a 40-ton load coming down on top of him, but until that supervisor released his thumb, he just had to sit there and wait. For this reason, conversations were kept short—in diver talk—and to the point. No rambling on about last week's beer fest.

Hand over hand Richard followed the rope down into the gloom. The growing roar in the water was coming from a cloud billowing neon phosphorescence in the distance and the rope he clung to led right to the heart of it. Richard kicked and pulled. The roar kept getting louder. Seconds later he reached the seabed at the edge of the cloud. For a moment he stopped and looked back to check his umbilical. The image he saw was dreamlike. Against a curtain of black space, the bell hung like a glowing orb in the night sky with an almost divine presence. At 270 feet, it was Richard's only safe haven if something should go wrong.

Adrenaline was now pumping furiously throughout his body. The roar was the roar of a thousand lions. The sled couldn't be seen, but he felt its rumbling power through the vibrations in the water. It felt like an earthquake machine and Richard began flashing on the photographs of his objective, its components, the checkpoints, and the areas he was told to avoid. Somewhere within this black blizzard, the claw was slicing the seabed with streams of high pressure water while powerful suction pipes vacuumed up the spoil and blew it out to the side.

In the back of his mind, he felt the supervisor's clock ticking. Holding tightly onto the rope, he gathered up his courage and with one deep breath plunged into the swirling tornado, his image disintegrating behind him.

For a moment Richard was lost, fumbling through the cloud with his arms outstretched, and then he collided with something hard and metallic. Over the clamor, he shouted into his mask, "On the job. All stop on the diver's umbilical."

Faintly, almost imperceptibly, he received a response from topside. "Roger, on the job."

Inside the swirling dust cloud, Richard groped around on what he would later describe as "a giant mess of tubular struts, hydraulic lines, and various and sundry protrusions."[17] Moving carefully and deliberately, he measured distances, collected soil samples, checked the towing bridle, the tension on the supply hoses, the sled's orientation to the pipeline, and reported to topside on the efficiency of the claw, whether or not it was cutting the trench as deep as it was supposed to. Checking all these things by "Braille" was dangerous work. Richard had to be extremely careful not to get himself or his umbilical near the suction pipes or the jetting nozzles around which he took his measurements. This was 1973, and barge captains did not always shut down the sleds while divers inspected. The desire to set "footage-per-day" records kept these seismic chariots blowin' and goin'. Sadly, it would take more than one fatality in the Gulf of Mexico to provide the kind of epiphany that required hardheaded swaggering barge captains to wake up to the horror of a diver being sucked into a 14-inch pipe.[18]

After completing his checks, Richard emerged from the tempest and stood below the hovering bell. In the heat of the dive, there had been no time to evaluate his surroundings or assess its beauty; but away from the sled, the water was much cleaner and his eyes were adjusted now to the darkness. With the bell casting a soft glow onto the seabed like a lamp shade, Richard stood there staring at the vastness of his surroundings, letting the breathtaking strangeness of this alien world take possession of his senses. All alone and listening to the sound of his breathing exploding into a gray black cosmos, it seemed like such an odd place to be. There were no rainbow colored fish, no beautiful kelp forests, no spiny black sea urchins, no frolicking sea lions, no majestic coral reefs, no landmarks, no rocks, no boulders—it was just a barren wasteland as flat as the corn fields of Iowa.

No one had ever set foot here before; Richard was sure of that. The panoramic lack of beauty, the awesome reality of what he was witnessing—its size and dimension so much greater than expected—Richard felt as though he were looking directly into the face of Mother Nature and it humbled him profoundly and filled his soul with a deep sense of peace. *Had Neil Armstrong felt this way on the moon?* Richard didn't know, but he was overwhelmed by the universal power it represented. And in an instant, as he gazed through the boundless sea, he knew he had been transformed and that his birth on this new planet had been complete. Without terror or frame of reference, he was the first of his species to arrive. He blinked his eyes in wonder, and then he turned and climbed his umbilical, back to his mothership.

Once inside, Richard and Dave blew the water out of the trunking, closed both doors, and pressurized the bell another ten feet to seal it.

"Ready to leave bottom, topside."

"Roger, coming up on the bell."

Like an elevator, the bell was slowly hoisted back to the surface, and because it was sealed, the internal depth gauge remained frozen at 240 feet, while the needle on the external gauge began to reverse itself. At the same time, the supervisor began their decompression schedule by opening an exhaust valve on his control panel. The inside depth was gradually reduced until they arrived at the first decompression stop. The exhaust valve was then closed and the depth held for a prescribed number of minutes, after which the valve was reopened to bring them to the second stop. The advantage of decompressing the divers in the bell was that everything was controlled, the divers were dry, and they were inside the safety of the bell.

More scheduled stops followed and it was in this way—by not allowing the speed of the ascent to exceed the rate at which the blood and body tissues were "out gassing" through the lungs—that the danger of decompression sickness was avoided.

Meanwhile the deck crew had physically brought the bell out of the water and landed it on top of the entrance lock. The two chambers were then clamped together and the trunking pressurized. With no place to go, Richard and Dave just sat inside while the supervisor continued the decompression until they reached the depth at which the entrance lock was pressurized to. At this point one of them reached down and opened a small valve on the inside door allowing the pressure of the bell to equalize with the pressure of the entrance lock. After equalization, Richard and Dave pulled the inside door back, climbed down into the entrance lock, and then crawled through a circular hatchway into one of the living chambers to quickly don oral-nasal masks called Bibs.[a] These were used to vary the breathing mix to the divers in order to accelerate the decompression. Finally, after two and a half hours, Richard and Dave arrived back at surface pressure with bloodstreams still fizzing from the dive.

As the days passed, Richard was constantly learning something new. There was always some gadget or tool to be fabricated to keep the job moving along. Divers had to be jacks-of-all-trades, and Richard liked that because it gave him the opportunity to use his mind and his hands to solve problems. The diving system was simple enough that if anything broke down, usually someone knew just enough to repair it. If no one knew the answer, then it was generally a bunch of guys practicing Rube Goldberg methodology.

But whatever knowledge divers had, they shared it freely and without reservation because they knew that withholding information could potentially backfire on them. There was no advantage in diving with someone who had a

---

a Built-In-Breathing System.

puzzled look on his face. If a diver discovered a better and faster way of installing a piece of equipment in the water, he never jealously concealed his technique; he gave up the secret (usually with a prideful grin on his face). And this was an aspect of the diving culture that Richard appreciated and enjoyed. And so the experienced taught the less experienced, and in this way the whole team benefited while individual divers were still able to make a name for themselves.

The crew on board the *Cherokee* was a mixed lot, half British, half American. And Richard liked them, thought they were "all pros," especially the British divers who were constantly pestering him to play cards. He also found simple pleasures in speaking to the Spanish welders. In the galley he would smile to them and say, "Hola" and they would say "Hallo" back and speak "neat things" to him in Spanish such as: "Mucho trabajo, eh, Ricardo?"[19]

But there was another type of mentality out there that he couldn't ignore, nor understand, and throughout the North Sea they were giving Americans a bad name. In a letter to his friend, Mary Ellen Fletcher, Richard wrote:

> You should see the red-necks (American south) with their bell-bottoms and mutton-chops and all that. I caught one a couple of weeks ago trying to squish a tiny wren that had taken rest near his foot. The vein on my neck stood out, man, and I walked next to him and burned my nigger eyes into his temple. He quit. The little guy escaped. Typical of the kind of brains working here.[20]

On September 27, a gale force storm warning was posted. Everything was battened down, and for two days, as 30-foot swells marched down out of the north and washed over the barge, no one was allowed on deck. They were heading for Norway, and by the evening of the 29th, the barge was safely nestled in a fjord not far from the city lights of Stavanger. Richard cleaned himself up and boarded a small coastal tug to go ashore. As it chugged down the watery inlet passing hundreds of fishing boats lit up like Christmas trees, he marveled at the colored lights bleeding across the surface of the glassy harbor water.

For a long time he walked the streets alone, bought some post cards and magazines, and then stepped into a pub. He sat down and ordered a Tequila Sunrise and had to instruct the bartender on the "complex construction" of the American drink. Feeling a bit desolate and homesick, he began writing to friends while he listened to American pop music blaring from the sound system, his mind flashing on images of the city he loved. Later, Richard wrote to Mary Ellen and described what happened next:

> A young Nord strikes up a conversation, quickly realizing I'm American—and within five minutes says adamantly, "I do not like Norway," and proceeds in English, with an English accent, to tell me the same thing I had been flashing on, only from a different perspective. Something is to be said about the questionable virtue of organized, efficient, colorful sterility and its ensuing boredom—absolutely no tribal essence—a few smatterings of mockery. I file the impressions, bolstered by a tangible outside view.

I trucked out, a bit snockered, wander around, find a few of the diving crew, and make it on in to a hoopla dance place. Rock band, ladies, more to drink, and the old nose begins to open up. I dance. The Spanish guys from the barge have made it there—young, soulful, vital crew they are. We are out there doing the shingaloo and the boogaling with all these well manicured, long shag, flawless skinned beauties. There were a great proportion of classically beautiful girls, I must say.

What followed was the story of the ages—much to my surprise—and the old fleeting glimpse of my old self. 'And here's Walker, chasing Norwegian tail, and here he is, being so cool. And here he is, and—and' And I feel the blood of her next to me. Few words were spoken—this act of procreation—its weight on me—the need real—I realize the depth and scope of more needs—and know I've changed. I've broken the equation, feel born again to human dedication and suddenly filled with love. I realize how much I want children—the long—awaited perspective on the idea of my role coming together, faith coursing through me. I love all my people at home with such ferocious strength, I am sure you must have felt it. You probably can't understand—but everything is beginning to fall together for me.

I stagger to the tug after a long morning walk, still my brain free-flowing with no sleep in the warm sunshine of this fine day. Putt-putt back out to the barge, the Sunday morning strollers fading on the docks, past the ships, and down to my cabin to find two letters. Your poem brought quick tears to my eyes. Your beauty is so deep and strong—I don't think you realize your power, and still it comes together. The depths of my experience with you is amazing. You gently take me down corridors in your letters—cry, laugh, feel with such intensity, looking round your head from the inside.

Please create—the world needs you.

Richard[21]

Five days later, on October 6, Syrian and Egyptian forces set the Yom Kippur War into motion by attacking Israel on their holiest day of the year, the Day of Atonement. Within two weeks, six of the largest oil-producing countries in the Middle East would hike the price of crude oil from $3.01 to $5.11 per barrel—not as a weapon against the supporters of Israel (that would come later)—but because, after years of confrontation with the oil majors, OPEC Arab oil ministers were finally throwing off the yoke of Western economic imperialism by taking control of what came out of their ground.

In the North Sea, the *Cherokee* had only been able to work three days that cold October.[22] Severe weather had stopped all work, and with nothing to do, tossing up and down on an inexpressibly smelly barge was hell. There were cards, books, and blue movies, but at some point their entertainment value became used up. If the chopper couldn't land, there were no newspapers, no letters, and no word from back home. Cut off from civilization, and feeling a bit restless and forsaken, Richard would go on deck and gaze across a cheerless sea where there were no visible signs of life, just a swirl of what seemed to be a nuclear winter filling the sky. He didn't know how long he could take it, but like most guys, he hung on because it was his job and it was a way to make fast money.

On October 24, after Phillips Petroleum concluded that keeping the *Cherokee* on location didn't warrant the expense, the job was shut down and Richard stepped ashore at Edinburgh, Scotland. Excited about going home for the winter, he wrote to Mary Ellen and remarked on the growing Middle East dilemma and his feelings of awe about being a North Sea diver.

> Arab oil is about to be rescinded. What was the kid saying a while back? If I were looking for a career with a future, I'm in it right now—piss on everybody's disdain for the oil industry. It's gonna take years to get out of the mess we're in—civilization ain't gonna stop tomorrow...I'd be glad if there was no more oil at all. At least I got the chance to make my mark here on the pipe—incredible how so much responsibility is given the divers. 30 sections [of pipe] to be repaired next spring—at the kid's command! The unadulterated POWER!
>
> See you soon,
>
> Richard[23]

Chapter Two

Damn the Torpedoes, Full Speed Ahead!

There will be argument for many years as to whether or not Britain should have developed her North Sea resources as quickly as she did.

—— Guy Arnold, *Britain's Oil*, p. 133.

In Britain, the onset of the October War and the latest oil price increase couldn't have come at a worse time for Prime Minister Edward Heath. All year long the leader of the Conservative Party had been struggling with an economy reeling from the effects of runaway inflation and a growing balance-of-payments deficit. January had brought thousands of British civil servants, including clerks and secretaries from his Cabinet Office, into the streets in angry protest over his earlier decision to impose a 90-day wage and price freeze. The work stoppages were the first ever mounted by civil servants, and the action disrupted the very fabric of ordinary British life: travel through airports, disbursement of government subsistence checks, tax collection, registration of births and deaths, even the reading room at the British Museum was closed to the public.[1]

A month later, gas-industry workers, upset about the government's cap on wage increases, disrupted deliveries. Schools, factories and homes went without heat. Investors, demonstrating their lack of confidence in Heath's anti-inflation policies, began taking their money out of the London Stock Exchange to the tune of $5 billion.[2] Meanwhile, as wages remained frozen, food prices continued to soar because half Britain's food supply came from abroad.[3]

Added to Heath's economic woes was the country's growing dependency on foreign oil. Britain was consuming about 98 million metric tons of oil products each year, 80 percent of which was supplied by the Middle East and Libya.[a] And now, after a 70 percent rise in the barrel price, reports were circulating that the oil import bill, currently at $2.5 billion, was likely to grow by another billion dollars annually.[4] Anticipating a shortfall in gasoline supplies, Heath ordered the stockpiling of rationing coupons at post offices throughout the country.[5]

Arab oil ministers were obviously flexing their muscles, but the worst was yet to come. On October 17, the day after the price hike, OPEC began implementing a policy of monthly five percent cutbacks on the flow of oil to countries supporting Israel (primarily the US).[6] The cuts would continue, it was declared, "until Israel evacuated the lands taken in 1967 and made restoration to the Palestinian refugees."[7] "Friendly countries" that adopted this view would continue to receive their normal supplies. When President Nixon asked Congress to authorize a $2.2 billion military aid package for Israel on the 19th, the Arab states promptly announced a full scale embargo.[8]

Although the Gulf states designated Britain as one of the "friendly countries" and gave "assurances" that it would continue to receive its normal supplies,[9] those assurances were under threat if the oil companies managing the distribution of Persian Gulf exports decided to ignore the will of OPEC. The Arabs warned that the embargo was not to be violated or there would be repercussions, and no one doubted the seriousness of that threat. Both Mobil Oil and Exxon had lost their positions in Iraq just one day into the October War.[10]

Meanwhile, British coal miners were revving up for a fight with Prime Minister Heath over the size of their pay packets. Coal production was down and Heath knew that a full-blown strike in the midst of an oil shortage would be disastrous for the citizenry. Concerned that Britain might not get its normal shipments for the upcoming winter, he tried to strong-arm the chairmen of British Petroleum and Royal Dutch/Shell Group into keeping the oil flowing into the country, no matter what the threats, but both men refused to cooperate.[11]

On November 5, after deciding that their initial cutbacks were not severe enough, Arab oil ministers announced a further 25 percent reduction in oil shipments to Western Europe.[12] Five days later, London predicted a shortfall of oil supplies by 15 to 25 percent for the upcoming winter. That revelation sent shock waves reverberating throughout the business community. To make matters worse, October's trade deficit of $821 million was the worst in British history, nearly doubling the September figure.[13]

With an energy crisis at hand and militant mine workers reducing coal stocks in order to destabilize the government, Heath had to take drastic measures to restore public confidence in his leadership. On November 13, he put the country into a state of emergency with industry expected to reduce power and fuel consumption by 10 percent. As theater owners turned off neon display signs

---

a 60% from the Middle East, 20% from Libya. *Source*: *New York Times*, October 18, 1973.

in London's West End, the Bank of England cut the money supply, sending the minimum lending rate rocketing from 11.25 percent to a record 13 percent, which in turn provoked a $3.6 billion sell-off on the London Stock Exchange.[14]

A month later, with coal production down 40 percent, Heath extended the state of emergency and imposed new restrictions to cope with the energy crisis. Television broadcasting was banned after 10:30 p.m., phased blackouts were implemented, industry was put on a three-day-work week, and the public was asked to turn lights down and reduce heating to a single room.

In a televised address Heath tried to prepare the nation for hard times. "We shall have to postpone some of the hopes and aims we have set for ourselves for expansion and for our standard of living," he said.[15] All across the country, the winter of 1973-4 would soon be regarded as the worst ever since the end of the Second World War.[16]

A suffering British economy, however, was of no concern to the OPEC Arab oil ministers. Their goal was to force the Israelis to give back the lands captured during the 1967 Six-Day War and make reparations to the Palestinian refugees. To illustrate the seriousness of their demands, they met in Tehran that December and sparked a global energy crisis by raising the barrel price once again, this time more than doubling it to $11.65, effective January 1, 1974.[17]

With OPEC now using oil as a political weapon, and the British economy continuing to slide into debt, there was a growing sense of urgency for government authorities to secure whatever indigenous sources of oil they had as rapidly as possible. In January 1974, this sense of urgency had become so acute, that Aberdeen editorials were offering advice to London in an effort to intensify the view that the country's overall economic prospects were linked to North Sea oil. And it was "the new job of people in the North-east," wrote British economist Roger Nicholson, "to convince the Government of this change."

Nicholson suggested that companies supplying raw materials to the oil industry be given special exemptions from power restrictions, that the infrastructure necessary to accommodate the oil industry be fortified, and that planning procedures be "speeded up and strengthened...so that contractors can press on as quickly as possible with developments."[18]

The quicker the oil came out of the ground, the sooner Britain could inoculate itself from the vagaries of Middle East oil sheiks and thereby ease its balance-of-payments problems. And "hopefully," wrote *The Scotsman*, "...lack of money will not be permitted to stand in the way of exploiting this priceless natural energy resource."

And what was the mood toward environmentalists? There would be some "heartaches," *The Scotsman* declared, "where economic benefits far outweigh environmental losses."[19]

Aberdeen's sense of urgency had not been overlooked by London. Since 1972, government officials were well aware that massive quantities of high grade, low-sulfur oil lay beneath the British sector of the North Sea: the huge deposits in the Brent and Forties Fields were evidence of this. And while Norway would adopt a slower strategy for development, combined with state participation in order to curb the wrenching impact of foreign influence upon its people, culture, and local economies,[20] Britain would take the opposite view.

Even at the outset, authorities placed a high priority on speed which was made clear by the Minister of Power in 1965 when he addressed the House of Commons detailing the guiding factors by which production licenses were to be allocated to the oil companies:

> In the government's view, the over-riding objective is to secure the most thorough and rapid exploration and development of the oil and gas resources of the Continental Shelf in the national interest. I shall therefore consider all applications which can contribute to this end...[21]

Faced with this mandate, British industry suddenly found itself caught off guard and unable to cope with the daunting task of claiming its energy resources below a nasty sea. Due primarily to its "disappointing economic record" and "low level of investment in the manufacturing sector,"[22] Britain neither possessed the technology nor the labor force skilled enough to fulfill the government's objective at the desired pace. Just in terms of exploration alone, most of the equipment necessary to drill for oil was in the hands of the United States. Of the 29 rigs operating in the British sector of the North Sea in 1974, 79 percent were American owned or jointly owned by American/European ventures.[23] Only one, the *Sea Quest,* belonged to the British.

In 1973, the Scottish Development Department projected that over the next decade 20 to 50 production platforms would be required to maintain a steady flow of oil to shore.[24] The task of building and installing those platforms on the Continental Shelf would require large project management teams, an area which, again, US companies dominated because of their previous experience off the coast of California, and in the Gulf of Mexico.

Britain also found herself ill-equipped in certain core industries such as the manufacture of large diameter pipe. For example, oil from the Forties Field would ultimately flow to shore through Japanese steel because there were no domestic mills capable of manufacturing pipe in the required thickness,[25] or in diameters larger than 16 inches.[26] Thus, the miles of 24- to 36-inch pipe needed for Ekofisk, Piper, Ninian, and Frigg would come from abroad. Nor did the British have the barges to lay the pipe, so the Italians, Americans and Dutch were typically hired to do the job.

But beyond the sheer logistics of building an offshore infrastructure, there were the enormous financial costs which the British banking system was unable to cope with. Initial estimates for exploration and development were in the neighborhood of £9.1 billion, not including operating costs.[27] Very soon, this calculation would fall to cost overruns and delays due to weather conditions and inflation. British Petroleum, for example, would borrow £360 million in

1972 to begin the development of the Forties Field. But by 1976, that figure would balloon to £750 million.[28] Brent would cost £2,900 million.[29] These were staggering outlays that would have to be paid back with huge interest payments. And if the oil companies hoped to make a profit, and if North Sea oil was to provide a counterweight to OPEC as the government expected, then the nation's oil resources would have to be brought to shore as quickly as possible and before any potential drop in market price.[30]

Keen on a policy of "rapid exploration and development," British authorities couldn't have found a more willing partner than the American toolpusher whose speedy traditions had been a way of life for decades. Driven by profit, there was no philosophy more appreciated within the oil industry than the kind of behavior which got the job done yesterday. By the time Richard landed on board the *Choctaw* in the fall of 1973, the industry was so imbued by this single most important priority that it set the tempo on the work deck and would ultimately play a role in determining his fate.

While the Minister's statement in 1965 was expressed on behalf of national interest, it also represented the cornerstone of official sanction allowing the oil companies to carry on with their normal and accepted practices. And so they did. Contracts were signed, construction orders placed, and fabrication begun on platforms even before the construction yards were finished.[31] Oil companies pressed for the earliest possible deliveries of the production platforms. And despite the fact that reliable data on wind, waves, and currents to determine their effect upon such structures did not exist, design and construction went ahead anyway.[32]

Author Bella Bathurst writes that for centuries Britain needed the sea, "fed on it, took employment from it, and profited by it."[33] Now she was in a hurry to find her destiny as an oil producing nation. And if 1973 could be regarded as the calm before the storm, 1974 would be the year where the rush for oil unleashed construction projects on an unprecedented scale, the diving population would double,[34] the government would create an entire Department to sponsor the oil industry, Prime Minister Heath would be forced from power, and classmates Tom Belcher, Burt Davis, and Craig Roberts would join Richard in Scotland.

1974 would also produce the greatest number of diving fatalities in North Sea history.

Chapter Three

The Promised Land

Tomorrow, I go into saturation—the work here has become heavy.

—— Richard Walker, 1974

As Prime Minister Heath battled a disintegrating economy, Richard returned home that winter to a nation in the middle of a gasoline crisis and President Nixon embroiled in the Watergate scandal. To earn money, he studied abalone habitats off Anacapa Island on a Sea Grant project funded by the University of California at Santa Barbara. It wasn't big time bell diving and it didn't pay well, but marine science projects were something Richard sought out whenever he was home because he enjoyed the science and it kept him connected to the community he loved.

By April, however, he was back in Aberdeen at 3 King Street in the office Ocean Systems kept above a bank when classmate Burt Davis suddenly walked through the door. Since graduation, Burt had worked in Peru for Ocean Systems and then briefly in the Gulf of Mexico where he learned firsthand why the Gulf had a bad reputation. As Richard listened, Burt described an industry far different from the one developing in the North Sea. The people, the philosophy of diving, the level of teamwork—it was everything he'd learned otherwise at City College. And to make matters worse, there wasn't much work going on.

"You'd maybe get a couple days offshore," Burt recalls, "and then you'd spend a week in the shop sweeping floors and doing shitty work for $1.75 per hour. That was stupid."

And then there was the attitude of the divers. "The Gulf guys," Burt says, "didn't like the California guys."

There was no team spirit, no shared values, no sense of camaraderie. Divers saw themselves as individuals, not as members of a tight unit. They formed self-serving alliances and carved out their turf for personal gain. It was a "what's in it for me" culture where information was held close to one's chest, not freely exchanged. But what annoyed Burt the most was the way the crew handled problems on the job. There was no unified approach to solving them.

"You'd go out on a job and if things got screwed up," Burt recalls, "they'd never sit around and talk about where it screwed up....Once after a certain job that they were having a little meeting of the divers and tenders...I said, 'Well, why don't you just sit down afterwards and talk about where'—and they all looked at me like I was a heretic because they all had their little personal empires, and they sure didn't want to let down any walls from that. It's a different mentality down there."[1]

And so it didn't take long for word to get around that the Gulf was a place to be avoided, that it was "old school, with old divers set in their ways," and that you had to put up with "a lot less safety." One American diver who quit the Gulf for the North Sea gave this reason for leaving:

> It was really just the integrity and the quality of the people. It was like a big Navy game....All these ex-Navy chiefs got together and they were all supervisors out on all these McDermott rigs and on top of that they loved Chino hands. They had these guys from Chino prison who graduated from dive school, and you know, you want to talk about intimidation, man, that's all these guys did was pump weight and grow muscles between their ears.[2]

Fed up with the Gulf game, Burt called up Oceans Systems and said, "I'm going back to Santa Barbara." And the man on the other end of the phone started to laugh. "Why is that Burt?"

"This place sucks!"

More laughter. "That's okay—there's going to be something coming up not before too long," he was told. "We'll, give you a call."

Six weeks later Burt was sent to Brazil. Unable to look down from a third-story window, he'd found his element making 500-foot bounce dives from a drill ship called the *Discoverer II*.[3] When the job ended six months later he decided his next venue would be the North Sea. Even at City College, he'd heard that the industry was booming there and thought that if he went to Scotland his chances of gaining more experience and working with people who understood the meaning of teamwork would be high. And Burt was right.

After that conversation, Richard was sent back to the *Cherokee* in the Ekofisk Field where there was more pipe to bury. But Phillips Petroleum was also keen on clearing the seabed of the thirty sections of pipe that Richard spoke of in his letter to Mary Ellen. Unlike the inspection of the jet sled, this recovery operation was going to take considerable time, certainly weeks if the weather kicked up. Bell bounce techniques therefore were out of the question. Instead, Richard was going into saturation, and "sat" was what divers strived for, where the money was, and where Richard wanted to be. At $500 per day, it was The Promised Land.

Saturation diving was also relatively new. It was only in the late nineteen fifties and early sixties when men like George Bond, Edwin Link, Jacques Cousteau and others began working independently to discover if it was even possible for human beings to live and work underwater for long periods of time. From that desire emerged the underwater experiments of Genesis (1957-1963), Man in the Sea (1962-1963), Conshelf (1962-1963, 1965), and Sealab (1964-1965, 1969) which proved that it was. Instead of bringing divers back to surface pressure after every dive, it made more sense—from the standpoint of decompression time and safety—to simply keep the divers under pressure. Once their bodies became *saturated* with the gas they breathed, no more decompression penalty would be incurred. They could spend days, weeks, or even months at the working depth until the superintendent wanted them out. Then, on a one-time basis, their bodies would give back the inert gas during a single, long and slow decompression to the surface.

And this is what Richard experienced that spring. He would lock out of the bell and swim to the job site as he normally did. But unlike his previous bounce dives, there was no stopwatch ticking in the dive shack, no pell-mell rush to get the job done, no decompression appointment to meet. He simply worked on the bottom and paced himself until the supervisor ordered him back to bell, usually after about four or five hours. Then he traded places with his bell partner who worked for the same amount of time. When the dive was completed, the two men sealed the bell which was brought out of the water and locked onto a multi-chamber complex pressurized close to the working depth. There, Richard remained cocooned until his next dive, in about a day's time.

Meanwhile, as he and his partner were cleaning themselves up, two other divers in sat had already climbed into the bell for the next "bell run." A new set of external and internal bell checks were performed and then the bell was away again with the fresh team continuing the work program. This "bell turnaround" process usually took about thirty to forty minutes and provided round-the-clock diving, stopping only for maintenance or if the weather blew up.

Inside the system, space was scarce. Ocean Systems normally employed small saturation chambers which stood in the vertical position and were easy to transport and set up on barges and rigs. These "verticals" measured only about eight feet in diameter and ten feet tall and accommodations within them were Spartan. Living in one was like camping out in an elevator. Everything was made of cold steel and aluminum, which sucked the heat out of the atmosphere and intensified the whirling noise of the CO_2 scrubbers. Hearing them for the first time made you wonder, *How am I going to live with that?*

Dividing the chamber were four half-moon shaped bunks attached to the bulkhead which forced even a medium-sized diver to sleep in the fetal position. Light penetrated through six-inch portholes which were scattered about and covered over with cardboard and duct tape during periods of rest.

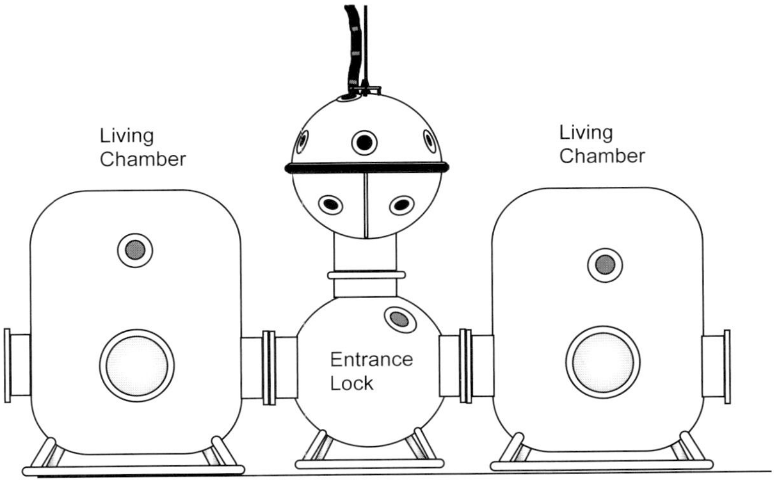

When Richard went into sat there were no regulations specifying the size, shape, or level of comfort for these steel igloos; and the helium used to pressurize them was expensive. To keep costs down, it was routine for diving contractors to economize the job wherever they could, and in the Ekofisk Field, Ocean Systems did just that. That spring, Richard spent his first sat in a deck chamber shaped like a large cigar tube not normally used for saturation. It measured nine feet long, and like the company diving bell, prevented him from standing up. "The work is dangerous," he later wrote to Mary Ellen, "and the living situation slightly better than staying in a cold, rusty sardine can."[4]

But even when new regulations came into effect at the beginning of 1975, the living standard in sat was only mildly improved due in part to the way in which the law had been written. Diving contractors were only required to provide a diving plant that was of "sound construction and suitable material," "in good working order," and adequate for the purpose for which it was intended. The allocation of space was never mentioned. Chambers merely had to enable the diver to lie down "without difficulty," and if he was going to remain inside for more than twelve hours, then by law, he had to be able to stand upright "without difficulty."[5] And throughout much of the 1970s, that was about all the space that was afforded the men who lived in them.

Stationed at the edge of the barge or beside a hole in the rig's deck called the moonpool, sat systems were exposed to the elements, and sometimes connected together in odd configurations which forced other divers to trespass through "your chamber" to get to the entrance lock for the next dive. The space between opposing bunks was merely two feet wide so elbows and knees had to be minded. The incessant noise of the scrubbers, the close quarters—it all made for a scenario of disorder and quarrel. And yet, even under these depressing conditions, tempers rarely flared. On the contrary, divers in sat were generally in good humor because there was nothing like a snootful of cash to clear the sinuses and make one forget the tiny box he was living in. Sat also got the diver off a cold windblown deck and down onto the seabed where he wanted to be. Claustrophobic bells and cramped chambers therefore were just accepted as part of the job description.

Pressurized to three or four hundred feet, of course, was no place for Richard to find out that he didn't like confined spaces. Once the chamber hatches clanged shut and the body saturated, there was nothing topside could do for him if he suddenly wanted out; the laws of physics prevented that. He could scream, holler, pull on the doors, but they were not going to budge. Not because they had locks on them, which they didn't, but because even at ten feet the amount of pressure against a typical two-foot diameter hatch was more than 2,000 pounds, at 560 feet, more than 112,000 pounds!

But Richard had long ago come to terms with the confinement. He knew he was expected to remain inside for the time the superintendent decided, which depended upon the workload and how well he performed. Moreover, Richard wouldn't have been told the number of days or weeks he could expect to be in sat, nor would he have foolishly asked. The whole purpose for going offshore was *to get into sat* where the money was, and it didn't improve his chances if he began asking such silly questions. As a matter of fact, once divers got into sat with the dollar signs dancing in front of their eyes, it was difficult to get them to come out voluntarily. In any case, the customary duration was thirty days, although forty- to fifty-day sats were not unheard of.

One thing Richard noticed about sat diving was that he had to learn to pace himself on these heavy construction projects because sat lockouts were long and exhausting; and if he made the mistake of blowing all of his stamina during the first hour of the dive, then he'd very quickly find his arms dangling down like blocks of stone, numb and completely useless. The image of a diver on the seabed muscling the job with gloved hands was a myth, especially in the North Sea. Water resistance, cold, and the sheer size of some of the objects he had to move made that impossible. So Richard had to use his ingenuity and resourcefulness (and the barge crane) to shift things around on the seabed. And the strange and wonderful thing was, he always found a way to do it because deep-sea diving was really a thinking man's game.

One other thing kept Richard working comfortably for hours in these frigid waters, and that was the use of hot water. The barge had a hot-water machine which pumped superheated saltwater down to the bell and out to Richard through a hose in his umbilical. The end of the hose plugged into a brass fitting mounted on the right hip of his hot-water suit, a kind of heavy duty neoprene rubber coverall that zipped up the front. From the hip connection the hot water flowed to different parts of his body through small distribution tubes embedded in the suit. This was an open-circuit suit so the water eventually flushed itself into open water through the hands, feet and neck area.

And when this water came down to Richard it was hot, so hot that he wore "woolly bears" underneath as insulation to protect himself from being scalded. Mixed with the surrounding ice water, the marinade kept Richard warm. But there were problems with these early hot-water machines, and divers were notorious for tinkering with them. Some caught fire or flamed out during the bell run. And when Richard lost his supply he usually knew about it before his supervisor because the change was almost immediate, and without hot water in the North Sea, he was cold, and he was cold *right now*, not tomorrow or the next day. Sometimes the situation was the reverse: the water flooding his suit would be boiling hot. In that event, he would simply turn a bypass valve on his hip which dumped the blistering flow into the surrounding sea, then order topside to come down on the temperature one or two degrees. But when the temperature was just right, it was like taking a warm bath. Without question, the use of hot water was perhaps the greatest technological advancement ever made in deep-sea diving, and without it saturation diving in the North Sea would have been impossible.

For Richard, the first couple of weeks in saturation tended to drag by. The physical adjustment was quick—the walls of his universe were right there, just three feet away—but the mental adjustment took time. Not yet settled and hermetically sealed, his mind naturally bounced between the world of freedom and the world of voluntary internment which forced him to depend upon topside for everything—towels, sheets, food, soap, books, mail, newspapers, everything. If he wanted the light turned on he had to ask topside to do it. If he wanted the toilet flushed, topside was involved. If the temperature in the chamber was too cold, Richard would shout into the speaker: "Topside."

"Yeah?"

"Come up on the temperature two degrees."

"Roger, up two degrees."

Even the act of eating involved the outside crew. The daily menu was announced over the horn speaker. Richard placed his order, and then someone on the outside went to the galley, served up the food, brought it back and loaded it into a small, one-foot diameter "medical lock" mounted on the side of the chamber which had a door on either end.

Courtesy of Marcus Taylor Ltd.

Typical Medical Lock.

Once the food was in the lock, the attendant closed the outside door and clamped it shut. Then he took a metal object (usually a wrench) and rapped twice on the side of the lock. This "take-the-lock" signal told the divers inside that the outside door was secure. One of the divers on the inside would then *s-l-o-w-l-y* open a small valve to pressurize the lock. Open it too fast and dinner would end up looking like it had been through a garbage disposal. Closed containers such as bottles of catsup and steak sauce had to be opened before being compressed, otherwise the divers would never be able to unscrew the tops. Sometimes the bottles simply imploded during pressurization, leaving a huge mess. Compressible items such as Styrofoam cups were useless because, by the time they reached storage depth, they were the size of a shot glass which made for amusing moments when someone actually made tea in one and blew it down to you.

Gradually, life in sat became routine for Richard. He would leave on his bell runs, come back, and fall exhausted into a deep dreamless sleep. Fourteen hours later he'd wake up still feeling bone-tired, muscles aching, and glance at his watch. Then he'd pull the cardboard back from a porthole to see if it was night or day. Outside, the other members of the crew were running around freezing their buns off for one-fifth the pay, while inside Richard was warm and dry, earning serious money, and giving the guys on the outside the thumbs up and that wish-you-were-here look.

And then one day Richard woke up and discovered that the walls of his chamber had stretched like taffy and that his brain had somehow disconnected itself from the outside world. Fresh air, freedom, time—their glorious essence had subconsciously slipped away. That annoying scrubber hum was gone, too. His voice on helium was the way it had always been. Food was now a form of entertainment and work on the seabed even more spectacular than ever. From the bell stage, he would glide down to the seabed through the watery darkness like a night owl descending on its prey, then trudge across the barren landscape, cut up his fill of pipe, and send it to the surface by crane. Hours later he was back inside his steel cocoon. This new world was strange and wonderful and oddly turned inside out. He was now operating from inside a transitional bubble, moving from barge to sea floor, then back again, day after day. It was as though he were being transported to work in some dust-filled mine on Pluto. The other world, the one that he had come from, didn't exist anymore.

There was no outside.

Day after day, this routine went on for Richard. Dive, eat, sleep. Dive, eat, sleep. Dive, eat, sleep. His hands, bathed in saltwater for hours, became rough and callous. He slathered antiseptic ointment on them like hand cream to keep the infections down. When the weather turned foul, operations were suspended. Storms lasted anywhere from several days to several weeks, so Richard slept and read and told funny stories, and spent a goodly amount of time thinking about his future while he waited on the weather. Sat was a state of mind, and it was where Richard discovered his inner strengths and weaknesses.

And on it went—dive, eat, sleep—dive, eat, sleep. Each day dissolved effortlessly into the next. Words like Monday and Tuesday became irrelevant. Under the yellow glare of incandescent light, Richard devoured newspapers and magazines from cover to cover; even the most innocuous advertisements seemed interesting. In his diary he drew cartoons and little sketches and wrote to everyone—friends, family, even "old flames," drinking in the words they mailed back as so much luscious poetry, causing him to laugh and cry, always breaking the lonely spell he was in and reminding him that the other life, the one he dearly loved, "in the land of the lotus," was sadly passing him by.

And then one day the speaker blared out an order for Richard and his partner to get their gear into the chamber used for decompression; they were coming out. Richard's mood suddenly shifted. His trips to Pluto were over. Unlike bounce decompression, sat decompression was a snail ride in diminishing stages: six-feet per hour, then five, then four, and finally the excruciating pace of three feet per hour during the last 50 feet to the surface. Richard's voice along the way got deeper, more resonant. He didn't have to try as hard to speak clearly to his partner, and as the days dragged on, he began to feel weaker and weaker as though vampires had been sucking the blood out of his veins.

Finally, just when he thought it would never end, he heard a *thunk*, like a can of coffee getting punctured. The chamber door popped open. His yellow canvas duffel bag was already packed and waiting on his bunk. He crammed one leg through the hatch, and then his head. A blast of frigid air smacked his face and suddenly his mind warped forward into the cold outside world where everything seemed strangely quiet—the whirling scrubbers weren't with him anymore. He was free.

It had been a long sat. He'd lost weight and his face was white as a ghost. He made a lot of money, but he'd earned it the hard way. Gaunt-faced, Richard walked to the edge of the barge and stood under an alabaster sky with his eyes closed and sucked the cold wind into his lungs, certain that the North Sea was where God made fresh air.

Statistics show that if someone is involved in an offshore construction, drilling, crane, or boating accident, the likelihood is he'll live to tell the tale. But in diving, an accident is almost always fatal.[6] Indeed, one North Sea researcher would later write that deep-sea diving was "by far the most dangerous activity on the Continental Shelf and possibly the most dangerous occupation in Britain." He calculated that the chances of dying in the North Sea as a diver between 1974 and 1976 were "twenty-six times its equivalent for quarrying, thirty-eight times the average for mining, and fifty times the comparable figure for the construction industry."[7]

For Richard, such statistics would simply have undergirded his feelings about the constant danger he witnessed offshore. He knew that the dive crew was as much a part of the safety equation as the equipment they used and the procedures they followed. And he also knew that you could take a bad diving system and crew it with expert, conscientious divers, and you'd never see an accident. In contrast, crew a state-of-the-art system with "wankers" and you wind up with disaster. Richard knew that the key to safety was people, and in an industry where throwing the wrong valve could cost someone his life, the vast majority of divers were extremely serious and no-nonsense about their work. Like soldiers in a foxhole, they watched each other's backs because their first priority was—nobody dies.

Nevertheless, serious screw-ups did occur. Misunderstandings have caused chamber doors to blow off their seals, forgetfulness to douse divers with dangerous levels of oxygen, and "bad gas" to be sent down the diver's hose. This was an aspect of the diving industry that Richard was fully aware of, not just because he understood the risks that were involved, but because, throughout 1974, divers were dying at an alarming rate. Ten fatalities within twelve months, with three deaths underscoring the inherent dangers of deep-sea diving, while two cases demonstrated that the death sentence could come from patently implausible sources.

Richard undoubtedly learned the details of the first accident that year because it claimed the lives of two fellow Ocean Systems divers, Pier Skipness and Robert Smyth, whose manner of death was so horrific that it would send nightmarish images rippling through the industry for years to come and ultimately sway how some superintendents and supervisors dealt with certain safety aspects of bell diving.

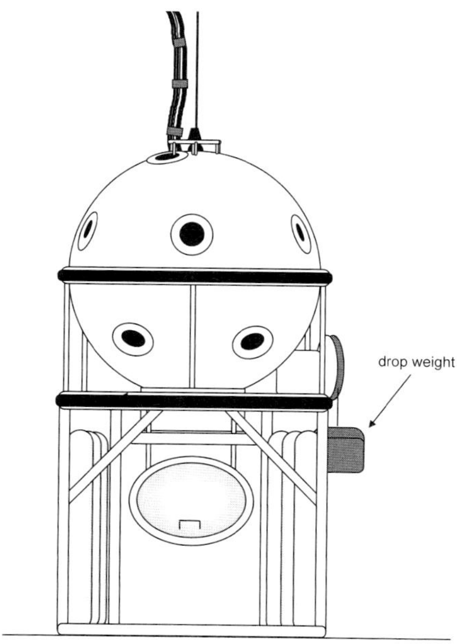

It was January 17, 1974, when the crew on board the *Drill Master* lowered Skipness and Smyth to 320 feet. The bell they were in, like all bells of that era, was equipped with a last-resort emergency ascent system: in this instance, a 1,000 pound drop weight mounted on the exterior of the otherwise buoyant bell to keep it submerged. In the event that Skipness and Smyth became stranded on the seabed, they could seal themselves inside, activate the release mechanism to drop the weight, and the bell would rise to the surface like a cork.

But something went wrong that day, and while one of the men was outside working, the drop weight accidentally released.[8] Within a minute, the bell shot to the surface, dragging the lock-out diver through the water. And while drowning was listed as the official cause of death, Richard knew that the effects of rapid decompression were equally as deadly.

At the end of August that year, one of the most insidious threats to the security of a diver killed Peter Kelly in the Norwegian sector. The supervisor neglected to analyze Kelly's breathing gas and instead of sending a mixture of helium-oxygen to the bell, he put pure helium on-line. Pure helium is instant death, and delivering it down a diver's hose almost always results in a fatality.[a]

Then there was the constant danger associated with keeping track of one's umbilical in the dark. On a lock-out dive at 265 feet in the Celtic Sea, British diver David Keane drowned while working on a wellhead. No one was absolutely certain how the accident happened—government records are so poor as to make the accident almost non-existent—but when Keane reported a "shortage of gas" to his supervisor,[9] the bellman was deployed only to find Keane's umbilical severed.

"Bad gas," drop weight failure, severed umbilicals— these were all perils well within the imagination of the common diver. But 1974 also produced two cases where death should not have been lurking.

It was the first day of March and Frenchman Mark Barthelemy and his Scandinavian partner were bell diving in the southern Irish Sea when something went wrong. Barthelemy locked out of the bell, but never made it back inside. According to Commander "Jackie" Warner, the government safety inspector who investigated the accident, this is what happened:

> The men were working at 300 ft., on the first operational dive, and there is little doubt they were out to prove themselves. The diver had left the bell and was at work, working very hard and observed by the diving supervisor on a television monitor in the dive control room topside.
>
> His respiration rate was high and the supervisor noted that he was working too hard. The diver was instructed to slow down his pace but, perhaps because of over-enthusiasm, he ignored the request and worked to the point where he encountered respiratory fatigue.
>
> He became unhappy and returned to the bell, the correct procedure in the event, but in his haste he failed to take the clear route back and passed on the wrong side of one of the bell guide weight wires.
>
> The bellman, in his efforts to pull his colleague back to the interior of the bell, and safety, was in fact pulling the diver underwater, because the umbilical was wrapped around the bell weight wire.

---

a In 1983, Sub Sea Offshore diver Mike Sherman survived such an incident while diving from the *MV Tharos*. Sherman was out of the bell when he suddenly became unconscious in the water. Only through quick action was he saved by his bell partner, Jim Harrison, who pulled him back inside and revived him.

> Panic ensued, both diver and bellman reverted to their native tongues, and the squabble ended with the bellman cutting the diver's umbilical, shutting the bottom door and telling the dive control to recover the bell to the surface.
>
> When the bell was brought out of the water the body of the drowned diver was found draped over the bell weights.[10]

Four months later, death came from another unexpected direction. On June 30, divers Ian Beaton and John Dimmer were being decompressed on board the *Sedco 135F* after making a short bounce dive to 490 feet, when, at about the 90-foot level, Dimmer suddenly complained of shortness of breath. He told his diving supervisor, Mike Spencer, that his chest felt like it was "full of air."[11] Spencer followed procedure and recompressed Dimmer back down to 100 feet where his pain was completely relieved. But when his symptoms continued to reoccur during a series of unsuccessful attempts to get him above 85 feet, a doctor with an alleged expertise in diving medicine was brought out to the rig.[12] Strangely he failed to diagnose and prescribe the proper treatment even after diving expert Al Krasberg—a former Westinghouse engineer with a Master's degree in Astrophysics from Harvard University and more than twenty years in the diving industry, including work on the American Sealab experiments of the 1960s[13]—correctly indicated to the doctor that Dimmer was suffering from a burst lung (pneumothorax).

Believing instead that Dimmer had a case of pneumonia, the doctor prescribed antibiotics, a slower decompression schedule, and an increase in the oxygen level in the chamber atmosphere. At this point Krasberg wondered what kind of diving doctor this was. He seemed to be unable to draw a correlation between the incidence of pain and the reduction in pressure.[14] Pneumonia was not susceptible to pressure differentials where as pneumothorax was. He did not take Dimmer's temperature, which might have produced some evidence to support his diagnosis of pneumonia. And there was that curious order to raise the oxygen level. As Krasberg would later explain at an official inquiry, "That seemed contrary to everything I knew. He said oxygen was sometimes used for treating pneumonia in hospitals, but on the other hand in diving oxygen is known to cause pneumonia."[15]

As for diving Supervisor Mike Spencer, he was "shocked" by the doctor's diagnosis and argued for a different course of action.[16] Dimmer was currently at 85 feet and experiencing pain. That pain had already proved to be relieved through recompression and Spencer wondered: why not recompress Dimmer, give him the relief he needed, hold him in storage, treat him with medications, then bring him out after the danger had passed? That might take days, or even weeks, but Spencer was unconcerned about the time factor, and he advocated that the patient should be cured within the chamber rather than subject him to more decompression because, at this point, one thing was undeniably clear, it was the reduction of pressure that was causing Dimmer harm.[17]

But the doctor objected to this suggestion and said that if anything was detrimental to the patient's condition it was to leave him in sat. Therefore it was imperative that they "get him out of the chamber as soon as possible and get him into hospital to treat the pneumonia."[18]

In the meantime, the doctor decided not to remain with his patient. Immediately after diagnosis he arranged for a helicopter to return him to Aberdeen, which was hours away. But what if an emergency were to arise? Both Spencer and Krasberg thought the doctor's departure unwise and tried to persuade him to stay,[19] but he assured them that this was a routine case, that he had one very similar to it recently, and that he could be consulted by phone or called back from shore "if there were any complications,"[20] which there were.

As the decompression continued, a pronounced swelling developed on Dimmer's chest.[21] The doctor made the long journey back to the rig, entered the chamber, and found Dimmer "showing some disorientation and a degree of distress and restlessness."[22] Then, perhaps out of a sense of frustration, he administered one of the most pathetic examples of bedside manner ever bestowed upon an ailing patient. According to Spencer, who overheard the conversation, the doctor put his stethoscope to Dimmer's chest and gave his patient a lecture, telling the injured man "to pull himself together and just to generally play the white man, sort of thing."[23]

Still, with mounting evidence pointing to a classic case of pneumothorax, the doctor continued to stubbornly adhere to his original diagnosis of pneumonia. He prescribed pain killers (Pethidine),[24] tranquilizers (Valium),[25] and sleeping pills (Mogadon),[26] an action that alarmed both Krasberg and Spencer, given that current literature on hyperbaric medicine advised not to give medications that could possibly mask a diver's symptoms and prevent him from telling anyone what he was feeling. Only later would it be revealed in court that the doctor had never attended any course on decompression or diving accidents.[27]

Finally, on the fifth day of decompression, the expanding gas in Dimmer's chest collapsed his lungs. At 0920 he was pronounced dead just 30 feet from the surface from complications due to a pneumothorax. He was survived by a young wife and child. A jury later found that his death was "caused by incorrect diagnosis and treatment by the doctor."[28]

Ten fatalities in twelve months. Ten reasons for a diver to wonder what he's gotten himself into and unless he's been around death a lot, the first report is the one he'll remember for the rest of his life.

Out there on a crane barge the wind is whipping his face, it's freezing cold, and he's been working eleven hours. By chance he passes the radio room and overhears two men deep in conversation. They are grim-faced and one of them mentions that there's been a bell accident up north. Two divers are dead. All of a sudden, the death notice becomes this diver's introduction to the industry and everything around him freezes for a split second. Every nerve ending and sensation in his body has been deadened by the paralytic realization that it's no longer a case of getting injured out here, but that death is an integral part of what he does.

What have I gotten myself into here? he wonders. But before he can ponder this question, his mind suddenly shifts to thoughts of his own mortality and whether this could happen to him, a prospect he decides not to dwell on—not perhaps because he's afraid of dying at sea, but because he's more afraid of killing someone else and having to live with the guilt.

These thoughts lie in the back of this diver's mind. But he's new to the industry and he hasn't yet realized that it's the strangeness and wonder of this new life that obscures the presence of danger. In time, the excitement and fascination will fade. He'll go to sea and do his job and most of the time nothing bad will happen. But the more time he spends offshore the more danger he sees. As the years pass, his life will oscillate between the exhilaration of working in deep dark waters and getting past a few moments of sheer terror. He'll become accustomed to facing down the fear and the danger no matter how great. A sense of invincibility will grow and he'll come to believe that no matter what happens, he'll be able to handle the challenge. The adrenaline, the camaraderie, the intoxicating sums of money, and this sense of stored invincibility will course through his veins like a heroin addiction. He'll love the work, but if he has any desire to lead a normal life, he'll always be in conflict with the demands it imposes on his life.

At home, he'll re-enter the atmosphere of everyday life by blocking out the fear and the danger. He does not discuss his brushes with death, but ends up keeping them to himself. As the weeks slide by, the drug in his veins slowly wears off and the urge comes back. His other life begins to strobe through his mind: ships pitching on white-capped seas, helicopter pilots struggling to set the chopper down, water flooding his mask, the dark barren sea floor—the challenge heralds him back. As his departure date nears, a cold chill runs down the back of his neck. Suddenly a premonition grips his mind and turns his imagination against him. He knows that out there everything is not under his control and that there are plenty of ways to die a horrible death, that if the diving doesn't kill him, the helicopter flights will. Still, he goes back anyway because he's willing to trade his fears and his nightmares for the money and the excitement of working in the dark. He goes back because he knows that only when he's actually on the job and inside the eye of the storm, he feels calm and focused. Only then, when he sees the sharp outline of danger, does the fear and the uncertainty go away.

All of this lies ahead for this new diver. It's a distant future destined to be marked by the survival of close calls, shaking hands, and the knowledge that every time he enters the bell—like the circus performer sticking his head into the lion's mouth—he expects to come out alive, but is never sure that he will. Only then will he ask himself: *Do I really want to be doing this?* But for now, at the beginning of this unimaginable journey, he carries on and vows not to make any mistakes because he's young and he's strong and he's confident that nothing bad will happen to him.

As 1974 was about to close, one further event marked the year. With a new Labour government in power, the oil companies were growing increasingly apprehensive about Prime Minister Harold Wilson's plan to ensure that North Sea oil be brought under "full public ownership." The current licensing arrangements had been criticized as a giveaway. As one former Energy Secretary would later write, the Heath administration had given "a large number of licences to oil companies to develop the North Sea without any provision that any oil be directed to meet United Kingdom needs. There was no proper statutory framework and no petroleum revenue tax."[29]

The price of oil had quadrupled over the past 18 months. And if it continued to remain high, administration authorities realized that the oil companies were going to make "enormous and uncovenanted profits" and were likely to transmit "half or more" of their post-tax profits to overseas coffers, a proposition that the Labour government viewed as "intolerable."[30]

To remedy the situation Wilson intended to restructure the Continental Shelf policy by doing two things: claim a bigger piece of the profit pie through taxation and exert more control over the country's offshore assets. That prompted then Secretary of Energy Eric Varley to declare that he *might* institute depletion controls and curb production, summoning frightful visions of state participation which Norway was pursuing. The oil companies cried foul. They had borrowed heavily and needed the "maximum degree of 'certainty' for the large investment decisions involved."[31] Depletion controls would upset that certainty. This was not the economic vision that Heath had painted, and now that the game was on, Wilson was changing the rules.

The oil companies needn't have worried. On December 6, Varley announced that there would be "no delay on discoveries made up to the end of 1975," that "production from such discoveries would not be cut-back until 1982," that "production from fields discovered after 1975 could recoup 150% of investment before a cut would be imposed," and that if cuts were to be imposed, they would be generally limited to "20% at most."

These statements, which became known as the "Varley Assurances," would effectively keep the government's rapid exploitation policy on course for at least the next seven years.[32]

Chapter Four

Forties Gold

We had a fire below decks yesterday, with life jackets, fire hoses, panic and heroism on the high seas—all is well now—quite exciting.

—— Richard Walker, 1975

At the start of 1975 OPEC showed no sign of lifting its steel-tipped boot off the neck of the western world. In January it boosted the barrel price another ten percent, increasing the financial strain on the British economy. The country's oil import bill, which was £800 million before the Yom Kippur War, was now approaching £3.5 billion.[1] But ironically OPEC's price increases also greatly enhanced the financial prospects for developing North Sea oil. More than ever, all eyes focused on the resources of the Continental Shelf as the key to a brighter future. And that future now had a more optimistic target date. Where best estimates for attaining self-sufficiency had been the mid-1980s,[2] the new Labour government had recently advanced the date to 1980.[3]

In the middle of this oil rush, Richard's future couldn't have been more secure. He was beginning his third season with Ocean Systems at a time when the construction boom was escalating at a feverish pitch. Meanwhile the total diving population had risen by only 20 percent over the previous year,[4] and skilled divers were in short supply. It was rumored that contractors were pirating men from other companies as they disembarked from the helicopters at Dyce Airport. And Richard was just the sort of diver they desperately needed. He had the training and his experience had grown steadily since August 1973. From surface diving to bell bounce to sat, he had inspected jet sleds, connected subsea pipelines together, repaired tanker loading facilities, supervised dives, and for a brief time worked on a Condeep project in Stavanger, Norway.

And while work in Norway had always been at a somewhat tempered pace, Richard was about to find it positively subdued in contrast to the intense rampage taking place in the UK sector, particularly in the Forties Field where British Petroleum was currently mounting a £750 million campaign to develop one of Britain's largest and most important finds. There, the workload was so massive it became the magnet that drew not only Richard and Burt to the field, but also fellow classmates Tom Belcher and Craig Roberts.

If anyone loved diving more than Richard it was Craig. At age nine he made his first dive in the tropical waters of the Caribbean. By age 16 he was teaching scuba out of a dive shop in the Virgin Islands.[5] Diving had seeped into the marrow of his bones and after graduating high school he entered the University of Miami with dreams of becoming a marine biologist. But in 1969 that vision was suddenly derailed when he took a job on General Electric's Tektite project in Lameshur Bay off St. John Island. The goal of the program was to conduct biomedical studies on a group of scientists undergoing long saturations inside an undersea habitat anchored to the bottom of the bay in 40 feet of water. NASA wanted the information to prepare its astronauts for future space flights. Once Craig got a taste of working with the equipment, and realized that the industry was still fresh and untamed, all thoughts of studying marine biology were over. Hearing about the Marine Technology program in California, he quickly shifted his energies to Santa Barbara City College.

After graduation he accepted the first job he was offered in the Gulf of Mexico with a rising young company called Sub Sea International. For the next eight months he tended diving and put sat systems together. Then, opportunity knocked that radically altered Craig's future. Sub Sea was awarded the "Forties" contract by British Petroleum in the North Sea. In March 1974, he arrived in Aberdeen with all the enthusiasm and vigor of a man about to land on the moon. He reported to Sub Sea's office and was subsequently sent out to the Forties Field where he began diving off a derrick barge called the *DB Hercules*. Four months later, on this same remote patch of water, classmate Tom Belcher unexpectedly showed up.

Short and stocky, Tom had the kind of cautious, analytical mind that would eventually put him in the superintendent's seat. The son of a Stockton, California lawyer, he entered San Jose State College and, in between political science courses, took a class in scuba diving with a couple of pals. He found it "interesting," so much so, that instead of doing what political science majors do after graduation, he immediately signed up for the Marine Technology program in Santa Barbara, where he roomed with Craig and became close friends with Richard.

When Richard left for the frontiers of the North Sea, Tom departed for the Gulf of Mexico where he spent nearly a year working for J Ray McDermott, then finally came to the conclusion that the Gulf was a "dead-end place" with no prospects for saturation diving. Meanwhile, the grapevine was alive with stories about the North Sea with its booming industry, high technology, and fast track to The Promised Land. That July, Tom "escaped" to Aberdeen, walked into Sub Sea's Office on Greenwell Road and was sent off to the *Hercules* where he bumped into Craig with a nodding smile on his face.

If the Forties Field was anything to British Petroleum it was the pearl of their North Sea operations and so named because it lay just off the Long Forties Bank 110 miles east of Aberdeen. Even by world standards its estimated recoverable reserves of 1.8 billion barrels of crude was considered to be a significant find. The oval-shaped field covered about 35 square miles, and because of its size, and in order to facilitate a more rapid extraction program, BP decided to install four fixed platforms (designated Alpha, Bravo, Charlie, and Delta) to provide a permanent base for drilling additional wells (about two dozen each) into the oil-bearing rock over eight thousand feet below the sea. When the project was completed, 400,000 barrels of crude would be siphoned off each day and exported through a 32-inch pipeline to landfall at Cruden Bay north of Aberdeen.[6] To say that Britain had high hopes for Forties oil would be an understatement. At peak production, this single field was expected to provide the nation with about one-fifth of its total annual requirements by 1978.[7]

In June, the base for Alpha was towed out to sea on its side from Teeside, England. To the uninitiated, the latticed steel tower, called a "jacket," looked like something King Kong had fashioned out of a gigantic Tinkertoy set. It was enormous, and by a process of controlled flooding it was upended and rotated 90 degrees in mid water to land on the seabed in 370 feet of water.[8] This was an historic event because Alpha was the first jacket to be set in the deeper waters of the northern basin. Two months later, Forties Charlie was towed out from Nigg Bay in northern Scotland and upended. Bravo and Delta were not due out until the following year. Resting on the Continental Shelf, only the top 75 feet of Alpha and Charlie protruded from the surface of the water. They were the Eiffel towers of the North Sea and in due time their stubby necks would be adorned with prefabricated modules to support the effort of bringing to shore that which Britain so desperately needed.

Once Alpha and Charlie were upright they had to be prevented from shifting on the seabed. The jacket for Forties Charlie alone weighed 30,000 tons; the modules would add another 17,000 tons.[9] To keep these steel monstrosities in place, BP hired the Houston, Texas firm of Brown and Root to effectively "nail" the structures to the seabed with a series of steel piles driven up to 250 feet into the earth's crust. To do that, Brown and Root brought in the *DB Hercules* with its giant 1600-ton Clyde crane mounted on the aft end. In the shadow of the crane, Sub Sea stationed its sat system where the crew began making dives up and down the jacket, which at night presented a dark looming figure, and during the day, a maze of confusing horizontal, vertical, and diagonal highways upon which a diver could easily get lost.

All during the summer of 1974, Craig and Tom performed a multitude of tasks. There were dives to establish down-lines, dives to open and close flood valves, dives to attach lifting slings, dives to help stab the piles through their guides, dives to check for damage, dives to video record and photograph, and dives to clear the seabed of debris to protect the diver from entanglement. All of these dives were dives of exhilaration and great joy, but they could not compare with the sheer pleasure of burning off all of the paraphernalia which enabled the float-out and piling of the jackets—including huge 50-foot diameter, 700-ton floatation spheres and dozens of pile guides. These had to be removed because they were no longer of any value to the jacket, but were now a liability to the structure for reasons of dynamic loading and corrosion protection. Burning was fundamental to a diver. It was straightforward, and the rewards were immediate: the more you cut, the more bragging rights you earned.

Thousands of dives and years of work lay ahead, but Sub Sea was not the only diving contractor BP had hired. Comex Diving was in the field on board the *Thor*, a 551-foot long derrick ship owned by the Dutch firm of Heerema.[10] As the summer schedule progressed, Belcher says there were periods when the *Thor* was on one side of the jacket and the *Hercules* on the other, when a competition erupted between the two companies to see who could get the most work done. The contest had developed innocently, but away from all the good-natured joke-making, BP was watching and evaluating the two teams with a ruthless eye. Within a year it would exercise its supreme power and impose severe consequences on the loser.

Like many crews, the *Hercules* divers were a real mixed bag of screwballs, social dorks and misfits. Some were brooding loners, a few were flamboyantly gregarious, but all were gung-ho and mission oriented. They could be eccentric and characteristically stubborn, but they were ardently professional and highly disciplined. They were not group thinkers, and yet they bowed to the absolute necessity of teamwork. In terms of their integration into the general offshore population, they considered themselves a force unto themselves and operated according to their own safety culture and moral code. If they needed something from the vessel—a tool or a piece of steel—they frequently stole it. They were flawed men, but they were hard-boiled individualists to the core. Their independent, non-conformist streak was reflected in the way they worked, and even in the way they reacted to a crisis.

On July 6, 1988, the largest offshore catastrophe in history occurred when a platform operated by Occidental Petroleum blew up in the Piper Field. Of the 226 men on board, only 61 survived. Occidental had written procedures on what to do during emergencies. Lifeboat stations and mustering points were predetermined. But no one could have imagined the chaos and confusion of a calamity that would kill so many and send Piper Alpha to the bottom of the sea. Later, authors Charles Woolfson, John Foster and Matthias Beck would write an account of the disaster and cite a study conducted by Chris Wright who analyzed the survival patterns of the workers of Piper Alpha. Referring to Wright's study, the authors summarized his findings:

> Some groups, he found, had perished almost entirely. Others, like the divers, largely survived. Wright finds that these different survival rates were closely correlated with two characteristics. One was the degree to which the group itself was, as a working collective, closely integrated. The other was the degree to which it was not incorporated into a wider system of control. The divers survived because they had previously worked out their own informal routines as to how to respond to a crisis and disregarded official procedures. Others perished because they followed the official emergency instructions, made their way to the galley accommodation and then obediently waited for instructions which never came.[11]

By October 1974, the piling of Forties Charlie was running behind schedule which put British Petroleum in a bind because it wanted Forties oil flowing to shore within a year to begin recovering its massive investment outlays. BP's choices at this point were simple: either close down the job for the winter and suffer the loss of revenues incurred by delaying the opening of the export line, or extend its development costs by keeping the ships and barges and dive crews on location over the winter, which is what it decided to do. The gamble would be expensive, costing an extra £20 million, but the company was betting that it would get the 50 days of work it required to make the bet pay off.[12]

Christmas and New Years passed, and by the end of January 1975, BP had recorded only 2.5 working days since November. For divers like Tom Belcher, the months spent riding out the winter storms were bitterly cold, boring, and extremely dangerous. The *Hercules* was a big barge, but it was not built for the North Sea where wave heights can reach 40 feet. The freeboard, or distance from the waterline to the main deck, was only about 15 feet. Storms frequently came out of nowhere and within a few hours the sky would turn black, 100-knot winds would begin whistling, and white streaks of foam would blow across mountainous waves that completely swamped the main deck. The *Hercules* was not self-propelled and had to be either towed to shore or across the field to keep it from foundering. After a number of hours, the tugs would make a U-turn, then go another 50 miles or so to try and hang on until the weather cleared up so the crew could get back to work.

In these storms the *Hercules* would climb horrendous swells then plunge violently down into the wave trough, creaking and moaning, only to repeat the dreadful cycle all over again. During one memorable storm, Belcher recalls that conditions became so bad that water started coming through the deck plates setting off alarm bells in the galley. At this point, some of the more demented divers seized the opportunity to begin taunting the Spanish welders.

"We were awash," Tom says...

> I mean it was just water coming over the deck here in this storm, and the Spaniards—the welders—were just flipping out. I mean they were all doing Hail Mary's and the divers were all walking around in their wet suits, going to dinner in their wet suits and there were, like, fights beginning to erupt...you know, one of the divers would say, 'Well, look it, I'll make you a deal. I'll sell you my wet suit and you can live.'
>
> This is the joke, the humor, the Americans had. The barge captain hated us for inciting the Spaniards.[13]

In February, weather conditions improved and BP logged 19 working days. By the end of March, the company managed to get the 50 days it wanted but at a whopping cost of £400,000 for each of those days.[14]

With weather conditions stabilizing, work in the field resumed in full swing. The jackets for Bravo and Delta were brought out and upended,[15] and Comex was summarily kicked off the *Thor* for failing to keep up with Sub Sea. Sub Sea then assigned a second crew and sat system to the vessel. It now had close to forty divers and support personnel working in the field.

Less than a few miles away, Richard and Burt were performing some of the most enjoyable and interesting work of their careers from an Italian pipelay barge called the *Castoro II*. Years earlier, BP considered two main proposals for getting Forties oil to market: either build a tanker loading facility in the field and ship the crude to market, or build a more costly £100 million pipeline to export the oil to shore. BP chose the latter course reasoning that typical bad North Sea winter weather would disrupt tanker loading operations and cause production delays. Of the four platforms, Charlie was the hub or "master platform" that accepted the petroleum contributions of Alpha, Bravo, and Delta through 20-inch flow-lines which were currently lying on the seabed between the platforms. Each platform in turn had a corresponding 20-inch "riser" which ran from the base of the platform up to the production facilities above. But the flow-lines were not yet connected to the risers. That summer, BP directed Richard and the other divers to begin connecting them. This involved bolting up sections of pipe called spoolpieces between the risers and the flowlines. Each end of a spoolpiece had a thick heavy collar, or flange, with bolt holes that matched the bolt holes in the corresponding flange of the riser and flow-line.

For Richard, the object was to avoid any unseen hazards in the dark and land the spoolpiece on the seabed between the riser and the flowline without getting hurt or crushed to death. With each pair of opposing flanges, there was one that could be rotated (swivel flange). With a bit of creative rigging, Richard turned it until the matching holes were roughly aligned. Then he used a special set of hydraulic jacks, drift pins, and high tolerance pins to fine-tune the alignment so everything could be bolted, then pressure tested for leaks. This sounded easy enough, but these were big pipelines, and the work was exhausting. The bolts were heavy. The jacks were heavy. Everything was heavy, which made it necessary for Richard to fill a five-gallon plastic bucket with gas and use it as a buoyancy compensator to move some of these objects around underwater.

When everything was running smoothly, his mood couldn't have been more euphoric. Flanging up a pipeline was arguably the best work there was. But when problems arose, his lockout could easily turn into six frustrating hours of basting in a hot water suit with a nose full of snot, and the only thought on his mind: *What the hell an I doing down here?* Sometimes topside sent down the wrong gear or rigged it wrong. Sometimes the equipment didn't work. Sometimes the hot water machine quit sending him darting back to bell. Sometimes the water was so hot, that by the time he diverted his attention away from his work, he had a huge blister on his leg. Sometimes the handheld torch he carried failed, and if the bell couldn't be positioned over the job site, he had to work in the dark, feeling his way around like a blind man. And, because he was breathing mixed gas, and talking through a device that marginally unscrambled what he said, he constantly had to spend the extra effort to articulate his communications as though his supervisor were a deaf and demented old man.

And there were moments when Richard must have been saying to himself, *Okay, keep it together,* because after hours of hard physical work, with no time for a drink of water, and with bits of saltwater shuffling in through his leaky mask, he'd become so dehydrated that it seemed absurd to be surrounded by an entire ocean only to have the same dreams as a dying man in a desert. There were also frequent moments when a drop of sweat would slowly crawl down his forehead on a torture trail to his upper lip. Without reprieve, Richard had no way to wipe this "liquid fly" away or shake it off; all he could do was stay focused and imagine that it wasn't there. If he didn't go temporarily insane, it was only because he forgot to.

Staying focused twelve hours a day, seven days a week, for 84 hours a week, week after week, under very intense, very serious conditions was mentally exhausting. And just when Richard was starting to "like it out there," he would find himself back in Aberdeen walking down Union Street hopped up on adrenaline and grinning bug-eyed for no apparent reason. The 360 degree revolving head he'd grown offshore to keep himself alive was no longer needed. Now there were no crane loads trying to crush the life out of him, no one turning his gas off, no monstrous wolf eels laughing at him, no swinging headache balls whizzing past his head, no 10-ton lift bags trying to drag him to the surface, and no loops of wire threatening to drag him overboard. The constant stress and the need to be one hundred percent focused was suddenly gone, leaving him to look back and laugh at all the danger, if only to keep from crying.

Battle fatigued and with a pocket full of cash, the adjustment back to civilian life was sometimes fraught with hiccups. Most foreign divers like Richard stayed offshore for as long as they could take it; and for the first few days onshore, Burt says he and Richard were "complete failures." To make matters worse, rain-soaked Aberdeen was not the definition of a Klondike boomtown. It was a conservative city with few amusements. Decent restaurants were difficult to find and the general décor of the city was cold and impersonal. Homes, banks, businesses, churches—everything was built of stone. Walking around the windblown city was like walking between two graystone cliffs, and for the majority of the year the weather was cold and wet, and the act of staying warm, a constant reminder of the ongoing energy crisis. Central heating was rare in boarding houses which were typically drafty. At night, if Richard was not out drinking in the pubs and chasing down women before the 10 o'clock closing hour, he was warming his ankles around a small electric heater in his room. Hot water for bathing was dispensed at specified times or through coin boxes like parking meters. Put in too little money and he found himself all soaped up under a drizzle of freezing water. His boarding house of choice was the Wilmar Guest House on Crown Street. A sign in the community bath read, "baths by arrangement," which he found so hilarious that he wrote home about it. The owner of the house was a sprightly woman by the name of Minny Gall who had a playful streak in her and was always laughing. She was also a pretty tough egg and wanted her young male boarders to know that she was not running a honky-tonk. Unspoken-for women would not be allowed to come into the rented rooms. Minny would not abide any spontaneous feelings of "wanting to get your leg over." She had the reputation of her house to look after.[16]

The summer of 1975 gave way to autumn, and then, in October, after almost three years of enduring the vicious hot and cold cycles of offshore-onshore life, the constant exposure to danger, living out of a seabag and walking the streets of a town that was not his own, Richard took a hard look at the direction of his life. He knew where he'd been and where he wanted to be. Suddenly the words just popped out of his mouth, "What the hell am I doing here?"

He knew the answer to that question; he was there for the money. But the allure and the mystique of being a diver had disappeared. He was sick of the constant danger. He missed his friends and wished he'd gone to school to do something "a bit less ridiculous."[17] Richard, like so many divers before him, had finally burned out.

There were glimpses that he'd questioned his future in this industry from the beginning. A month after he arrived in 1973, he wrote to Mary Ellen: "Jesus Christ, I can't adapt to this life too well. No telling how long it's going to last..." And then 10 days later: "Lasting through six months of this seems increasingly impossible." Most likely these words were a reaction to the initial shock of the harsh life, but it wouldn't have taken Richard long to realize that there was nothing romantic about spending the rest of his life migrating from one dangerous filthy barge to another. Richard had the soul of an artist and the intense reality of the North Sea did nothing to reinforce the deep personal feelings he had for the ocean. He loved sat diving because it gave him the opportunity to explore a part of the earth he never could have experienced otherwise, but he hated the North Sea for the cold, the danger, the grueling work. And it was this love-hate relationship—fueled by the big money and the constant need to earn a living—that lured him back every year.

"Goddamn fucking money!" he declared in one of his letters to Mary Ellen. "If I could stick it out, in the end it won't seem like all that much a sacrifice."

But the sacrifices had become too great. Richard couldn't stick his head in the sand and forget that his heart was in Santa Barbara. He wanted to stabilize his life and become part of his "land of the lotus," and he wanted his friends close by, to go back to school, and to be absorbed by an environment that challenged his mind intellectually.

"I want to paint, sculpt, write, play music, and do it with the feeling of not having to worry about security or having to modify the way I want to live for the sake of fucking practicality," he declared.

He also wanted children and living offshore was no way to attain that goal. In the end, it was the artist in Richard, not the explorer, that finally won out, and so he left. And for the next three years no one in the North Sea knew of, or ever heard from, Richard Walker again.

That year, 1975, the North Sea claimed the lives of nine more divers; the bodies of three were never found.[18]

Chapter Five

Skip Guiel

If it was medieval times he'd be a king.

—— Roger Guiel[1]

Courtesy of Sheila Borsari

Skip at age 3.

In the tall grass the two boys squealed like newborn piglets. Joey Graveline was five, his neighbor, Victor Guiel, was three and they were building a fort. Not just any fort, but a Viking fort to protect themselves from the invaders that were sure to come and steal the jewels they had hidden away. It was a warm day and the West Springfield, Massachusetts air had the smell of disturbed grass. Joey and Victor were covered in fresh dirt but took no notice and played on. Finally the fort was completed according to an imaginary blueprint known only to the boy-architects. They crawled inside, and then they waited for the impending battle.

These and other memories would remain with Joey Graveline for the rest of his life. But he would never be able to recall when he first met Victor, only that, like some genetically embedded dream, the little boy with the infectious laugh and sidewall-haircut had always been part of his life.

Victor Francis Guiel Jr. was born on August 21, 1950. He was the son of Victor Francis Guiel Sr. and Shirley Nascimbeni. Their big green clapboard house on Lowell Street had a chicken coop and two car garage out back and was owned by Shirley's parents, Art and Edna Nascimbeni, who lived on the ground floor while Victor's family occupied the second floor. Three years after Victor's birth, a second son, Roger, arrived, followed by baby sister Sheila in 1954.

The Guiels lived in an area of West Springfield known as Al's Corner which sat at the confluence of the Connecticut and Westfield Rivers and directly opposite the Big E, the Eastern States Exposition. Named after one of those old time soda fountains where you could buy a cherry smash or vanilla coke, Al's Corner was blue-collar, largely Italian and built on a flood plain protected by a dike. The backyards of the large two and three story homes were spacious with clotheslines, vegetable gardens, fruit trees, and lawns infected with crab grass.

Joey and Skip played and built their forts in the vacant lot behind Skip's house. As they got older, the forts became more sophisticated. They scavenged for pallets and cardboard boxes in the alleyways, whatever they could haul away. When some of the other boys joined in they eventually formed a small gang and called themselves the Dayriders. Several blocks over, Ricky Borsari and the Fontana brothers did the same. They became the Nightriders. Eventually the two gangs began making raids on one another. While the Dayriders were off pinching cherries or apples or vegetables from a neighbor's garden, the Nightriders would sneak in and smash their fortress to smithereens just to show the Dayriders that their defenses were weak. Retaliation was swift and in kind. Eventually the two sides began to build secret underground forts which the other side knew about, but couldn't find. Skip was a Dayrider, but his brother Roger was a Nightrider. When Skip tried to pry the secret location from his younger brother at the dinner table, their father would cry out, "Stop it! None of this! Leave that outside."

During the summer, the forts were pummeled by dirt clods, and in the winter, by snowballs. No one got hurt during these contests, but they didn't dare ride their bikes into the other gang's territory. When Skip got a job delivering the afternoon paper it forced him into "occupied territory." Nightriders hung from tree limbs like monkeys, and rained water balloons down on him. When one of the other gang members rode down Skip's street one day with a big smirk on his face, someone picked up a broom and threw it like a javelin, locking up the front wheel and tossing the kid onto his head. There was never a dull moment in Al's Corner.

In 1961, the Guiels purchased the house across the street. A year later, Skip started junior high school and joined Joey on his three-quarter-mile hike across the train tracks towards Park Street. Coming home they rummaged through rail cars hunting for anything that wasn't bolted down—batteries, flares, whatever they could find, even if it was of no use to them. Later they would figure out what to do with it.

On Saturdays, when Joey and Skip weren't leading their siblings down Memorial Avenue to the weekly matinee like mother geese, they walked to Johnson's Bookstore and bought small bottles of chemicals just to see what they could mix up. In the minds of these two adolescents, it was the contents of these bottles, with their unusual sounding names, that held the promise of something new and more powerful than making rockets with match heads. And that power was nitroglycerin. They wanted to make nitroglycerin.

On a workbench in his cellar, Skip established his chemistry lab while Joey did the same under his home. When something either foamed or smoked or changed color, they conferenced like mad scientists comparing notes. In retrospect, Joey says it was amazing they never poisoned each other. And then one day Skip was working with his test tubes and beakers when a frightening explosion erupted: "It blew the place apart," Roger says. "Soot and everything, everywhere. That was the end of the chemistry set."

When Joey was twelve his uncle gave him a broken car and everyday Skip would come over and they would try to make something work or take it apart. As Skip grew older he began to realize the importance of books as a powerful source of new ideas. His interests ranged from history to science fiction to books that told him how to do it because Joey says Skip "always had to know the reason why." And he regarded all of his books as personal treasures. Fearful that they might get lost, he often scribbled his address on the inside cover with the added forewarning: "Send it back sucker!"

Every summer the Guiels rented a small cottage in Maine on the edge of Lake Megunticook near Lincolnville. There was no running water and the outhouse was *out* by the road. Drinking water was retrieved with a bucket on the end of a stick from a well in the yard. At the crack of dawn, Skip's father would shake the family, hitch up the boat, load everyone in the car, and drive to Duck Trap where Roger and Skip would skin dive in a flowing channel, while Sheila looked for shells.

Courtesy of Sheila Borsari

Skin diving in Maine at age 15.

He was more than just a friend.

—— Conrad Rivest

Six miles north of West Springfield, in the small community of Chicopee, Conrad Rivest taught marine biology at the local high school. At thirty-three and fluent in five languages, this no-nonsense, self-described flower-child brandished a pigtail, a set of earrings, and refused to follow the administration's dress code.

He liked wearing jeans and sneakers better than a coat and tie and preferred teaching high school rather than college because high school kids are "malleable, and the college kids are not."

When he wasn't filling malleable heads with marine biology, Conrad taught scuba diving for various YMCAs in the area and for Al Pop, the owner of Holyoke Underwater Supply. Tall and powerfully built, Conrad was popular with his students because he exuded confidence and knew how to make the everyday "hoi polloi" understand the physics of diving. It was the look in his eye and the sound of his voice that made him the best scuba instructor around. The other instructors in the area didn't like Conrad because they were mostly jocks or factory workers who knew how to dive, but didn't know how to present the material. As a result, they didn't have many classes and Conrad did.

And then one June day in 1966, he was teaching a class at the local YMCA in Springfield when he noticed he had a fifteen-year-old kid in his class who was a likable young man, but "didn't like to take orders." Glancing down at his clipboard, he looked at the name on the roster. It read Victor Guiel.

At the beginning of every course Conrad administered a swim test to cull the "weakies" from the crowd: twenty laps of the pool. And where everyone else finished with their tongues hanging out, Skip swam the last four laps in butterfly just to show Conrad what he thought of his test. Then he hopped out of the pool and said, "What's next?" When Conrad ordered the class to tread water with their fingers out of the water, Skip held his wrists high with a big smile on his face. Conrad stared back with a blank face, but behind his expressionless eyes a red flag went up in his head. Clearly this was a kid that if he told him to go thirty miles an hour he was going to go fifty.

Conrad knew he had a rebel on his hands, but as the course progressed he discovered that Skip's maverick attitude had limits. This became evident when he was explaining the rules of diving. If the decompression schedule said to decompress for five minutes, the barrel-chested instructor boomed, "YOU GO FIVE MINUTES, NOT FOUR, NOT SIX—YOU GO FIVE MINUTES," not because he was telling them to do it, but because "you don't mess with Mother Nature!" And Skip was impressed by that logic. He was a rebel in one sense, but when he realized he was dealing with a higher power, he obeyed.

After certification, Skip became Conrad's dive assistant for the next four years, hauling scuba tanks and gear around, and taking students out for their check-out dives.

"As far as I'm concerned he was the best diver I ever met," Conrad says of his former pupil. He also thought Skip had an uncanny appreciation of his underwater surroundings and could be trusted implicitly, even while diving in black water. "If I said, 'Vic, be here.' I'd reach out and he was there. I don't care what I did, he was there. He had a sense of where I was going, when I turned; I didn't have to tell him, and I didn't use my hands. I'd turn this way and Vic was there."

Conrad's position on safety had always been clear and absolute, and Skip remained loyal to the rules. When his younger brother pestered him to take him diving, Skip told him no, that he had to take the course. Diving was not something to fool around with, and as Conrad learned, if Skip thought the rules of safety had been crossed, he was not afraid to punish the rule breaker.

On October 1, 1970, Skip enlisted in the Navy, partly because he needed direction in his life, and partly to diffuse the tension building at home with a father who was feared for his fits of anger and stormy mood swings.

Courtesy of Sheila Borsari

Petty Officer Victor Guiel did well in the service, although he would later warn his brother that if he ever followed in his footsteps, he would "bruise him up."[2] The military kept meticulous records on Skip's career and filed a NAVPERS 792 performance evaluation a year after his enlistment. In the report, his commanding officer describes a young man who is "effective and reliable," "requires a minimum of supervision" and "willingly follows commands and regulations," "promotes good morale," and "handles men very effectively." He concluded the evaluation with these remarks:

> He is an industrious worker who is aggressive and cooperative which produces timely and accurate results. He is always willing to help others which is best demonstrated when assigned to other duties within the division. He gets along extremely well with his fellow workers and has a friendly attitude at all times. This attitude increases his effectiveness in the assignment of his responsibilities.[3]

And every evaluation thereafter carried the same common thread: that Skip had an outgoing personality, was dedicated in his work, and frequently helped others in need. It was a pattern he followed even in his off-duty life. When his ship sailed into port, he barely gave his mother notice he was bringing home a couple of crew mates to have a home-cooked meal. Their race, religion, politics and personal hang-ups meant nothing to him. What mattered was that they didn't have their families around them, and Skip couldn't stand to see that. So he would bring them into the family fold.

On March 15, 1974, after three years, five months, and fifteen days of active duty, Skip was transferred to the U.S. Naval Reserve and issued an honorable discharge. At home in Massachusetts, he resumed his friendship with Conrad Rivest.

From the start of their relationship Conrad had been Skip's mentor, the eighteen-year gap in their ages conferred as much, but Conrad discovered that the impetuous rebel he once knew no longer existed. Skip was older, wiser, and had gained a more cosmopolitan view of life. Conrad now saw Skip as his equal.

A year slipped by and in West Springfield Skip's sister was making plans to get married. Having grown up virtually the only girl in Al's Corner, Sheila was no shrinking violet: she was tough and strong-willed. Throughout her childhood she had followed in Skip's footsteps: taking over his paper route, winning trophies as a sharpshooter, and learning to scuba dive. She'd watched her brother and Joey work on cars, stood up to Skip when he picked on Roger, then defended him when Roger became the bully. Sheila always took care of the underdog.

In high school, she made her own dresses, drove around in a car named Bambi, and was a member of the cheerleading squad. She was feminine, but ballet lessons were not her style. With long, dark brown hair and a wholesome unrestricted laugh, she had a softness to her that easily attracted men. When she dated, Skip and Roger did their best to scare off the bad prospects. After graduation, Sheila worked at the Breck factory on the production floor, then at Hampden Electric as a bookkeeper. And then one day someone from her past suddenly woke up and realized he had been living just a few blocks from this goddess his entire life.

Former neighborhood Nightrider Ricky Borsari was talking with Skip one day when Sheila unexpectedly drove by. Apparently Rick hadn't seen her for several years because he suddenly blurted out, "Is that your little sister Sheila?"

Growing up in Al's Corner, Rick had never paid much attention to her. She was there, but she wasn't there. In fact, he didn't think of her as a girl. But now Rick began to re-evaluate the person he thought he knew. Besides Sheila's good looks, he knew she was strong-willed and loyal. Anytime her brothers needed anything she was there for them. She was athletic, outgoing, and she could cook!

Completely smitten, Rick discovered that Sheila was "kind of engaged" to a tall Polish kid. When he approached Skip to "make approval," he was given the green light: "Don't let her marry that Polack," Skip told him.

And, indeed, he didn't.

His courtship of Sheila ultimately yielded a wedding announcement but not without a bothersome snag. Rick was an ardent deer hunter. And because deer season began on the first of October, he wanted to tie the knot in September. But the priest had other obligations and told him he couldn't perform the ceremony until the third of October. Undeterred, Rick continued to press for a September date, until the exasperated priest finally shouted back, "THE THIRD IS IT PAL!"

So two days after deer season opened, Rick and Sheila were married in St. Ann's Church and immediately thereafter Rick presented his new bride with a brand-new compound bow and off they went on their honeymoon to a log cabin in the White Mountains of New Hampshire.

By now, Skip had moved to the New Hampshire coast and was living with a ragged bunch of ex-Vietnam vets in a big house on the boardwalk at Hampton Beach. And from time to time, Conrad would drive down to go diving, usually at a place called Pulpit Rock near the town of Rye. The guys Skip lived with had magnificent motorcycles and the best diving gear money could buy. But other than that, Conrad thought they looked like bums.

"Bums in the sense they were neat, but long hair at the time and all that. And whenever they would have trouble with the police, there were swamps out in the back; and these were all vets; they knew how to handle the swamps and the local cops never caught them."

For ten years Conrad had grown to love and understand his former pupil. He knew there were aspects of Skip's life that he didn't approve of, but at the same time he didn't condemn him. He learned that once Skip trusted you, it was total trust, unless you did something not to deserve that trust. And because Skip made every person feel important, Conrad came to understand his true worth.

"If he was with me and my brother at a restaurant, he knew his manners. And if he was with a bunch of guys slugging down some beers and having a Bar-B-Q, he acted that way. He never made a person feel uncomfortable because of their education. In other words, if you were well educated, he knew how to fit in. If you were poorly educated he didn't flaunt himself. There is an expression in the Bible that Paul uses: 'When I was with the rich, I acted like a rich man. When I was with the poor, I acted like the poor.' And that's the way Vic was."

And there was something else Conrad found remarkable: Skip never presented his problems to anyone but himself. When Conrad discovered one day that Skip had had a problem but kept it quiet, he asked: "Why didn't you tell me about that, I could have corrected it," and Skip answered, "I can handle myself."

And then one day in late 1975, Skip announced to Conrad that he was going to become a commercial diver.

High pay and adventure—sounded pretty good to me.

—— Bob Gemmell

On January 23, 1976, Skip was standing in line on orientation day at Commercial Diving Center in Wilmington, California. With shoulder-length dark brown hair, patches on his jeans and Fu Manchu mustache, he was talking to someone about Air/Mixed Gas Course D-68, when the guy in front of him suddenly turned and asked, "You're in D-68?"

"Yeah," Skip answered.

"Well so am I."

When the young man introduced himself, he said his name was Bob Gemmell. At twenty-three, married, and with the responsibility of two small boys, Bob was tired of running a metal lathe at L.J. Manufacturing in Eugene, Oregon. He wanted to break away and do something extraordinary with his life. He loved scuba diving and had read the ads in *Skin Diver Magazine* about becoming a professional diver, a career thoroughly cloaked in glamour and mystery. After checking out Commercial Diving Center, he flew home, told his wife Maggie

Courtesy of Sheila Borsari

Bob Gemmell

that commercial diving was what he wanted to do even though he had no idea what divers earned.

Now, standing in line that first day, Bob thought Skip looked like a biker, and as things turned out, this self-professed Oregon redneck and Massachusetts biker would become the best of friends and perfect roommates.

Commercial Diving Center had a global reputation for being the fast track into the diving industry. It was a good school, but in some respects, it was the complete opposite of Santa Barbara City College. It wasn't perched high on Cliff Drive in one of the most desirable communities of Southern California, nor did it have a spectacular view of the Pacific Ocean. It was in the heart of L.A.'s industrial section with an oil-refinery horizon. White plumes of smoke spiraled into a chemical-grade sky above acres of tubular steel. Freight trucks seemed to forever rumble across the railroad tracks leading down to a wharf where an occasional hairless, bloated cat would float by in the filthy waters of the harbor.

For four months Bob and Skip had classes in diving physics, physiology, decompression procedures for air and bounce diving. There was a lot of discussion about partial pressures of oxygen and helium, Sur-D O_2 tables, and symptoms of decompression sickness. They made dives in the training tanks wearing various dive hats and ran the decompression chamber. They learned how to tie knots—the bowline, the clove hitch, the fisherman's bend, the rolling hitch—and how to burn and weld steel underwater. They learned just enough to avoid making the grossest of errors for where they were diving. The reality of 40-foot waves and 90-mile-an-hour winds would come later.

On the weekends Skip and Bob scoured the sea floor for scallops and abalone from Laguna to Redondo Beach. At their apartment on Bayview, Skip was in charge of cooking because, when they first met, Bob immediately brought up the subject. "Skip, man, I can't cook; I don't like to cook; I don't want anything to do with cooking." And that suited Skip just fine. Having grown up under the tutelage of his Italian mother, cooking was one of his passions.

As the months passed, the two roommates grew closer and Bob looked upon Skip as one of the most honest, loyal, all-around kind of guys who deliberately reached out to people. Where Bob would walk into a room and look for a wall to lean against, Skip would mix with the crowd to meet everyone. He wasn't loud; he wasn't someone who had to draw a lot of attention to himself, but Bob says Skip wasn't separated from a lack of ego either.

On May 20, 1976, the two divers received their diplomas. Skip placed second in his class narrowly missing out on the top score to Tim Brookings from Darwin, Australia, while Bob wasn't far behind in fifth position.

Adrenaline was now flowing in Skip's blood. He was leaving the sanctuary of school for the greatest challenge of his life. Sensing he was on the precipice of an adventure, he wrote to his girlfriend, Debbie Bulewick, shortly before he graduated:

> I'm getting all psyched for a change. East for the blessing or west to work, who knows where. But I need a change somehow. I'm getting homesick every time I think about you and Springfield. But school's almost over and something will break. The gods must be getting ready to roll Skip's dice.

Eleven days later this tall, good-looking twenty-five-year-old Italian-American from West Springfield, Massachusetts walked past the Immigration counter at London's Heathrow Airport carrying a seabag over his shoulder.

No one noticed, but he was making a beeline to the North Sea.

Courtesy of Sheila Borsari

Skip (far left) at dive school, April 1976.

Chapter Six

You're Only As Good As Your Last Dive

There's only two can'ts—if you can't do it, you can't stay.

—— Charles Woolfson *et al.*
Paying for the Piper, p. 47.

By the time Skip arrived in Scotland in late May, five divers had already died that year while working on oil and gas projects. One case in particular bears special attention because it demonstrated how the role of safety could be subjugated by the interests of a powerful oil company and that diving from a heaving ship was infinitely more perilous than diving from a stable barge.

For a second time the incident involved drop weights, with the fatality resulting from another accidental bell surfacing of the kind that killed the two Ocean Systems divers in 1974, and by a chilling coincidence it occurred on exactly the same day, January 17.

Under a contract to provide diving services to Hamilton Brothers in the Argyll Field, Comex Diving Limited installed a portable saturation system on the stern of a small supply vessel known as the *Smit Lloyd 112*. The stern was a convenient location from which to launch the bell, but it was also an area most susceptible to heave. By late August 1975, the ship had arrived in the field to commence work,[1] and as the winter months approached, rough seas repeatedly caused diving operations to be aborted due to the difficulty of getting the bell through the splash zone as well as it bouncing on the end of its lift wire like a yo-yo.[2] To get the bell through the air/sea interface more rapidly, Comex officials authorized the dive crew to add additional weight, but this, too, proved ineffective.[3]

Unhappy that their work schedule was being adversely affected, Hamilton Brothers began to apply pressure upon Comex to remedy the situation.[4] A meeting was called to discuss the situation, and according to the Comex Diving Safety Manager, the oil company "demanded" that the diving system be adapted to allow operations to continue in more marginal sea conditions.[5] The client's basic concern, the project manager would later testify, "was to get a system with increased efficiency as quickly as possible."[6] To Comex, the message was clear: either fix the problem or get off the job.

With the threat of losing a lucrative contract looming, Comex decided to take a completely unorthodox approach to solve the problem of the yo-yo effect. Instead of suspending the bell over the job site, the company decided to allow the crew to modify the bell to set it directly onto the seabed, a procedure never used in the North Sea before. To counteract the effect of a heaving ship on a slack bell wire, Hamilton Brothers instructed another company to build and install a heave compensator to act as a shock absorber for the system.[7]

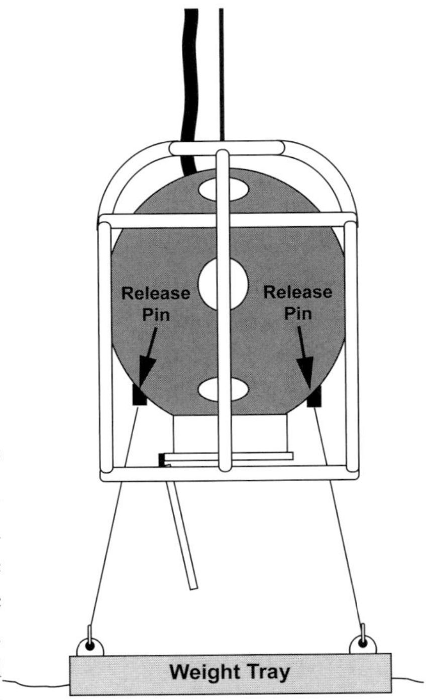

Like most diving bells of that era, the one on board the *Smit Lloyd* was fitted with a drop weight which was flat and doughnut-shaped and rested like a steel collar around the perimeter of the trunking just inside a protective bumper guard. When work began on the modifications, the doughnut weight was removed and replaced with a heavier 2.2 metric ton weight tray which was slung beneath the bell by two wire strops. These strops were attached by lengths of chain to a pair of drop weight release pins mounted on the hull of the bell.[8] The new weight tray now functioned as the drop weight, and when the system was in operation, it would rest upon the sea floor while providing ballast for the buoyant bell floating about six feet above. Unfortunately, when Comex constructed the lead-lined tray, it neglected to provide holes in the bottom to negate whatever suction force the seabed might impose upon it as well as the more vulnerable release pins.[9]

By October, the tray was installed, but the heave compensator had not yet been fabricated. Nevertheless, Comex was prepared to dive the bell without it, and did so. On a routine dive that month, the ship was pitching up and down like a hobby horse as the crew began lowering the bell with Derek Bannister and a man named Brorurd inside.[10] At about 30 feet or 30 meters from the surface, Bannister couldn't be sure, he felt a "sudden jolt."[11] The next thing he remembered, the bell was on the surface and he was looking through one of the lower portholes at a pair of broken chains. The weight tray had come free. No one was injured apart from "the odd bruise," Bannister said, because the bell had remained sealed with the inside door closed.

Shaken by the accident, Bannister went on leave while Comex built an identical weight tray (the original was abandoned on the seabed) and equipped it with stronger, more fortified sling components. By the time Bannister returned to the ship in mid-January, the heave compensator had been installed and tested, at which time Hamilton Brothers discovered that a pipeline connection near one of their subsea satellite wellheads was leaking oil. They instructed Comex to repair it. A young Texan by the name of Clay Ellis and John Mallin were originally slated to make the dive, but by the afternoon of the 17th, Bannister volunteered to take Mallin's place.[12]

One of the guiding principals of safe diving is to keep the bell above seabed obstructions at all times, particularly subsea wellheads which are box-kite-shaped structures that tower 20 to 30 feet off the mudline. For this reason, the decision by Comex to conduct a dive that put the bell into direct contact with the seabed and in close proximity to a wellhead was a dangerously ill-conceived plan. It may have served the interests of Hamilton Brothers, but it deliberately and unjustifiably put Bannister and Ellis at greater risk. If the *Smit Lloyd* suddenly dragged her anchors or was mistakenly moved in the wrong direction by an order from the diver, the bell could impact and snag on the wellhead with potentially disastrous results.

At 1722 the launching of the bell went according to plan and Bannister and Ellis came to rest in 256 feet of water approximately 10 to 15 feet from the wellhead.[13] Argyll oil at that time had only been flowing for about six months so the surrounding seafloor was still covered in drill mud, a thick, gooey substance used during drilling operations. Twenty-year-old Ellis exited the bell, traveled down the pipeline at the full length of his umbilical but could not reach the leaky connection. He could see it, but he couldn't reach it.[14]

He had to move the bell.

Communications through his mask was poor so Ellis returned to the bell and stood on the bottom door with his head inside the trunking. He took his mask off and instructed topside in which direction the ship should move. Topside then disengaged the heave compensator and began hoisting the bell. It took some force to break the weight tray out of the mud—one engineer would later calculate as much as "20 times the maximum force that the release mechanism could withstand."[15]

This action, along with the shock applied to the bell wire from the heaving ship, sheared one of the bronze release pins and completely bent the other connecting the weight tray to the bell.[16] There was a loud bang and Bannister called out for topside to stop what they were doing. Seconds later there was another bang. With no ballast securing the bell to the bottom, it started to shoot to the surface, main lift wire and umbilical trailing behind.

As the bell soared to the surface, expanding gas bubbled noisily out of the trunking. Ellis lost his footing and would have fallen out had Bannister not managed to grab him and pull him inside. With the atmosphere fogging from the rapid decompression, Bannister was still struggling furiously to close the inside door when the two men arrived on the surface forty-five seconds later. Eventually he got the door closed and started to re-pressurize the capsule, but in the desperation of the moment something had fallen across the face of the hatch preventing a seal.[a] Moments later the two men blacked out from the lack of oxygen in the now-depressurized atmosphere. When Bannister regained consciousness hours later, he was paralyzed from the chest down and Ellis was dead.[17]

Clay Don Ellis was "mad about diving" his mother would say to reporters,[18] and he was the 27th diver to die in the North Sea since the government began collecting statistics on fatalities. Immediately after the accident, Comex reinstalled the doughnut weight.[19]

Strangely, the Department of Energy, the agency responsible for investigating diving accidents, did not make an official inquiry into Ellis's death, having been persuaded by Comex that current safety regulations did not cover diving operations on subsea pipelines.[20] To close this apparent gap in the legislation, new regulations were quickly drafted and by July 1976, the Submarine Pipe-lines (Diving Operations) Regulations were put into effect. For the next five years, they would set the minimum standard of safety for diving in the North Sea.

By this time, Skip had been hired by Sub Sea International. There were now about 1,500 divers working in the North Sea, almost three times the number since Richard's arrival in 1973. In terms of contrasting management styles, Sub Sea put a less benign face on the industry for Skip than the one Richard first encountered with Ocean Systems. The pay scale, the level of competition, and Sub Sea's view of divers as a human resource was drastically different.

Unlike Ocean Systems, where no formal lines of demarcation separated divers, Sub Sea divided its troops into a rank and file system which was largely influenced by the way the Anglo-American company ran its operations in the Gulf of Mexico. Beginning divers were called tenders, followed by diver/tenders, and then divers. Fresh out of dive school, Skip began his climb to the stratosphere of commercial diving from the bottom rung.

Depending upon the urgency of the situation, Sub Sea's normal practice was to work a new tender like Skip in the shop for two or three weeks before

---

a Bannister suggested in court that it might have been a glove.

sending him offshore. Shop pay was $20 for eight hours of work plumbing sat systems, welding, scraping paint, whatever the foreman ordered him to do. Offshore, his rate doubled to $40 for a twelve-hour shift. Diver/tenders earned between $60 and $80 a day, and full-grade divers, $100 to $140 a day. If the diver went into sat, an extra $360 per day was added to his pay packet. Payment was issued to any designated bank and the amount was "neat:" no taxes were deducted. Thus, if a diver earning $140 a day went into sat for exactly thirty days, he would receive $15,000.

Unlike normal onshore employment, Sub Sea did not extend its divers amenities such as holiday pay, sick leave, comp time, redundancy pay, or overtime pay. If there was a sudden panic to mobilize a new job, Skip was expected to work until the panic was over, no questions asked, and with no added bonus. In America, overtime pay is generally associated with labor unions, but in the North Sea there were no unions for divers, and it was perhaps inevitable that a story would begin circulating at Sub Sea about a young American tender—newly hired—whose eyes lit up with dollar signs when he was informed that he was expected to work overtime.

"Overtime—great! I hope I get some of that," he was heard to say.

His supervisor just turned to him and smiled, "Oh, yeah, we've got overtime."

At Sub Sea, Skip's march towards economic nirvana was an uncertain process, full of pitfalls, operating more by Darwinian principles of natural selection than by any organized system of evaluation and training. There were no well-marked trails or guidance counselors to show him the way. The company had come to the North Sea with the redneck, yell-and-shout Gulf mentality. Its motto over the doorway could have easily read, "Every Man For Himself," and those who showed even the slightest ambivalence towards hard, physical, dirty work (or worse, complained about it), were unlikely to make it. Skip had to either sink or swim in the whirlpool of competition, and many new tenders, shocked by the harsh new life, simply walked away from their "dive tour" after a couple of months.

Unsure of how long he was going to last in this gladiatorial environment, it became almost a moral imperative for Skip to go about the business of getting into sat as fast as possible, a philosophy that he quickly embraced right from the start. After working all summer on various jobs around the North Sea, he was sent to the Thistle Field about 80 miles northeast of the Shetland Isles. Thistle was where he would earn nearly all of his money, and where his fortunes would ultimately change. The field was discovered in September 1972 by Signal Oil and Gas Company.[21] Its reservoir of Middle Jurassic ooze was more than 160 million years old and lay simmering within in a sandstone deposit trapped under nine thousand feet of rock. It was estimated to contain 824 million barrels of oil, but only 453 million were expected to be recovered. Not a huge find, but rich enough for the Halibut Group, the consortium of companies that had an interest in the field, to go ahead with development. For Signal Oil, however, the strike would never pay off. The company ran into financial problems and was eventually sold to Burmah Oil Company, including its 24 percent stake in Thistle. Halibut then put Burmah in charge of operations. But sometime in 1974, Burmah also fell into financial difficulties, and by early January 1975, the company had completely collapsed.

Meanwhile, the new Labour government, in its desire to take a more active role in controlling the fate of North Sea oil, passed the Petroleum and Submarine Pipe-lines Act 1975 which created its own state-run oil company, British National Oil Corporation. In January 1976, BNOC (*Bee-nock*) took over all the offshore petroleum assets of the National Coal Board, and after paying about £103 million in July, acquired most of Burmah Oil's North Sea holdings, including management of the Thistle Field.[22]

In August, Thistle's jacket was floated out and upended in one of the most hostile regions of the North Sea where 100-year storms could theoretically produce 150 mph winds and nearly 100-foot waves.[23]

By the end of September, Skip had taken only five days leave since joining Sub Sea, received his raise to diver/tender, and witnessed a near fatal bell diving accident on the *Active Duke* while it was tied up to the jacket. On the third day of October he was transferred to the *Orca*, a Dutch barge laying pipe in the field. As the mechanized drum of pile drivers echoed in the gray distance, he began rotating in the water, swimming the stinger every six to eight hours to check the pipeline as it rolled off the stern of the vessel.

"It's a good opportunity if everything goes good," he wrote to his parents, "and it's worth points towards me making diver pay rate."[24]

For Skip, witnessing the unimaginable spectacle of assembling the world's largest platform against gale-force winds in a white-capped sea, was an awesome experience and one that he often shared with his family back home. In a letter to his mother and father, he detailed the demands of putting in 16 to 18 hour days in between the storms that suddenly blew up.[25]

To his brother Roger, he wrote there was not much to gripe about except that day shift thinks night shift screws off and vice-versa, that life offshore presented "none of the complexities of the city," and that he didn't look too deeply into the flaws of his crew mates because "You can't afford enemies."

He said he'd learned to tie some "pretty neat knots," splice rope, weld, and burn steel a bit better. He knew more about diesel engines, boats, and diving. He described how the barge had to run from a storm for a week, and when he spoke about the North Sea, he did so with affection.[26]

October turned to November which gave way to December, and as each cold winter day surrendered more of itself to darkness, Skip spent most of his time reading, sleeping, and writing letters. In his entire life he'd missed only one Christmas at home, when he was in the Navy, but now he was making plans to herald the New Year from a lonely outpost in the middle of a desolate sea.[27]

But Skip didn't spend Christmas offshore that year: he flew home to surprise his family.

"You'd never know when Victor would come home," Rick and Sheila recall. "He wouldn't call you and tell you he was coming in. He'd call you from the Cloud Nine, smashed...after he had five or six Jack Daniel's just to make sure he had a good buzz on."

Once he was home there were two things Skip readily attended to. "He'd come in and he'd swallow my refrigerator," Rick says, and he'd get an update on his family. And if he discovered that Roger was not doing what was expected of him, and didn't have a good explanation, Skip would get angry and remind him "You take care of this; this is your responsibility; this is what you have to do." Roger says it was his brother's way of putting "all his ducks back in a row."

Just after New Years, Skip telephoned Bob Gemmell in Oregon and told him to get over to Aberdeen as soon as possible. The diving companies were going to hire early, he said, and he wanted his friend to get in on the action. And then, on January 10, 1977, he flew back to Scotland.

In London, the Department of Energy reported to Parliament that the prospects for 1977 looked "bright" for the North Sea. Two more oil fields and one gas field were expected to come on-line. Predictions were that oil production by the end of the year would be at "a rate equal to half or more of the UK consumption," and that the 1976 oil import bill, which was £4 billion, would be "eliminated by 1980."[28]

Meanwhile, Bob Gemmell kissed his wife Maggie and sons Jason and Adam good-bye, flew to Aberdeen, landed a job with Richard's former employer (now called *Solus* Ocean Systems), and heard that a house was available in the small village of Stonehaven about thirty minutes south of Aberdeen. One month later he checked the place out, secured it with the landlord, and phoned Maggie. "I got a house," he said to her. "Come," and with that Maggie stored their belongings, organized the kids, uprooted her life from the Pacific Northwest, and flew to Glasgow with the boys. At the airport, Bob was waiting for them. But as luck would happen, he was home for less than a week when the office sent him back offshore. "And that was Maggie's introduction to diving," Bob says with a quick laugh.

Eventually, Skip began staying with the Gemmells when he was on the beach. He became a member of the family, played with Jason and Adam, took them for walks and rides on his bright red BMW motorcycle. It was a bike Skip spent hours on, driving down narrow country roads with yellow broom and purple thistle splashing the wayside. Scotland's treeless scapes fascinated Skip, with hills of peat and black rock, and flocks of white sheep dotting dark green pastures like puffs of cotton. He made friends wherever he went, tried the local food, and drank the beer.

That winter, Skip went into saturation for the first time on the west coast at Loch Linnhe to conduct underwater trials for British Petroleum.[29] It was a short sat, but it was a big step in the right direction and it demonstrated that Sub Sea was boosting his career. A month later he was back in the North Sea surface diving. They were good dives, easy work with good visibility, but not without some alarming moments. In a letter to Roger he wrote:

> I got another dive in a few days ago. What a goof. I was working clamping some pipes together at 40 feet on the rig legs and they leaked some 'mud' into the water around me. Anyway, I come to the surface 2 hours later exhausted and wanting a breath of fresh air. So, I signal one of the tenders to pull off my helmet and the boss yells, 'Nobody touch him!'...Then some funny little guy comes charging at me with some crazy little gadget that looks like a Geiger counter. About this time I can't figure out whether to shit my pants or kill someone. Well, everything turned out OK, but that 30 seconds sure checked out my heart. Seems they thought they had dumped some radioactive mud out on me. But they didn't.[30]

There was another short sat in March, and then in May, Skip received his promotion to full grade diver. His pay rate was now $100 a day. This was good news, but he was having second thoughts about remaining in the North Sea and working for Sub Sea. Throughout his career, Skip had always maintained an optimistic attitude. But with the constant isolation, the cold, and the competition to get into sat fierce and unrelenting, he began to doubt if all the sacrifice was worth it. In a letter to Sheila at the end of May, he wrote: "Saturation looks a bit far away. I'm on a boat full of hungry divers and most of them have plenty of experience."

Since the Forties project, Sub Sea had accumulated a tight cadre of experienced divers that formed a seemingly impenetrable wall of talent that required not just hard work and patience to breach, but a lot of luck. There was no quota system for the allocation of sat; in theory the system was based upon merit with the most experienced divers going into sat first, and the less experienced queued up at the chamber door. Sometimes it worked that way, sometimes it didn't. Sometimes politics and favoritism crept into the formula. Sometimes a helicopter full of supervisors, the office had promised sat to, would unexpectedly show up. Suddenly Skip's thirty-day queue up got pushed back another thirty days.

That kind of character building experience had to be looked upon philosophically. Even though the merit system had broken down, there was nothing Skip could do. He couldn't whine about it because in the diving industry the complaint department and the unemployment department were one and the same. The best he could do was to quietly leave the job and try to find another opportunity.

Were these conditions to persist unabated year after year, then few divers would enter the business. But the diving industry had a revolving door to it and opportunities opened up all the time. Superintendents constantly looked for new talent, and there was always the chance that the next diver scheduled to be blown down into sat might unexpectedly become injured on deck, go on leave, or simply retire from the business. And as one diver left, everyone else moved one step closer to the chamber door.

But Skip's frustration with Sub Sea ran deeper than his uncertainty about getting into sat. Sub Sea's behavior was beginning to have a negative impact upon his morale. In another letter that same month, he severely cursed the company for the way it treated its divers.[31] He had good reason to. Whether it was slapping a fresh coat of paint on a piece of broken equipment and sending it offshore or failing to supply enough tools to get the job done, Sub Sea constantly did things to annoy the men who executed the contracts offshore. And the one area where company managers created the most strife, above all else, was their ability to make promises then fail to deliver on them.

Sometimes they would promise the diver a raise, but continue to pay him at the lower rate, then, when the error was discovered at a later date, feign ignorance of the mistake.[32] Good-natured divers, unwilling to provoke a confrontation, would simply let the matter slide. Never compensated for the blunder, the only reason they were unable to discount the affair as an "oversight" was because the practice had been repeated with other divers.

In one incident in the Ninian Field, Chevron Oil agreed to pay Sub Sea a bonus every time one of its divers got wet (wet pay), plus a graduated bonus for the depth of the dive (depth pay). According to the superintendent who ran the job, Chevron wanted to provide good incentive for good performance. But after the contract was signed, Sub Sea kept the details of the agreement to itself and began pocketing the wet-pay bonuses that should have gone to the divers. When the fraud was eventually uncovered, Sub Sea began releasing the bonuses to the divers but refused to pay them retroactively.[33]

In late 1980, the company sent a team of divers to Stavanger, Norway to undertake the dangerous salvage of a capsized rig called the *Alexander Kielland* that had been towed into a nearby fjord. Assured that high Norwegian taxes would not eat up their paychecks, the divers spent three grueling months working in subfreezing weather only to discover that 44 percent of their wages had suddenly vanished via an alternate tax maneuver.

Normally the error would have been discovered in a month's time when paychecks were disbursed, but in this instance Sub Sea withheld the wages during the job. When the company finally announced that payment was imminent, it apparently anticipated a backlash because it dispatched its operations manager to the job site to personally deliver the long-awaited checks and smooth out ruffled feathers. The excuses he offered, however, were less than convincing and many of the original crew, feeling deceived, asked for replacements within a week.[34]

These were the kinds of shenanigans that Skip knew about and had to be wary of. But pay manipulations were insignificant when compared to the way company managers used the promise of sat to leverage control over its divers. The North Sea oil industry was booming and Sub Sea was growing by leaps and bounds. Jobs cropped up overnight, and oil companies expected contractors to move resources and personnel around at a moment's notice. Faced with a slew of contracts to service, Sub Sea brandished the only weapon it knew would produce results: it promised the diver sat. Trouble was, the promises were issued in such an uncoordinated, haphazard manner that no one had any real idea whether those promises could be fulfilled, and many were not.

Still, the promise of sat pushed a lot of buttons with divers; after all, getting into sat was their prime objective. So divers spent weeks and even months building sat systems on the promise of sat, took supervisory positions on the promise of sat, interrupted their leave and flew back to the North Sea on the promise of sat. They could have said, "No, I'd rather not supervise," but refusing to take a job was dangerous to one's career because office managers cultivated grudges and kept them on ice for years. As Skip accurately pointed out in his letter to Roger, "You can't afford enemies."

At the same time, Sub Sea pushed the need for loyalty. Loyalty would bring rewards. But time and again the company hired new divers and paid them higher wages than those who'd been loyal for years. Eventually, divers used to joke that if you wanted a raise, the best way to get one was to leave Sub Sea and go to work for another company.

Another pitfall Skip had to mind was attaching himself too firmly to a topside specialty. At dive school, he was told that the more skills he had, the more attractive he'd appear to his employer. That logic made perfect sense, but at Sub Sea, there were instances when it had the opposite effect. One well known case involved a British sat diver with an exceptional expertise in hydraulics. He was so good at what he did, Sub Sea sent him from rig to rig fixing broken systems. Unfortunately, he rarely got wet and eventually everyone forgot he even was a diver! Still, he refused to abandon Sub Sea because he was loyal and the company kept making the usual promises to him. The only way he was able to break out of this straitjacket and into sat was to threaten to walk off the job, and then Sub Sea relented.

Energetic, hardworking divers with a welding specialty were also vulnerable. Sometimes they were deliberately kept on the beach for weeks burning rod in the shop, and then, when they were allowed offshore, welded for superintendents who valued their fabrication skills, but were reluctant to let them out of their sight to dive. Over time, they came to realize the best strategy to get into sat was to never ever declare that they could do anything else but dive.

Why did Sub Sea continually push the system to the limit? To a large extent, its conduct was a refection of the way the oil industry operated in general. The government's rapid exploitation policy and the massive costs at stake fueled a cut-throat environment where enormous pressures flowed downhill. Oil companies such as Chevron, British Petroleum, and Mobile Oil, for example, carried the financial risks, but they did not perform the diving work. All of the work during the phases of exploration, construction, and maintenance was farmed out to subcontractors like Sub Sea and Comex who were constantly reminded of their duties and obligations, as well as the negative financial consequences if the clients' expectations were not met. Implied threats were common and taken seriously, and subcontractors bent over backwards to yield to them. This was what happened in the Clay Ellis case when Hamilton Brothers applied just enough pressure on Comex to induce it to modify its bell and implement a completely unorthodox and dangerous bell diving procedure to keep the *Smit Lloyd 112* working in marginal seas and beyond its capabilities. It was a decision that ultimately cost Ellis his life and Bannister the use of his legs.

At Sub Sea, the same dog-eat-dog mentality that existed offshore, existed in the office. Many of the people in positions of authority—the company president, the operation's manager, and most of the project managers—were former divers with strong personalities. They understood the offshore environment, knew how to size up a new recruit and calculate gas consumption figures. But they were not businessmen with any formal management training; they were divers in three-piece suits. Consequently when it came to resolving administrative problems, they simply resorted to the same ruthless tactics they'd learned offshore.

While upper management was full of former divers, lower management, particularly the personnel office, was staffed with people who had no offshore experience and no appreciation of how the common diver contributed to the fortunes of the company. Inevitably, most of Sub Sea's upper echelon of divers avoided the personnel office, preferring instead to lobby for the prime jobs directly from the operation's manager, while relatively new divers such as Skip depended upon the next phone call. Unfortunately, if he stepped away to buy a bag of chips, and Sub Sea rang his house, he rarely got a second call. The secretary simply went down the call-out list to the next man, which one British diver feels was a strategic mistake from the point of view of keeping the diver loyal to the company:

> ...next, next, next, next, you know, not interested who it is. But you see, it falls down on them companies because if you get no loyalty one way, you don't get it the other way. The guys would just say, 'Well, all right, I've got another job for another few quid. I'm away.'

As a tender and diver/tender, Skip did what was expected of him because he was new to the industry and his strategy was to take any job, anywhere, to get as much experience on deck and in the water as possible. But in May 1977, Skip was a sat diver, and he realized that if he had any hope of breaching the wall of talent, he had to push for those jobs which would enhance his opportunities for sat, while tactfully avoiding those which would not. That meant he had to accept the political reality that it did him no good to believe the people in the personnel office—who were not competent at assessing his skills as a diver—had his goals and interests at heart. On the contrary, if he passively allowed them to dictate his career path, there was the very real probability that he would find himself lost in Sub Sea's Gulag relegated exclusively to air diving. And air diving was not why Skip went to the North Sea. Air diving was financial starvation. This is why the vast majority of sat divers adopted the strictly mercenary view that, with only a limited number of dives in their bodies, why waste them? Thus, sat divers wanted to sat dive and nothing more, not work on deck or surface dive for one-fifth the money.

With saturation looking a bit far away, and the company repeatedly dishonoring its promises, Skip had several options. He could stay and continue to grumble over the situation, or he could leave and start over with a new company, something he considered and spoke of in several of his letters to his girlfriend Debby:

> After days like today, I wonder what the hell I'm doing here. Put in 15 hrs of banging your head against the wall trying to fix problems that shouldn't have been there to begin with without proper tools or parts. It's so fucking frustrating. And to know it's probably going to happen again tomorrow. If everyday was like this there wouldn't be money enough to keep anyone here. But this is an exception. At least it better be or I'll be looking for another company.[35]

But leaving Sub Sea was a risky proposition for several reasons. First, it took time to build a relationship and earn a reputation in a new company. If Skip left now, he would lose the connection he had with those who knew his capabilities. Sometimes these moves worked out for the best, sometimes they backfired. Contracts shifted dramatically from year to year and the company you left last year might be the one with all the work this year.

Second, loyalty at Sub Sea ran in one direction. Trying to get back into the company came with a price tag. Mean-spirited managers didn't just make threats, they carried them out. Examples abound where a topflight diver was branded a "day-rate whore" and expected to serve penance by taking a pay cut or supervise an air job. Explanations that he had bills to pay or a family to support went in one ear and out the other. It didn't matter how long he'd worked for the company, that he'd never missed a helicopter flight, never caused an accident—history meant nothing to Sub Sea. He was always given the sense he was only as good as his last dive.

It usually took a couple of years before Sub Sea washed any romantic notions of company loyalty out of a diver's head. By early 1978, the cold, the boredom, as well as Sub Sea's disingenuous behavior, finally led Skip to question his relationship with the company that had hired him into the business. As Conrad Rivest said of Skip: "Once he trusted you, it was total trust, unless you did something not to deserve that trust." And in small ways and large, Sub Sea constantly demonstrated that it could not be trusted. As a consequence, the grapevine resonated with bitter feelings of cynicism and resentment as the company cultivated disloyalty within its ranks. Any sense of allegiance the diver felt was not to those who jerked his pay around or gave false promises, but to the diving superintendent who put him into sat. And in this "wild west" atmosphere, all that was needed to change the fortunes of a discontented diver was the appearance of a more favorable circumstance, and in the spring of 1978, Ken Duell provided Skip with just that opportunity.

Chapter Seven

Infabco Diving Services, Ltd.

It was a startup company with a lot of potential.

—— Roger Palmer, Infabco diver.

Like most divers, Ken Duell was a risk taker. But unlike most divers, he thought more broadly about how the oil industry worked, the role politics played, and the importance of relationships. He knew early on that slogging around on a dark muddy seabed with someone else's hand on his gas supply valve was not his cup of tea. The young American had been a member of Sub Sea's *Hercules* crew in the Forties Field, and whenever Sub Sea pulled a new chair out for him, he readily sat in it because he was ambitious and looked for opportunity. As a consequence, his career rose sharply from diver to superintendent to eventually becoming Sub Sea's project manager in the Thistle Field,[1] a position that brought him in contact with the movers and shakers of industry and with a young English engineer by the name of Brian Masterson.

In his early thirties, Masterson owned a London based company called Infabco Group Ltd. which was providing engineering support and maintenance services to British National Oil Corporation. Duell liked Masterson. He found him friendly, personable, and with a passion for business. And not just one line of business. He delved into all sorts of things which appealed to Duell and would eventually lead him into a partnership with him.

For more than a year Sub Sea had been the dominant force in the Thistle Field. It had captured the diving work on the drill rigs, the pipelay barges, and the platform where it established a very expensive, and some say, overly complicated diving system. In March 1977, Sub Sea divers were also involved in the installation of a tanker loading buoy about a mile from the platform. For all practical purposes, Sub Sea considered Thistle part of its domain. But in mid-

1977, its supremacy began to erode when BNOC by-passed the American-run company and awarded the maintenance contract for the loading buoy to a British diving firm called Wharton & Williams. Sub Sea still retained the platform work, but in early 1978, Duell heard rumors that BNOC was not going to renew Sub Sea's contract. At some point he discussed the matter with Masterson. Both men knew the information they held was not only significant, it might be worth a fortune. There was a lot of money in diving contracts, and with Masterson interested in becoming more prosperous and Duell dissatisfied with Sub Sea's leadership, they decided to launch their own company—Infabco Diving Services Ltd.—to fill the vacuum that would be created by Sub Sea's exit.

With all the history Sub Sea had in the Thistle Field, Duell thought it curious that BNOC would run them off the job. Ultimately he concluded that the relationship between the two companies must have soured, and that BNOC had grown tired of Sub Sea's temperamental diving system on the platform. But there is a fair amount of evidence to show that Sub Sea's impending departure was more likely due to the government's ongoing drive to give British businesses "full and fair opportunity" to compete in an offshore industrial market dominated by the US. Indeed, there was even a special branch of the Department of Energy devoted to encouraging the oil and gas industry to "buy British." It was called the Offshore Supplies Office and it required oil companies to submit quarterly returns to indicate where the contracts were going.[2] Moreover, the oil companies, through their lobbying group, the United Kingdom Offshore Operators' Association, had signed a "Memorandum of Understanding"[a] with the Department of Energy to "use goods and services of British origin...whenever they are competitive in regard to specification, service, delivery and price."[3]

In 1975, the Offshore Supplies Office intervened in the affairs of Burmah Oil Development Limited (BODL) when it became clear that Burmah wanted to award the contract to build the Thistle tanker loading facility to a Monaco-based company called Single Buoy Moorings Inc., rather than to the competing British firm of David Brown Vosper. After some lengthy negotiations, OSO performed a shotgun marriage between the two companies as revealed in the project engineer's "Assessment Report."

> During the bid stage it developed that the SBM design of a [tanker loading facility] was most favoured by BODL, and there was good justification for awarding the contract to SBM alone. However, the Offshore Supplies Office of the U.K. Government applied pressure that a U.K. company should get this contract. Thus a joint venture company comprised of David Brown Vosper (Offshore) Inc. and Single Buoy Moorings Inc. was seen as the compromise solution....The contract was therefore awarded to DBV, but was to be assigned to the joint venture DBV-SBM, which was to be formed. This forced marriage of direct competitors produced a strained relationship which eventually led to a complete breakdown of joint venture plans.[4]

a On November 3, 1975.

At the beginning of 1978, BNOC had an interest in 13 of the 17 fields currently in production or under development,[5] and as author Guy Arnold suggests, was in a position to "use its strength to help OSO and the Department of Energy get their way." In early 1977, Arnold says BNOC "made certain" that a contract for the supply of steel bound for the Murchison Field was awarded to British Steel Corporation.[6] Thus, BNOC's decision to purge Sub Sea from the Thistle platform was probably not due to any major complaints about Sub Sea's performance, but because BNOC was looking for a British diving firm to align itself with OSO objectives.

While Duell may not have known about the government's policy to give preferential treatment to British firms, Masterson almost certainly did. He was British, he was a savvy businessman, and he would have known about the official policies of his government that would affect his chances of winning a contract in the highly competitive field of deep-sea diving. Although Masterson knew nothing about diving, Duell was confident that, when the time came, he would be able to produce the necessary manpower to form a working company. Sub Sea International, after all, was mass producing disenchanted divers every day.

One other thing probably played a small part in the decision to go forward with Infabco Diving Services. Duell knew that gambles of this kind had been pulled off before. About 18 months earlier, in the autumn of 1976, Ric Wharton and Malcolm Williams had left Comex Diving Ltd., established a place of business in an old meat packing factory near Dyce Airport,[7] and from a small inshore contract with Total Oil, were now aggressively expanding their empire. It was their British-owned company that made the first inroads into Sub Sea's territory. They were the ones who picked up the maintenance contract on the oil tanker loading buoy in the Thistle Field. And if Duell and Masterson didn't get moving, it probably would be Wharton & Williams, and not them, that would pick up the pieces of Sub Sea's lost dominion.

One of Duell's first moves was to register Infabco Diving Services in Dublin, Ireland.[8] Then he approached three Sub Sea Superintendents—Ed Goltz, John Nadeau, and Craig Roberts—with an offer of co-partnership in the new company. They had worked together in the Forties Field as part of the *Hercules* team and as Duell would later say, "we all grew up together, you know, it was Roberts, Goltz, myself, Nadeau—I mean, we were the mainstay of Sub Sea."[9]

Dissatisfied with Sub Sea's leadership and looking for greener pastures, Duell says "Roberts was the first to come across," then he "coaxed Goltz into it, because we had all had enough of Ted McDowell [Sub Sea's President] at that time. And then we coaxed Nadeau to come across."

For Craig Roberts, leaving Sub Sea was a chance to build a company from scratch with people he could trust and count on. He'd witnessed what Sub Sea's rapid expansion had done to lower the quality of the people he had to work with and he didn't like it.[10]

Masterson, Duell, Goltz, Nadeau, and Roberts—they would be shareholders in the new company with the four superintendents providing the diving expertise while Masterson would bankroll the venture.[11]

Initially, Duell ran the business out of the Skean Dhu Hotel,[12] then moved the operation south to a workshop in the small village of Montrose. He purchased a portable ScanDive sat system from Norway, then he and Masterson submitted a bid to BNOC for the new Thistle contract.

Meanwhile, word quickly spread throughout Aberdeen about Duell's new company. By mid-April, Sub Sea divers began showing up at Infabco's doorstep looking for work. The day rates had been set slightly above the industry standard to attract divers. But more than the few extra dollars they were offering, it was the reputations of the superintendents that created a mass exodus of Sub Sea divers across company lines. At the end of May, Canadian diver Chris Roper left Sub Sea specifically to follow Duell. "Ken was the wise old man, and we were all nineteen and twenty-year-olds and kind of naturally pissed off with Sub Sea."[13]

A few weeks later Skip followed. By then Infabco had captured the Thistle contract. Roper remembers Duell walking into the shop that day to deliver the announcement: "The good news is we got the contract; the bad new is we're going back out to Thistle."[14]

The sat system was made ready, loaded onto a supply vessel, and transported to the Thistle Field where it was installed, not on the platform where Sub Sea had originally been working, but on the *Gulnare* parked alongside the jacket.

Courtesy of Marcus Taylor Ltd.

The *Gulnare*.

The *Gulnare* was a five-legged, self-propelled drill rig elevated about eighty feet off the waterline. BNOC had hired the floating motel to accommodate the extra labor required to complete the final construction of the Thistle platform. Viewed from a distance, it resembled an old kitchen table floating in the sea, but close up its imposing mass was that of a coliseum. It had a helicopter pad, a four-story block of accommodation quarters stacked like shipping containers next to the oil derrick, and on the port bow, under the forward crane, Ed Goltz installed the ScanDive system as Infabco's beachhead in an industry fraught with danger.

On September 10, Skip and Chris Roper went into sat as bell partners. For their first couple of dives, they snapped photographs and inspected the 16-inch Dunlin riser for damage. And then a seven-day storm blew up. When they

Courtesy of Gary Legere

The ScanDive system situated under the crane.

went back to diving on the 19th they spent the next 18 days working along the southwest face conducting surveys and clearing the seabed of debris at 530 feet.

This was the deepest oil producing field in the world and every time Skip locked out and stood at the base of the jacket, with its massive steel crossbraces zigzagging through the liquid ether and fading into the gloom, he was filled with a irrepressible sense of awe. It was impossible for him to see the entire structure, the distance between each footpad was over 260 feet, but he could feel its overwhelming presence nonetheless.

Occasionally he would glance over his shoulder to check his umbilical and catch a glimpse of the bell hanging in the night sky like a burning star, illuminating the gray wasteland below. It was a surrealistic vision made more striking by the gauzy colloidal suspension of detritus slowly showering down upon him. When he looked up, he saw glistening bubbles mushrooming from his mask on their voyage to the world above. It was a fantastic sight and one he never got tired of. Submerged in the breathtaking serenity of the universal sea, Skip was in the womb of Mother Nature, a place where he felt as much at home as the fish that swam around him.

After more than two years in the industry, the creatures he saw were still wondrous and mystifying. Schools of pollack and cod swam by with huge inquisitive eyes while the occasional octopus and giant crab scampered along the seabed in search of prey. Sometimes the water surrounding the bell would come alive with pink swirling snow storms of small shrimp and krill attracted by the lights. They, in turn, would draw the larger fish that would move in to take their fill. In a letter to Rick and Sheila, Skip described what happened next:

> After an hour or so, their bellies were bulging from the feed. What I would give to run into a similar situation while sport diving. You'd have enough fish to feed the town for a week. The average size fish was between 2-4 feet with a couple bigger or smaller. There were a few thousand fish minimum.[15]

Sometimes these fish schools would move in like hail storms, their numbers so great they extinguished all light. For the diver who got in their way, there was nothing he could do but stop work and stand there motionless while the swarming mob poked and prodded every square inch of his body.

To an observer the mere spectacle of a man sucking cold gas through a thin rubber hose at 530 feet might seem ridiculous and insane—motivated only by the prospect of earning big money. But deep-sea diving was more than just a fat paycheck to Skip: it was in his DNA. It was about the mental and physical challenges of working in deep water. It was about the technology, the camaraderie, and the fact that his crew depended upon him. Topside work was infected by the cold, the filth, and the long hours of boring work. But below the surface, Skip transcended himself and found great joy in leaping and bounding along the seafloor in slow motion, somersaulting off the top of the bell, and witnessing firsthand the unbelievable creatures that most people only see in books. He was down there in a world so dark and mystifying that the creatures don't see their prey, they sense them.

In a letter written to his girlfriend, Skip explained his attraction to the sea, and why he put up with the disagreeable aspects of his profession:

> ...I, too, am confused as to what I really want in life. My body and that part of my mind that deals with feelings, sensation, not numbers, wants to be away from here experiencing and sharing. But the other part of me needs to be here doing the things I do—picking up on a different emotion, one transmitted by the sea, and by the needs of the people I work with. It's not the same as back home....
>
> This life is something I'm not ready to give up. It's something that is necessary, simply as a way of making a living, of putting food in my mouth and a roof over my head, but also as something I desire and need, something [that] adds meaning to living....
>
> I sit here in a chair; I look up and see a wall of gauges and dials, regulators, digital readouts, and television monitors. Two days from now I'll be controlling the lives of four brothers from this same seat. I'll monitor and control everything in their environment so that they can live, so they can experience the sea as none else can, so they can become part of it.[16]

Whatever the task, working in the dark under highly focused conditions was like working with blinders on. Sometimes Skip wasn't aware he had "company" until he stumbled upon them. Hundreds of monk fish lay scattered about the seafloor like fan-shaped land mines with iridescent green crocodilian eyes sticking up through the silt. Normally they were fairly docile creatures until stepped on; and then all of a sudden, out of a big cloud of silt, a large vacuous mouth would come charging, flashing cat-like teeth. They were easily fought off, but the fact that everything underwater appeared 25 percent larger didn't help.

And Skip ran into wolf eels down there—an utterly repugnant monster with a bulbous head, hollow remorseless eyes, and a gob full of long vicious teeth. It seemed to be a creature resurrected from the dead. It burrowed under pipelines and piles of debris thrown from the platform. It was usually encountered with about four feet of its thick body half out of its lair, goosenecked, and undulating above the seabed with small dragon-like wings unfurled and fanning the water. There always seemed to be a pile of bones at the entrance to its hideout; and it frequently had a laughing, seductive grin on its face which seemed to say: "Come closer, just a little bit closer. I dare you."

This was not a subdued animal by any stretch of the imagination; it was a completely fearless hideosity and even the most macho diver instinctively understood not to challenge it, but to pass it by.

Back on the beach, the Infabco partnership was melting like winter snow on a hot spring day. Duell discovered too late that the contract with BNOC had been issued, not in the name of Infabco *Diving Services* Ltd. as expected, but to Masterson's parent company, Infabco *Group* Ltd. Consequently, payments were pouring into an account over which he had no control.

When he confronted Masterson about it, he says Masterson claimed the money had to go to him because he was carrying the financial burden. But whatever the true dimensions of the unexpected shuffle, they set alarm bells off in the mind of Ed Goltz.[17]

Since 1969, Goltz had one of the strongest reputations in the diving industry. He was respected not because he had this kind of John Wayne persona, but because he spoke his mind and defended his positions regardless of what the consequences were. And Ed had no respect for people who conducted less than straightforward business deals of the kind that Masterson had pulled off on Duell. Nor did he like the things he saw going on between Masterson and BNOC, some of which he says "didn't look too Kosher."[18]

Shares in the company, the partnership—none of the original enticements were being fulfilled. Goltz says he never received an answer regarding what his stake would be, never signed papers for incorporation of the company, or knew who would hold what office or delegate what responsibilities.[19]

"There didn't seem to be any money behind it," Goltz recalls, "although there was always talk about the millions of pounds that Brian had available from other investors and this sort of thing. But Ken was still paying for all the [hotel] rooms, all the expenses off his credit card."[20]

The business venture, once thought to be a good idea in Goltz's mind, now seemed to be "shaky" and poorly capitalized. And when more questionable business manipulations began to surface weeks later, Duell says that's when Ed decided, "this just isn't on."[21]

One night the ex-Navy diver went to see Duell and the two men had a long talk. There was an uneasiness in the air as they discussed the dimensions of a muddle that had clearly gotten out of control. At the end of the conversation Goltz finally said, "Ken, I'm off."[22]

"Well, I can't blame you, Ed," Duell replied.

And Goltz left, leaving behind his interest in the company to anyone who wanted it.

For Duell, Ed's departure was the last straw. Demoralized, and always feeling he had to get on bended knee for money—he realized too late that Masterson's plan all along was probably to gain control of the company.

Several weeks later, on a Saturday afternoon, he and his wife, Yvonne, met with Masterson and tried to persuade him to abandon the path he was leading them down. They argued over personal philosophies and business ethics, but Duell says Masterson wouldn't give in. The next day, in the Montrose office, Ken and Yvonne signed the legal forms giving their 40 percent share over to Masterson. "They're yours," Duell said to Masterson, and with that, the couple walked out the door and never looked back.

It was a painful episode in Duell's life. He'd put his heart and soul into the venture; and he desperately wanted it to succeed. But in time, Ken would come to realize that divorcing himself from Brian Masterson would be one of the best decisions of his life.

With Duell's shares in hand, Masterson now had majority control of the company. But half his expertise had left and he couldn't afford to lose Roberts and Nadeau. He needed them to run the *Gulnare* contract. Duell says, in a conversation with Roberts and Nadeau, both men expressed their intention to leave Infabco as well, to which he responded, "Please yourselves. If you go, you go, if you don't, I'm not going to be pissed off with you."[23]

But Roberts and Nadeau stayed with Masterson. Later, Duell found out why.

"I remember John telling me one night when I met him up at the Skean Dhu, they had relinquished their shares in return for a higher day rate."[24]

There was no diving work here at all. He [went back to] working as a glass blower at Kilovac because there was no work here. And he went to Texas a couple of times. He would have occasional short jobs with Ecomar. He worked tearing down the barracks. And he wasn't the kind of person that could ever sit around...and he didn't really want me to work. I had Marisa. I worked up until I had her and then I wasn't working. And he wanted me to stay home. And that was fine, it was no big deal. And it was just no work, and it was getting towards Christmas time and there was still no work.

—— Jeanne Walker

Half a world away, Richard Walker couldn't have been happier enjoying life in Santa Barbara. The perils of the North Sea were far behind and out of mind, but only because he consciously resisted the lure of the big money. After returning home in late 1975, he was taking a silversmith class when Ocean Systems notified him that diving operations on board the *Castoro* would be resuming on the 28th of February. But Richard wouldn't go back.[25] He wouldn't allow himself to be consumed by the constant hunger for the sat dollar. And while he used to say, that "somehow, coldly enough," more money helped, ultimately the unrelenting pursuit of it did not satisfy his soul.[26] He was home now, far from the compressed life of the North Sea, and that's where he intended to stay.

A thousand feet above sea level, Richard was living in the small private community of San Marcos Trout Club perched on the side of the Santa Ynez mountains overlooking Santa Barbara. It had once been a vacation retreat for weekend fisherman, but now this sleepy little hamlet was a homeowner's association of about thirty dwellings where birds sang songs to the passing breeze and dragonflies zigzagged in the hot afternoon sun like miniature helicopters. Overhead, red-tail hawks circled a blue sky, while down below, a skinny-dipper's creek flowed through a steep ravine studded with acacia and scrub oak. It was quiet and peaceful here. And it was here, away from the hustle and bustle of city life, that Richard met the woman he wanted to marry.

Jeanne Michelle Stevenson had grown up in the Pacific Palisades area of north Los Angeles and attended UC Santa Barbara as an art major. In early 1974, she left the university at the age of nineteen intent on seeing Europe, but ended up traveling overland to India, flying over the closed borders of Burma, and landing in Singapore. Eventually she arrived in Australia just before Christmas. By the time she returned to Santa Barbara in the summer of 1976, two years had gone by.

Tall and slim, with long, straight, dark brown hair, Jeanne had a terrific sense of humor and embodied all the charm and beauty that appealed to Richard. Both were free spirits, both were artistic, and it was only natural they would catch each other's eye and fall in love. Richard's best friend, Doug Campbell, understood the bond they shared.

"Jeanne is a fabulously talented person; she's real exceptional. She was a ceramist and a sculptress, a very good artist, and they had a real connection. I think through that they had very similar minds in the way they could look at things and be very creative about it. And they were in love; they were just crazy about each other—absolutely crazy about each other."[27]

They married at the local courthouse then took up residence in the Trout Club in a small forest service lease house overlooking the canyon that faced the Pacific Ocean. Neither worshipped money. They lived a simple life with few possessions. Richard drove around in an old 1949 flat bed truck named Ethyl. It was rust colored and one of the fenders was dark green. He was still making a living as a commercial diver, picking up jobs with Oceaneering and Martech. But the bulk of his work came from a small company called Ecomar Inc., a marine biological consulting firm which studied the effects of human influence on the ocean such as offshore drilling, nuclear power plant discharges, sewer outfalls, and break water construction.

Jeanne was working as well, and together they began searching for a piece of property of their own. Their first thought was the Trout Club; it was isolated and private, yet close enough to Santa Barbara to enjoy the beach life and the things the city had to offer. But land was expensive and properties for sale scarce. Undeterred, they kept watch for someplace to anchor themselves to the community, and because they would need more space for a family. Jeanne was pregnant.

Photo by Author

Hollister Ranch

And then one day Doug came to them with information about an exclusive piece of property for sale north of Santa Barbara at Hollister Ranch. Once part of an old Spanish land grant, the 107-acre parcel encompassed an entire valley with a small stream, a waterfall, and access to about eight miles of private beach and world-class surf. It was ideal, and on April 1, 1978, Richard and Jeanne signed the papers to close escrow under a partnership agreement with Doug and his wife, and Doug's parents. For Richard and Jeanne, Hollister was the answer to a dream.

In May, Jeanne gave birth to the child that Richard always wanted. Burt Davis remembers the day it happened. Having quit the North Sea the previous winter, he was at the Trout Club visiting Richard when Jeanne started having contractions. Later that night he received a phone call from Richard shouting, "It's a girl! It's a girl!"[28]

They named her Marisa. And she awakened in Richard emotions he found surprising because he never thought of himself as a family man. But in the days that followed, Marisa became the light of his life, and with each passing day he doted over her, fed her, wiped her face, rocked her in his arms, kissed her chubby little cheeks, and told her fairy tales even though she was too young to understand. At the age of thirty-one, Richard's transition to fatherhood was total and without reservation. Doug saw the change in his friend even before Marisa arrived.

Courtesy of Jeanne Walker

Richard, Jeanne and Marisa at Hollister Ranch, January 1979.

"I remember he built this crib for Marisa before she was born. He just spent endless hours on this thing and it was just absolutely beautiful, wooden crib, all handmade and just polished and gorgeous. He just put a lot of love into it. He was so happy with Jeanne, and so happy with Marisa. It was really neat to see Richard like that. Kind of had the feeling he was going to settle down, that his life had made a big change, certainly for the better."[29]

But in 1978 a dark cloud suddenly began to hover over the Santa Barbara diving community. Work dried up. Tom Belcher—having returned from the North Sea with a case of burn out—remembers those difficult times. "We were both kind of surviving in Santa Barbara barely just with what was going on—and there was nothing going on."[0]

Richard picked up work wherever he could, but it was never enough to pay the bills. What's more, the forest service began sending him notices to vacate the house he and Jeanne were living in.

"We tried to delay and delay the forest service," Jeanne recalls. "And they finally said, 'No, we're tearing the house down.'"[31]

They had until after Christmas to move out.

And then Richard and Tom received a phone call from Craig Roberts offering them a job with Infabco.[32] With no work in Santa Barbara, and unrelenting mortgage payments, Richard decided to return to the North Sea. He loathed the idea, but he felt he had no other choice. He needed the money.

On October 13, Richard turned 32; less than a month later he kissed Jeanne and Marisa good-bye and boarded a plane for Aberdeen. On the flight, all the emotions relating to why he'd left the North Sea came flooding back. He was sacrificing his life to *that sea* again.

On November 6, Richard received his diving medical certificate from Offshore Medical Support. Within days he found himself donning an orange survival suit once again at Dyce Airport, smelling helicopter jet fuel, and flying over the gray sea he thought he'd left far behind. The sight of it, and its contrast to Santa Barbara, must have been too stark to acknowledge.

But Richard was determined to make his new job with Infabco work. He knew he had to compartmentalize his feelings and bear the emotional weight until he earned enough money to pay off his land and build a house for his family at Hollister. And then he would be free of the North Sea.[33]

As gray as the clouds seemed that day, they had at least one silver lining in them. His friend, Tom Belcher, was on board the *Gulnare*. When Craig Roberts offered him the position of shift supervisor at $285 per day, he accepted.

On November 19, Skip showed up. And it was on board this lonely outpost called the *Gulnare* that Richard and Skip's paths crossed for the first time. How well they became acquainted with each other is not known. But it is known that they didn't dive together on this tour. Richard would remain on deck until his return to Santa Barbara in time for Christmas, while Skip would spend Christmas and New Year's in sat.

In terms of diving fatalities, 1978 was a relatively quiet year. The rising tide—ten deaths in 1974, nine each in '75 and '76—had slowed to five in 1977, and now there had been only one reported fatality in January 1978 when a diver was killed on a 1,000-foot bell run in Norway. But on November 26, the ten-month lull came to an end when reports began to spread about a dramatic rescue attempt to save two divers trapped on the bottom of the Beryl Field, 110 miles south of Thistle.

On board the *Gulnare*, Richard and Skip almost certainly heard about the accident as news of diving fatalities tends to spread from flare stack to flare stack like wildfire. The next day the Aberdeen *Press and Journal* and *The Scotsman* were running front-page articles with banner headlines and graphic illustrations describing the event.[34] But what the reporters didn't know, and therefore couldn't report, was the extent to which the rescue failed: why the diving bell was rigged to go into the water in violation of the safety regulations, why it took so long to find it, and why the divers could not release the drop weight to save themselves.

These details, and their impact upon the outcome of the rescue, as well as their potential to improve lost-bell rescue procedures, would not be revealed to the public for at least another year. By that time, the lessons learned by the dual fatality would go largely unheeded by the diving industry and its government overseers.

Chapter Eight

The SALM

It was only discovered by somebody looking over and saying, 'That buoy doesn't look like it's standing up straight anymore.'

—— Chuck Alexander
Project Engineer for the Thistle SALM

At the beginning of 1979, the North Sea oil industry was at the crest of its heyday. The price of oil, hovering around $12 a barrel, was poised to rocket to $38 a barrel by the end of the year due to the impact of the Iranian revolution.[1] The total diving population had grown to 2,200, and where there had once been 14 oil platforms nailed to the UK Continental Shelf in late 1976, now there were 21. Where there had once been five producing fields, now there were twelve. More than 700 miles[2] of pipeline were delivering crude to shore through five submarine export trunk lines: Forties was linked to Cruden Bay north of Aberdeen, Piper to Flotta in the Orkneys, Ekofisk in the Norwegian sector to Teesside, England, and Ninian and South Cormorant to Sullom Voe in the Shetlands.[3] Not all of the producing fields, however, had access to this export network, and those that didn't had to rely upon tankers to ship their oil to market until additional feeder lines could be commissioned.[4] Initially, Thistle was one of these fields.

Long before the installation of the Thistle jacket, BNOC's predecessor, Burmah Oil, anticipated that the platform would begin producing oil before a pipeline connection could be established to Sullom Voe. In February 1975, Burmah made the decision to build a tanker loading facility in the Thistle Field, and in March of 1977, Sub Sea was paid $1.2 million to install it about a mile from the platform.[5]

The SALM, as it was called, was a Single Anchor Leg Mooring system and it was the first of its kind in the North Sea. It was conceived by Exxon Research and

Engineering of New Jersey, designed by Single Buoy Moorings, Inc. of Monaco, and built by Motherwell Bridge Offshore in Leith, Scotland. It cost $6.4 million to build and about $11.4 million to set into position.[6]

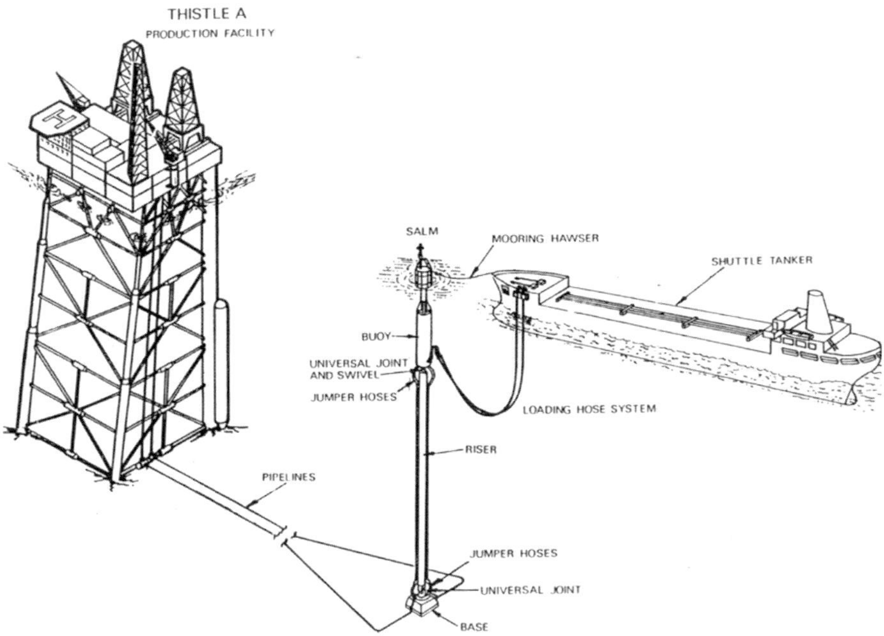

Courtesy of Chuck Alexander

The SALM functioned much like a "filling station" for oil tankers. It had three major components: a yellow-tipped upper section called the buoy, a middle section called the riser, and a gravity base which anchored the facility to the seabed to prevent it from drifting away. From head to toe it was 545 feet long. The 182-foot buoy was 13 feet in diameter, the riser was 335 feet long, and the hex-shaped SALM base was roughly the size of a small house (28 feet high, 47 feet in diameter).[7] Two articulated "universal joints" allowed the buoy and the riser to pivot in rough water.

During operation, Thistle Alpha fed oil to the SALM through a pair of 16-inch pipelines,[a] which approached the SALM from opposing sides. This hard piping ran up the base and terminated at the bottom of the riser near the first U-joint. From that point the oil bridged the articulation through a couple of very thick 16-inch flexible "jumper hoses," then continued up the riser through more hard piping, and over the riser-buoy articulation through another set of jumper hoses. From there, the oil flowed through a pair of product hoses to the shuttle tanker which was moored to the buoy.

---

a Initially, only one of the lines was used for oil, the crude line, while the other, the ballast line, was used to return dirty ballast water from the tanker to the platform.

Although the SALM was installed in March 1977, it did not go into operation until February 1978, when Thistle Alpha began producing oil.[8] It remained in operation for the next eight months, but then on October 15 a "pin hole leak" was discovered in some pipework on the inside of the buoy during a routine maintenance inspection. Preparations were undertaken to fix the leak, but poor weather conditions interrupted the repair work.[a]

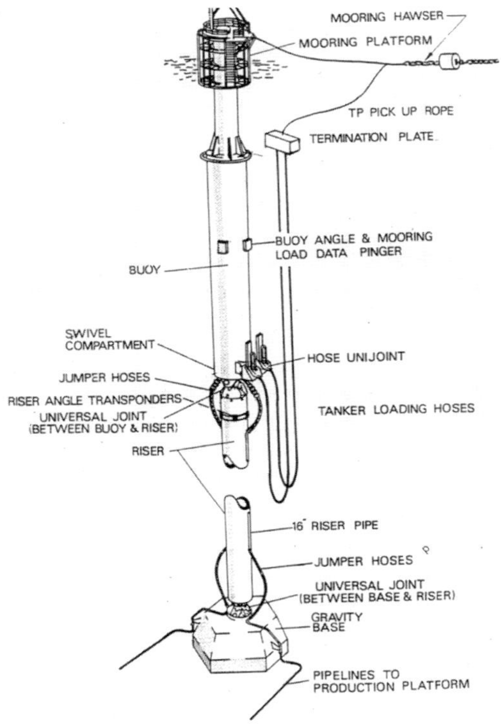

Courtesy of Chuck Alexander

SALM components.

Over the next three months, as the buoy tossed about in one of the roughest seas in the world, the pin hole leak developed into a split. At the same time, a crack in the 16-inch pipework outside the buoy allowed water and oil to flood the central trunkway.[9] This caused the buoy to lose more of its buoyancy throughout the winter months. The loss of buoyancy in turn caused the upper U-joint to go into compression which allowed the 2,404 pound U-joint pin to wrench itself free.[10]

On January 17, 1979, someone on board the Wharton & Williams diving vessel, *Stena Welder*, reported that the buoy was not standing up straight anymore. Attempts were made to dewater the buoy, but two days later, more water was found to have flooded the internal compartments. A diver was sent down to investigate the buoy-riser U-joint and discovered that the pin had fallen out.[11] The only thing keeping the buoy from floating away were the jumper hoses bridging the articulation.

On January 20, after the buoy sustained more flooding, a diver reported that it was now actually grounding on top of the riser. Alarmed by this news, BNOC immediately notified the London office of Chuck Alexander, the project engineer responsible for the 1977 installation, and ordered him to mount an emergency operation to rescue the 340 metric ton buoy before it sustained irreparable damage, or worse, broke free and became a shipping hazard.

---

a Although the SALM was taken off-line, BNOC was still able to get its oil to market because the export line that ran to shore via the Dunlin Field was now in operation.

"Easier said than done," Alexander recalls. London Weather Center was forecasting a big storm front closing in on Thistle.

That night Alexander flew to Aberdeen,[12] and by the next day, he was taken out to the diving support vessel, *Tender Carrier*. After further inspection, Anderson discovered that some of the buoy's tanks were flooded with oil which he thought was odd.

Meanwhile, back in Aberdeen, there was a mad scramble to locate a pair of specially modified, 400-ton[a] Green Pin® shackles that were lying in a storage yard somewhere. Alexander knew the shackles would be needed to salvage the buoy because his team had used them during the 1977 installation. When they were found BNOC had them loaded onto a supply vessel bound for Thistle. It also hired the huge crane ship *Pearl Marine 27* on an emergency basis and requested it to report to the Thistle Field to recover the buoy.

On board the *Tender Carrier*, Alexander asked the Wharton & Williams diving superintendent to disconnect the jumper hoses, but he says the divers refused to go near the buoy because it was violently bouncing up and down and too dangerous to work on. The jumper hoses were a bonded composite of steel wire and very thick rubber. They were flexible, but very stiff, and disconnecting them required that they be rigged and tediously unbolted from the buoy/riser assembly, not an easy job to do even in the best of weather conditions.

With a storm approaching and the buoy unmanageable, Alexander decided the next best thing to do was to to blow the hoses off with explosives. BNOC, however, didn't have any explosives experts, but Shell Oil did, and they were eventually brought out to the ship with their gear. When the problem was explained to them, they said, "Right, we know what we'll do," and they took detonating cord and fashioned two "half-moons" with a hinge to allow the diver to "clam shell" the explosive wreaths around each of the 16-inch hoses. The diver would have to swim 170 feet down to the job site, fit the wreaths, then lead the detonating cord back to the surface, which he did.[13]

On January 23, when the signal was finally given, Alexander heard a bang and saw bubbles come to the surface. A diver was sent down to have a look and discovered that, although both charges had gone off, only one of the hoses had been cut. No one knew why, but now the buoy was listing even more precariously than before.

About that time, the silhouette of the 669-foot-long *Pearl Marine*, with its 2,000-ton Clyde crane sticking up in the air, appeared on the horizon with dark storm clouds looming in the background. The situation was growing more tense by the minute. At the last moment the divers finally relented and agreed to unbolt the remaining hose. The 400-ton shackles were attached to the top of the buoy, and at 0700 on the 24th, it was raised out of the water and onto the deck of the

a Safe working load.

Pearl Marine.[14] And just in time. As the vessel headed for a fabrication yard in Amsterdam, the weather was now rotten; it was snowing; it was blowing, and the seas were starting to build.

After the storm subsided, BNOC ordered Wharton & Williams to inspect the top of the riser to see if it had sustained any damage. This time, instead of using a diver, an ROV (Remotely Operated Vehicle—a kind of mobile robot with a video camera) was sent down to make a video inspection, and the report came back that the riser had sustained no damage to it, and BNOC officials breathed a sigh of relief because the knuckle joint, where the pin had fallen out, was a crucial component of the SALM with precise dimensions. With the emergency over, the inspection tape was then mailed to Alexander in London and no one thought much more about it.

Several weeks later, Alexander decided to take a look at the tape "just for fun." He inserted it into a VCR and began watching it—casually at first—but then more closely with a lot of concern on his face.

"And I started to see parts of the riser that were not familiar to me," Alexander recalls. "I was involved with the building of the riser, and I knew what the thing looked like. All of sudden I was seeing unfamiliar joints I don't expect. I finally realize that the video was being taken *inside* [the riser]!"

Somehow the wreathed charge had blown a big hole in the top of the riser. The explosion had also rocked one of the 16-inch flow lines out of its clamps, displacing it 30 feet below its original position.[15] BNOC now had no choice but to remove the riser and take it to shore for repair.

That March, BNOC officials considered whether there was enough time in the 1979 season to complete the necessary repairs and re-install the SALM.[16] At one of the meetings, the question was raised: was it worth spending millions of dollars to repair and re-install the SALM, now that the pipeline from Thistle to Sullom Voe was in operation? According to Chuck Alexander, BNOC's managing director said, "Yes, we're going to do that because we need backup just in case something happens to the pipeline."[17] And so on March 20, 1979, BNOC authorized the repair and re-installation of the Thistle SALM.[18]

The following month, on April 15, the massive semi-submersible workshop, *Narwhal*, was brought in on a "no cure, no pay" basis to remove the riser.[19] Sub Sea divers stripped the riser of loose objects, cleared the base of debris, attached a pair of 500-ton shackles with six-inch diameter wire rope slings, then removed the pin connecting the riser to the base. By the 24th, the *Narwhal* was steaming southeast towards the Norwegian town of Bergen with the riser laying across her deck.[20]

I'm going again to the North Sea. I was really depressed yesterday.

—— Richard Walker, April 24, 1979

That same day Richard found himself flying to Scotland again. The previous December he had moved his family from the Trout Club to a small house on Alisos Street. He tried to find a better job—anything to keep from going back to the North Sea—but the better job never came. On April 27, he landed on board the *Gulnare* at 4 o'clock in the afternoon and was put on the nightshift under the supervision of Tom Almon.[21] Over the next several days the crew was conducting surface diving near the 80-foot level on the jacket, when a huge storm blew up with 70-mph winds and 30- to 50-foot seas.[22] On the first of May, the *Gulnare* pulled away from the platform and for the next five days, Richard worked the night shift in the freezing snow, re-taping burning gear and changing out the bell winch wire. One of the men who worked with him would later make an entry in his journal: "Everyone was soaked to the bone. Laughing at the silly situation we had all committed ourselves to."[23]

Two days later, Skip arrived and was put on the dayshift.

Forty-eight hours later the weather cleared and for the next two weeks the crew resumed bell diving. On the 22nd, Skip was put into sat with Bob Gemmell, his roommate from dive school. Bob had come to the North Sea on the advice of Skip, was initially hired by Solus Ocean Systems, then migrated to Wharton & Williams where he found the work spotty. When Skip joined Infabco, he told Bob about the *Gulnare* job and by December of 1978 he'd become a member of the regular crew.

Then, on the 25th, after spending nearly four weeks in the sat queue, Richard was blown down into sat with Howie Hodgson. The diving was mid-water, from 260 to 328 feet and primarily inspection work. Perched on one of the horizontal members while wire brushing welds and looking for corrosion, all Richard saw as he peered over the side was black fog. On several dives, the platform dumped drill mud on top of him cutting his visibility to zero.

On the 31st of May he picked up his diary and began writing:

> On board *Gulnare*, in sat w/Howie Hodgson. Supposed to dive soon. Been in 6 days now. I've been thinking clear. I do not enjoy or respect this business. The thrill is gone. I love my little girl so much. I want more chances to make a good, healthy family out of us...
>
> On the record, it would be very good for me to get the heck out of this mess. I don't know how—but this particular job is not for me. I really wish very hard for the thing I've chased for so long—the job that would maximize the use of my peculiar talents and thereby, would be enjoyable.
>
> This is now a rut, a money addiction—in debt and this is the only viable, visual way to pay and have time to do. I need my tiny hugs and evenings with Jeanne, my chosen one...

> What is this feeling—I need to let something go, to feel. I am tied to worry. Have become pragmatic. Have lost my adventure, rough and ready spirit...

For the next twenty days, Richard and Skip were on opposing bell teams, passing each other to and from the bell, swapping books and stories during downtime, listening to the latest developments in each other's lives and exchanging information about the job. They heard that Infabco was preparing to bid on the re-installation of the SALM. Richard was interested in the contract because, through his contacts, he learned that Exxon Corporation was planning to install a similar loading buoy off the coast of Santa Barbara in the Hondo Field. And he knew if he gained the experience in the North Sea, he'd have an excellent chance of being selected for the California job. This would put him closer to Jeanne and Marisa.

Financially, Infabco had been good for Skip. The sat work had been steady, there seemed to be a lot of it, and there was no indication that the contract with BNOC was going to end any time soon. In fact, BNOC had recently given Infabco permission to add a second sat system to the rig. With his bank account growing, Skip had already begun taking steps to buy a piece of land in Maine.

Three months earlier, he and Roger had gone in search of land, driving north on Highway 1 along the shores of the Penobscot Bay to Bucksport, and then east to the town of Ellsworth.[24] They continued as far north as Machiasport and stayed the night in a small drafty trailer that billed itself as a "hotel on the shore." To Roger, the surrounding desolation was so stark, it was frightening.

"I mean there's not even any hills," he says looking back. "The highest thing is a rock! The only plants that live up there are, like, blueberries. I'm like, 'No fucking way! Let's get out of here.'" So Machiasport fell quickly off their list.

Later, Skip found a 20-acre parcel on Leache's Point, sandwiched between the Orland and Penobscot Rivers. The asking price was $12,000. It was raw wilderness, thickly forested with a mixture of White Poplar, Northern Red Oak, Paper Birch, Beech, and Maple running down a five-degree slope to a wide frontage of black tidal mud and a few enormous rocks. The air was fresh and sweet, and it was quiet. The only normal sounds were from the wind rustling 100 feet above in the forest canopy, the crickets, the caw of an occasional crow, and the scream of an osprey diving for an Orland fish. The closest town was Bucksport, about four miles away. This was where Skip decided he and Roger would build a log cabin, and in the town of Ellsworth, he hired the law firm of Hale and Hamlin to begin a detailed examination of the property records.

On June 1 Infabco owner Brian Masterson received a telex from BNOC inviting him to bid on the re-installation of the Thistle SALM.[25] Masterson was keenly interested in the contract. In fact, a week before the telex, he'd sent his general manager to Norway to negotiate an exclusive right to a brand-new diving vessel which was then coming off the assembly line. Masterson wanted to use the vessel to re-install the SALM. What's more he was making plans to add a second sat system on the *Gulnare*, and the word was out on the streets that Infabco was hiring.

This expansion news, however, did not appeal to Craig Roberts who thought Masterson was making a big mistake. "I was against it," Craig says reflecting back.[26] He had spent nearly a year assembling the kind of crew he could trust to stay out of trouble, and now he felt Masterson's plans were going to disrupt the balance of the company. Craig's idea was to keep the company small and staffed with high quality people. But with the addition of more contracts, Craig knew the company would be forced to bring in a lot of "questionable entities," people he wouldn't know and couldn't count on. Dive crews were not just a bunch of bodies to Roberts. They spelled success or failure in the way they covered, or overlooked, each other's mistakes. Roberts, however, had no control over what Masterson did with his company.

Back on the *Gulnare*, Richard had been offshore more than six weeks when he reached for his diary again:

> June 6 - Waiting to dive again, melancholy again. Oh how I miss my family. I'm their guy. I should be home. I'm very anxious to see them. I ache inside for them, especially to just hug Marisa and be her friend and Daddy again.

On the 9th, he and Howie began decompressing back to the surface allowing Richard to turn his thoughts to Hollister Ranch and the daunting task of building a house in a remote location. Issues concerning power, water sources, the type of structure, and where it should be located were on his mind. He worried about building restrictions and codes. But even though he worried about these things, he never became overwhelmed by them. He was constantly making lists of things to do to solve the problems. And the more he thought about the Ranch, the more he became excited about the prospect of having a place where he and Jeanne could grow their own food, take up fishing, all in the effort to become self-sufficient. "But I know it will be hard work," he wrote in his diary. "After it's paid, and if taxes stay low, why we can just live. That is my goal."[27]

While Skip and Bob would stay in sat for another week, Richard and Howie's twenty-day sat came to an end on the 14th of June. When Richard climbed out of the chamber onto the windblown deck of the *Gulnare*, he'd been offshore for 49 days and earned almost $14,000. It was Thursday, and for safety reasons relating to decompression protocol, he still had to wait another day before flying off the rig. Through the grapevine he was aware that Infabco had bid on the re-installation of the SALM, that it was preparing to mobilize a new diving vessel for the contract, and that he might be drafted to become part of that effort. Richard, however, was anxious to collect his pay and get home:

> 15 June - 3:30 p.m.
>
> On the chopper. The last one. Craig's gonna want me to hang around for sea trials on this new diving boat...but I must go home. A necessity. No word from [the] office if there's any money there. Don't even know if I can get there in time. It's fucked. I'll get them to pop for a plane fare and take it out of my pay, but I should get a booking today. Almost impossible at this late hour. I hope I get home soon. If it's Monday, I'll be pissed off.

Richard needn't have worried about joining the new vessel. Roberts was already hundreds of miles away in Norway mobilizing what was about to become the most infamous diving ship in North Sea history.

Three days later, on June 18, BNOC awarded Infabco Diving Services Ltd. contract no. 0561 to re-install the SALM.[28]

Chapter Nine

Alarm Bells in the Night

Came on the ship, worked like a dog. There was a big push to get us out of port and nothing was coming together very quick. We'd hear from the upper end that this thing's supposed to be going out to do this job, and we're looking at, 'There's no parts!' So we worked like dogs, and the next thing, we're pressing people into sat wondering if this is going to work.

—— Gary Legere
Infabco Life Support Technician[1]

Smedvik Shipyard No. 66, Tjørvåg, Norway

Like a Titan's trampoline, the helideck sat perched atop the new vessel's towering snow-white forecastle. With her sleek international orange hull, imposing construction crane pointing skyward on the back deck, and twin smokestacks standing like goal posts beside the wheelhouse, the *M/S Wildrake* was one of the most impressive-looking dive ships ever built. No other vessel had her distinctive profile. She was brand-new; she was magnificent, and she radiated an aura of majesty and grandeur that symbolized the advent of a new era in North Sea diving.

Moored to a shipyard pier near the town of Ulsteinvik on the island of Hareid, the *Wildrake* had been built for the shipping firm of Anders Wilhelmsen & Company of Oslo. Her primary function was to conduct diving operations, but she was also classed as a tug supply ship and firefighter. This was no dirty barge or filthy oil rig where diving contractors had to install clumsy portable systems in awkward locations. Nor did they have to waste time marrying the diving plant to the vital systems of the vessel. This was a "purpose-built" diving support vessel with a built-in saturation system where the universe revolved around putting men on the bottom of the sea. She was faster, more versatile, and less costly to operate than a construction barge or rig, and Wilhelmsen, as landlord of the *Wildrake*, was eager to have Infabco Diving Services as its first tenant.

It was early June, and while Richard and Skip were in sat on board the *Gulnare*, Superintendents Craig Roberts and John Nadeau were in Norway with a mission to accomplish—mobilize the *Wildrake* diving system by the first week of July, BNOC's target date to begin the re-installation of the SALM.[2]

Preparing a saturation system, even one with a prior history, requires an enormous amount of work. Like musical instruments in a symphony orchestra—the bell, the chambers, the gas storage system, the umbilical, the environmental control units—everything has to work together in harmony to prove the system safe and reliable before diving operations begin. Mobilizing a brand-new system, right out of the box, presents an even greater challenge because the equipment has never been tested.

Not surprisingly, when Roberts and Nadeau arrived on board the *Wildrake*, they found a ship reeking of fresh paint and a diving system far from complete.[3] Throughout the vessel, welders were still welding, painters were still painting, and technicians and consultants were still ferreting out unforeseen teething problems. And although Wilhelmsen were the owners of the *Wildrake*, Roberts would later learn that the design of her virgin diving plant had been subcontracted to Møllerodden A/S of Haugesund, an obscure company that had never installed a diving system on a vessel before.[4]

Spread over four deck levels, the saturation system was crowned by the bell van overlooking the back deck. Inside, a series of eight windows gave the supervisor a clear view of the bell and its handling system (an enormous davit). Towards the center of the ship, there was a breach in the deck where it looked as though the builders had constructed a small rectangular swimming pool. This was the moonpool. In the same way a rock jetty protects a harbor, this bottomless hole in the vessel was designed to provide sheltered water in which to launch the bell.

Between the davit and the moonpool, the bell was docked to the upper portion of a two-foot diameter "tunnel" protruding above the deck. The lower end of this tunnel was mated to the top of the entrance lock one level below. Through this narrow passageway, the divers would be able to exit the bell after a dive and enter either chamber one with eight bunks inside, or chamber two with four. A few feet away, a small control room (sat van), packed with analyzers, closed circuit television monitors, and a large console studded with valves and gauges, regulated the environmental conditions within the saturation complex. And down a staircase directly below the chamber room was the gas storage room where half a million cubic feet of gas would be packed into 132 torpedo-shaped tubes.

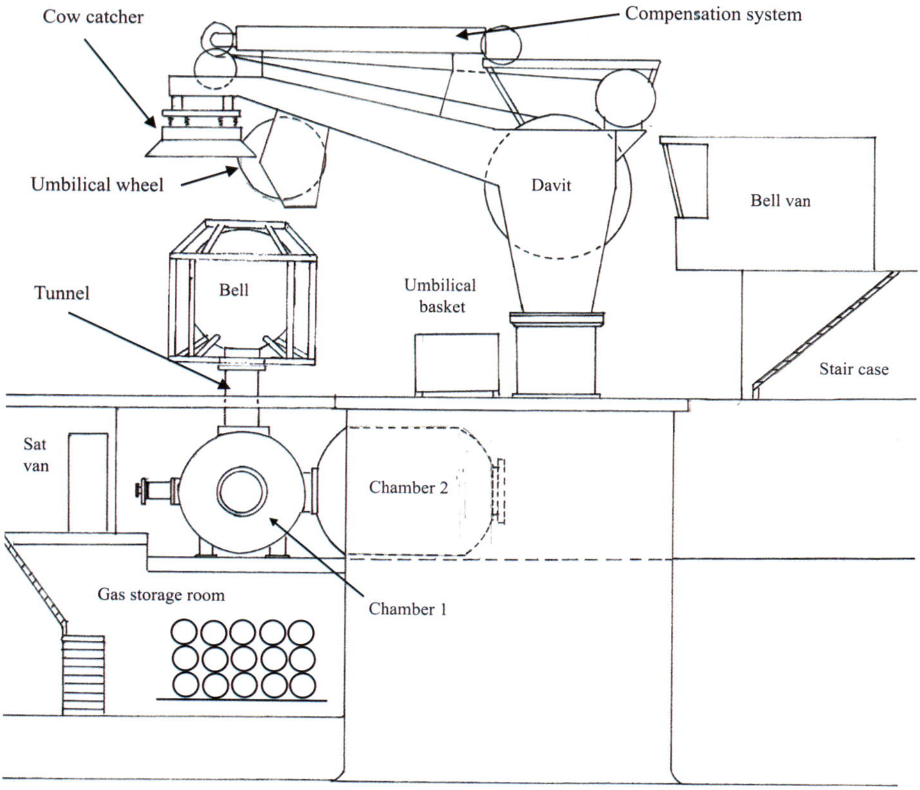

Courtesy of Marcus Taylor Ltd.

The *Wildrake* saturation complex below deck.

Among those assisting Roberts and Nadeau were diver Tim Newson, Supervisor Brian Reid, and Life-support Technicians Mark Hansen and Gary Legere. All had been recently hired by Infabco for the *Wildrake* project, and all were veterans of the diving industry.

Against a deadline to get the ship out of port, and with officials from Wilhelmsen anxious to know whether Møllerodden's system was "workable," Legere says everyone was "really flying at it" to get the system to come together. Twelve-hour shifts blurred to sixteen-hours calibrating depth gauges and gas analyzers, solving problems with the environmental control units, labeling hundreds of valves, trying to find spare parts which didn't exist, and repairing miles of gas tubing because Møllerodden's system leaked like a sieve. Locating these leaks was made more difficult because Legere discovered some of the Norwegian deck hands painting over the gas connections. At the same time, he had to spend countless hours transferring thousands of cubic feet of gas from delivery trucks on the quayside to the storage tubes below deck with a small, high-pressure pump because the ship's primary gas transfer system wasn't working.

Courtesy of Marcus Taylor Ltd.

The back deck of the *Wildrake*.

Legere had come from the ranks of Taylor Diving, one of the premier diving companies in the industry, and he was not impressed by what was being touted as "the most advanced diving system at that time." On the contrary, he concluded that it had been "just thrown together." And in time, that assessment would be shared by others.

As the weeks wore on, Roberts was asked by Infabco to look for a British superintendent to run the installation of the SALM.[5] And it was on board the *Wildrake*, that he became acquainted with an English diver working on behalf of Møllerodden to layout the gas storage room. His name was Peter Holmes.[6]

Tall and thin, with short dark hair and sharp features, Holmes carried himself well and projected an overconfident disposition. One former Infabco supervisor who worked under him says, "If you put Peter Holmes in an RAF uniform, he would have been the epitome of an RAF officer." His resume indicated that he'd been a diver, a lockout submarine pilot, and a diving superintendent. He also had a special talent for working on diving systems. When Roberts asked Holmes if he was interested in joining Infabco for the SALM project, Holmes said yes. Roberts then recommended him to Infabco's office in Aberdeen.[7]

By the end of June, Richard and Skip were on leave in the United States when the *Wildrake* sailed south to the port city of Stavanger to add finishing touches to the diving system and take on additional crew. From all over the world, more than two dozen divers and technicians began arriving for the maiden voyage. From Santa Barbara, Supervisor Ted Roche stepped from his taxi, took one look at the gleaming new vessel and said to himself, *"God, look at this beast. This is bitchin!"*[8]

On July 3, officials of Det Norske Veritas (DNV) stepped aboard the *Wildrake*, conducted tests, and certified the diving system in compliance with the current diving regulations.[9] While the *Wildrake* had met BNOC's deadline, the riser/buoy repair work taking place in Bergen and Amsterdam had fallen three weeks behind schedule.

BNOC's original plan was to reconnect the buoy to the riser in a sheltered fjord south of Bergen, then sail out to the Thistle Field, prepare the SALM base, then attach the SALM to the base. BNOC was allotting 44 days for the project.[10] But because of the buoy/riser refurbishment delays, the oil company ordered Infabco to prepare the SALM base first in the Thistle Field, then sail back to Norway to work on the SALM.

By the time the *Wildrake* began its two hundred mile journey to Thistle, Peter Holmes and Craig Roberts had left the vessel leaving John Nadeau in charge of diving operations.[11] There had been no shakedown cruise to test the new diving system against what the designers envisioned, or what nature would impose on the new vessel.[12]

In the Forties Field, Nadeau had made a name for himself as a talented diver working for Sub Sea International. His ability to burn steel underwater was near legendary. In every respect, this tall, serious man with a deep commanding voice epitomized what a superintendent should be: smart, decisive, hands-on, a good technician that could "keep the trains running on time." One British supervisor had this to say about him:

> I had a lot of respect for John Nadeau because he was a very, very competent guy. He's clever. He's shit-hot with rigging and whatever, and he always had an answer to a problem.

On July 5, the *Wildrake* arrived in the Thistle Field.[13] The captain backed the stern up to the *Gulnare* and Infabco divers Steve Lobban and Dave Lane were set down onto the back deck by basket transfer, a type of white-knuckle ride more hair-raising than anything Disneyland had to offer. The "basket" was a cone-shaped piece of cargo netting attached to the end of a crane wire. You tossed your bags into the center of it, then clung chalk-faced to the outside of the contraption while the crane driver swung you and your buddies through the air like a human tether ball hundreds of feet above the water. There were no safety belts or harnesses to wear. If you let go, you simply fell to your death.

The Thistle SALM base.

That evening Nadeau held a meeting to familiarize the divers with the objectives of the work program.[14] Photographs of the SALM base were displayed showing what looked like an enormous steel "tortoise shell" with 16-inch piping running up two of its six sides. On top of the base, a ten-foot-tall pyramid-shaped socket (called the yoke) provided the attachment point to anchor the SALM. The work program involved clearing the base of debris, changing out a pair of jumper hose connectors, trimming down some gusset plates, preparing the U-joint pin tray and installing a large jack (the "haul-down jack") for drawing the bottom of the riser into the yoke.[15] To complete the work, Nadeau selected four divers to go into sat: John Joly, Dave Rhodes, Butch Wright, and a diver named Adrian Moore.

On July 7, the *Wildrake* took up a position over the SALM base and diving operations began.[16] Adrian and Butch blew down into sat and were lowered to the work site. Ted Roche was supervising the first dive when a conflict developed between the two divers at the end of the bell run. Eventually, order was restored in the bell and Adrian was deemed to be the culprit. When the bell was brought up he was promptly decompressed alone and sent to shore.[17]

As Adrian's replacement, Nadeau picked Tim Newson, a veteran sat diver with a steel-trap memory and probably the only guy in the North Sea who ironed his work shirts offshore. Newson came from Cornwall, and in 1970, after seven years as a merchant marine running cargo between New York and Fort Lauderdale, the soft-spoken Englishman told his wife he was going to become a deep-sea diver. For the next nine years, he freelanced with all the major diving companies in the North Sea, the Middle East, the Caribbean, and West Africa; and in March 1977, he was part of the dive crew that originally installed the Thistle SALM.

For the next ten days the two teams of divers worked back-to-back bell runs preparing the base for the attachment of the SALM. The jumper hose connectors were replaced; the haul-down jack was installed; the pin tray was rigged, and the gusset plates were profiled. Only one job remained: the welds on the yoke had to be tested to ensure that they were sound. And because none of the guys currently in sat had the proper inspection qualification, Nadeau had to put another team inside.

At 0545 on the morning of July 18, British divers John Brook and Dave Lane blew down into sat.[18] Both men were new to Infabco, but not new to commercial diving. Lane had been a diver since 1969, Brook since 1977.

It was the final dive to prove that the SALM wasn't cracked, and as it turned out, it would be the longest dive of John Brook's career.

Like a June bug on the end of a string, Brook locked out of the bell and began setting up his equipment on top of the base to employ a technique called Magnetic Particle Inspection (MPI). To find a crack, he first had to apply a magnetic field across the weld with a horseshoe magnet, squirt "magic ink" on it from a squeezy bottle, then illuminate the area with an ultraviolet light. If there was a crack, it would clearly show up as a jagged fluorescent line to which Brook would report euphemistically that he had found an "indication," as oil company personnel did not like divers blurting out frank statements such as, "Hey, I found a crack down here!"

Centimeter by centimeter, Brook moved along each fillet weld with his inspection gear. It was tedious work, but made more so because he had to keep sending the small squeezy bottle to the surface to be topped up, and the black light kept fritzing out on him. Three, four, five hours passed and the former Royal Navy Aircrew Diver began to wonder when topside was going to order him back to the bell. While UK law limited surface dives to three hours, no such limit applied to sat dives.

Finally, after what seemed like an eternity, Brook asked if he could be changed out with his bell partner, Dave Lane.

"And after some lengthy negotiations to which I wasn't party to," Brook recalls, "I was informed that they discovered that Lane had not got the right qualification...and I would have to stay out and finish the job, to which I was not most pleased!"

By the end of the bell run, after finding nothing wrong with the yoke, Brook had been in the water seven long hours. The next day, July 19, as the *Wildrake* steamed towards Bergen, Mark Hansen opened a small bleed valve on the sat panel and started decompressing all six divers from their storage depth of 460 feet.[19]

There were problems with everything when we first started off.

—— *Wildrake* Supervisor Brian Reid[20]

It had taken 17 bell runs to prepare the SALM base. Nobody died. But those were not bell runs without obvious warning signs. The bell handling system, the moonpool, and the bell in particular proved to be a Pandora's Box of defects, both real and perceived. And because the decision had been made to commence diving operations without adequately field testing the system, it was only during actual working conditions, with live bodies in saturation, that Nadeau began to run into all sorts of sundry problems.

Courtesy of Marcus Taylor Ltd.

The *Wildrake* davit.

At the forefront of the troubles was the bell handling system which consisted of a huge davit, a kind of fixed boom mounted on a rotating pedestal, responsible for getting the bell in and out of the water safely. On the end of the davit, the manufacturers had constructed an inverted basket-like device called a "cow-catcher" whose main purpose was to exert sufficient force against the top of the bell to prevent it from swinging uncontrollably while it was being maneuvered on deck. But early attempts to "snub" the bell into the security of the cow-catcher failed because a vital link between the end of the lift wire and the pad-eye on the top of the bell was too long.[21] That crucial link was a 14-inch swivel.

To shorten the distance, Nadeau replaced the swivel with two heavy shackles.[22] Surprisingly, one was a screw-pin shackle. Screw-pin shackles had a reputation for dropping their load and experienced divers knew they were unsafe for applications where the pin could be subjected to side movement. The second shackle in the new arrangement was a safety shackle which remains safe because it allows the pin to rotate during loading without the threat of the pin "walking out."

Where the swivel had once made a solid connection to the bell, now there were two shackles with daylight between them that held hands with the open socket on the end of the lift wire and the pad-eye on top of the bell. Møllerodden's original design, based upon good rigging principles, and with close tolerances with the clevis pins securing the bell to the wire, had been dispensed with.

This was not "shit-hot" rigging practice. It was a Band-Aid fix involving a vital connection. Shackles are designed to lift loads under static conditions, not dynamic conditions, and for years, rigging manuals have advised not to subject them to shock loading. According to Dave Lane, he questioned Nadeau directly about replacing the swivel with these shackles and was told that it was "accepted practice."[23] And in the past, especially at Sub Sea where Nadeau had worked, shackles were employed in this manner, which worked reasonably well from a stable barge.

But Møllerodden's system was operating from a heaving ship. And every deep-sea diver knows that a bell on the end of a 500-foot wire behaves like a marble on the end of a bull whip, and that the much lighter bell does not always rise and fall in harmony with its heaving mothership. As a result, a condition called "snatching" occurs in rough weather with the bell subjected to sudden, violent jerks as the bell and ship move in opposite directions. In that event, the

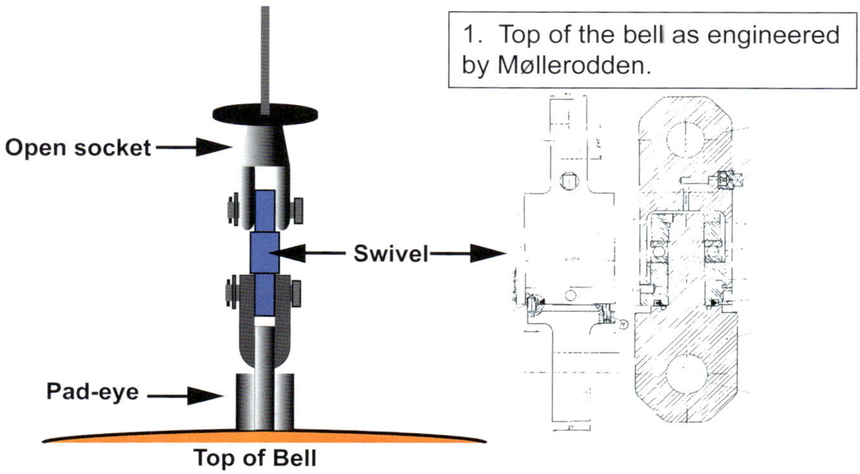

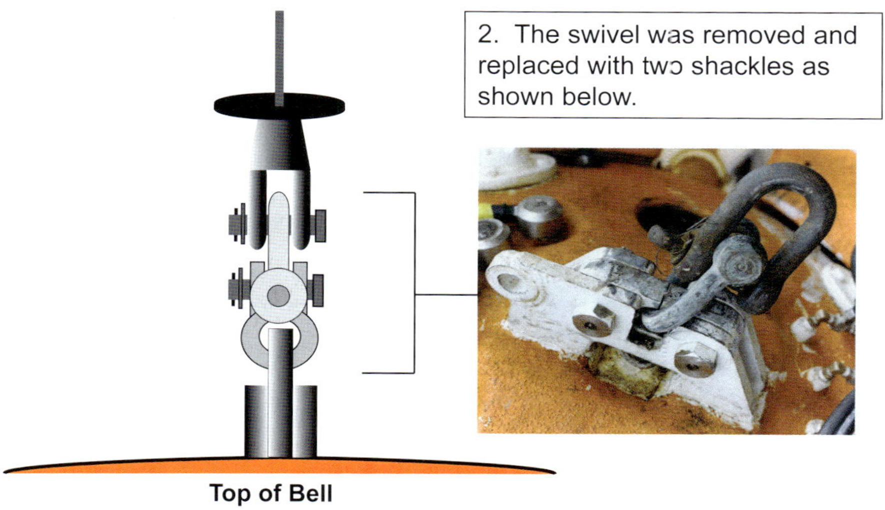

duty upon the shackle pins becomes more severe which would not otherwise exist if the swivel were used. Forces such as shock loading, bending, fatiguing, twisting, and shearing become amplified under these conditions.

And there was another reason why this shackle arrangement was a bad idea. Almost two years later, Craig Roberts would explain it to an Aberdeen Court:

> If there is a swell on the vessel, and the diving bell is heaving with the vessel... the shackles...will move, and where there should be a fairly close tolerance, you end up getting wear.[24]

As foreseeable as this problem was, no attempt was made to restore a solid connection between the bell and its lift wire during the eleven days of diving. On the final dive, John Brook said he could hear the shackles colliding against each other from inside the bell. To him, they sounded like "the ghost walking down the corridor beside you with the chains dragging."[25]

At the core of another problem was Møllerodden's apparent conviction that launching a seven-ton bell from a pitching deck should be done with a davit. Davits are, and were then, dinosaur technology. Sub Sea and Ocean Systems had been using them for years, which worked well on heavily anchored barges and rigs, but not on heaving ships. Instead, companies such as Comex, Hannes Keller, and Drass were employing more expensive and highly effective "cursor systems" which effectively gripped the bell in a "bear hug" and forced it safely through the splash zone under controlled conditions.

Even a diver with moderate experience would have been able to recognize that Møllerodden had borrowed the concept of diving from a barge and was attempting to apply it to a heaving ship. And even a diver with moderate experience would have been able to foresee that, once the *Wildrake* bell left the security of the cow-catcher, the laws of physics would cause it to swing like a wrecking ball, stopping only when it smashed into a solid object or hit the water 30 feet below. Indeed, diver John Joly said some of the crew expressed serious doubts about using the davit before operations began.[26] But Møllerodden did not have the foresight of a moderately experienced diver, and however unsound the davit decision was, it was compounded by a further complication.

For a number of years, manufacturers had been adding "clump weights" to diving bells as a safety feature to recover the bell in the event of an emergency. On a conventional system, the clump weight is a large horizontal bar lowered to the working depth by a pair of guide wires. As it hangs hundreds of feet from the ship like a trapeze bar, the bell is clamped between the wires, then run down to the job site. The supervisor stops the bell about 15 feet above the weight and diving operations begin. If the main winch or lift wire should fail, the clump weight can be raised like an elevator, retrieving the bell to the surface. It is a back-up recovery system with obvious benefits, and Møllerodden, intent on giving the new vessel state-of-the-art appeal, produced one for the *Wildrake* bell.

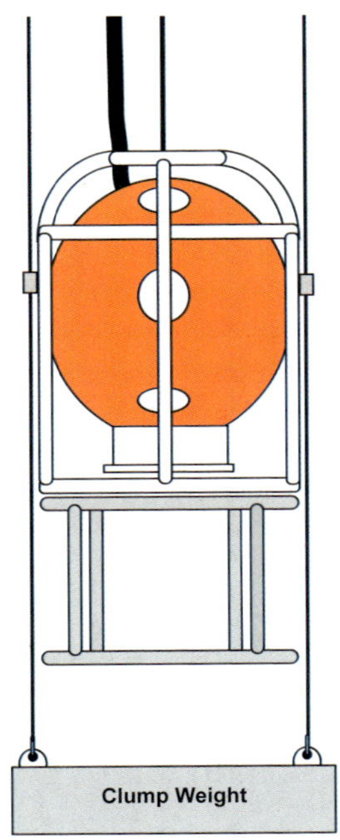

Conventional clump weight system.

But unlike a conventional system, which handles the bell and clump weight *separately*, Møllerodden economized its system by assigning the davit the responsibility of managing *both the bell and its oversized three-ton clump weight simultaneously*.[27]

Perhaps financial pressures created this completely unique, one-of-a-kind hybrid system. Or perhaps, as one diver pointed out later, Møllerodden's engineers didn't fully appreciate the conditions under which its system would be dived in the North Sea. But whatever mistakes Møllerodden made, they were made under the scrutiny of a third company which ironically turned out to be Nadeau's former employer, Sub Sea International.

In early 1978, Wilhelmsen hired the firm of Smedvik Mekaniske Verksted to build the *Wildrake*. Smedvik then subcontracted the provision and assembly of the diving system to Møllerodden, and according to the terms of the contract, Møllerodden's design specification was subject to review by Wilhelmsen consultants, Sub Sea International,[28] a company with considerable expertise in the use of davits but virtually none in the type of sophisticated bell-handling technology used by Comex *et al.*[a]—technology that almost certainly would have saved Nadeau the trouble of modifying the bell.

The *Wildrake*, with its top-mounted helideck acting like a sail in the wind, was so unstable that it rocked back and forth at sea like a ticking metronome. When diving operations began on July 7, the crew launched the bell/clump weight assembly through the moonpool as designed, and as predicted, both objects swung dangerously about threatening the safety of the crew. For the divers in the bell, it was like being inside a wrecking ball as it smashed against the walls of the moonpool. As John Joly recalls, it was "not a nice ride."[29]

Watching on deck, tender Steve Lobban thought the whole operation was a fiasco. "It was horrible," he says looking back. "Scary. It didn't work. There was a real heavy swell coming up in the moonpool and it would knock the shit out of the bell."

That "heavy swell" was created, in part, by the size of the moonpool. Here again, when it came to designing the system, Møllerodden did not follow conventional wisdom. Normally moonpools are built to allow just enough clearance for the bell to pass through the ship. The twenty-foot by ten-foot "swimming pool" in the middle of the *Wildrake* was big enough to accommodate two bells. Moreover, the walls were not baffled to dissipate the surge.[30] They were flat and smooth, and the aeration system, which was supposed to inject compressed air into the water to cause the bell to sink faster, wasn't working.[31] As a result, the moonpool generated a wave much larger than the surrounding sea. Water would wash up and spill onto the back deck then quickly recede like a powerful rip tide leaving a mist of salt spray in the air. Faced with this absurdity, the only thing the divers could do was shake their heads and joke about how they were bobbing on a Force Four sea, while a Force Eight raged on in the middle of the ship.[32]

---

a Within the industry, Sub Sea's reputation for low-technology solutions earned it the nickname "Sub Standard."

Møllerodden, however, designed their system with a trump card. The davit was ambidextrous. Not only could it swing left towards the moonpool, but it could swing right to launch the bell over the portside railing. By adjusting the *Wildrake*'s heading, the starboard side of the ship could be used to block the weather to allow the crew to launch the bell into the much calmer lee. Thus, the moonpool could be avoided altogether. But when the crew tried to launch the bell with the clump weight over the side, they discovered another flaw in the Møllerodden system. The clump weight would not clear the ship's railing,[a] and Wilhelmsen would not give Infabco permission to cut it down.[33]

Nadeau then made a critical decision that would have profound consequences. He decided to abandon the clump weight and launch the bell *naked*, without a stage, on the end of its lift wire. No dangerous clump weight to avoid on deck, no potential entanglements with clump weight guide wires, and no back-up safety system to recover the bell if something should go wrong. By the time Tim Newson took over for Adrian on the third bell run, the clump weight and guide wires were no longer in service.[34]

Nor was the drop weight system immune to tinkering. The emergency ascent system consisted of two flat, crescent-shaped slabs of steel cradled inside the lower bumper ring. The tips of each weight were poised within a pair of steel pockets while the teetering mid-sections were supported by chain/turnbuckle assemblies running up to two "clothes pin" type release mechanisms on the hull of the bell. These release mechanisms were activated from inside by spinning two stainless steel wheels counterclockwise. That action would open the clothes pins allowing the chains to slide free, the weights to drop out of their cradles, and the bell to rise to the surface.

Although Nadeau was not an engineer, what bothered him was the integrity of the release mechanisms. They didn't look strong enough to remain intact during diving operations. If the bell began snatching on the end of its lift wire, a sudden jar might cause the weights to break away. And it was well known what the consequences would be if that were to happen. The names of Pier Skipness, Robert Smyth, Clay Ellis, and Derek Bannister had long since faded from memory, but the tragic events of January 17, 1974/1976 still haunted the working diver, testimony that accidental release was no foolish fear, but that ballast systems worked all too well even when they weren't supposed to.

Nadeau then made another crucial decision. Rather than fix the release mechanisms or require Møllerodden to prove they were safe, he ordered the drop weights lashed to the bell frame with white nylon rope.[35] This was one of those bad seminal moments in history—a decision so brazen and feckless—because it effectively left an essential, last-resort emergency ascent system inoperable.

---

a Tim Newson and Steve Lobban said the clump weight did not have a conventional horizontal bar shape. It was tall and basket-shaped, like a Sub Sea stage without the emergency bottles.

Lashing the Drop Weights

1. This illustration shows the Wildrake drop weight before modification as designed by Møllerodden.

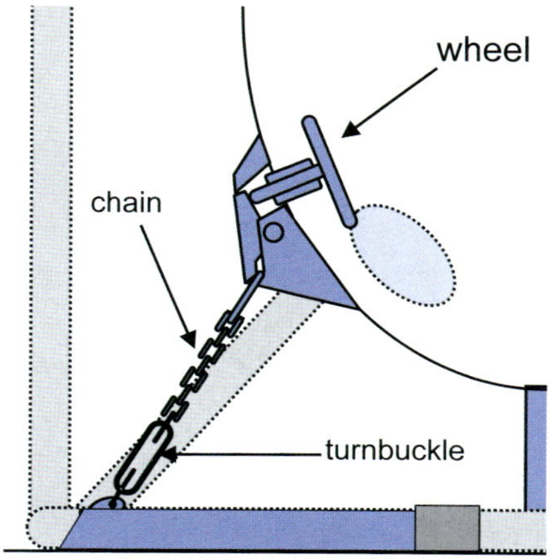

2. To release the weight, the diver turns a wheel inside the bell. This opens the release mechanism and the weight falls.

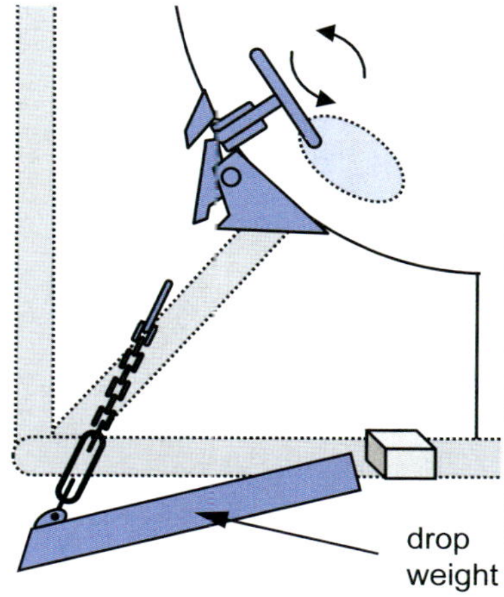

3. Modification: The turnbuckle/chain assembly was replaced by chain, and pad-eyes and rope were added.

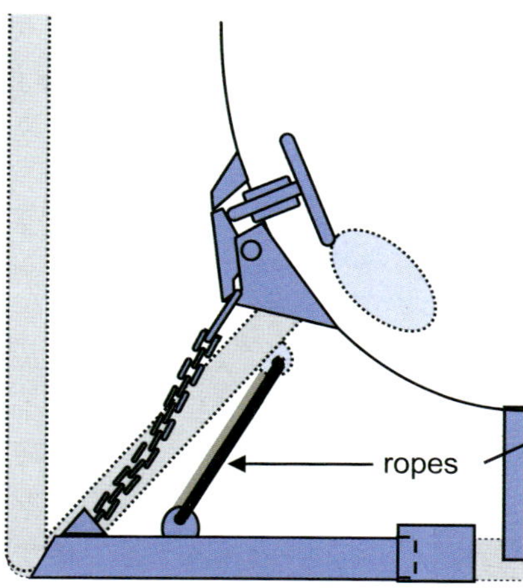

4. Before the accident: blowup of the bell reveals the modification.

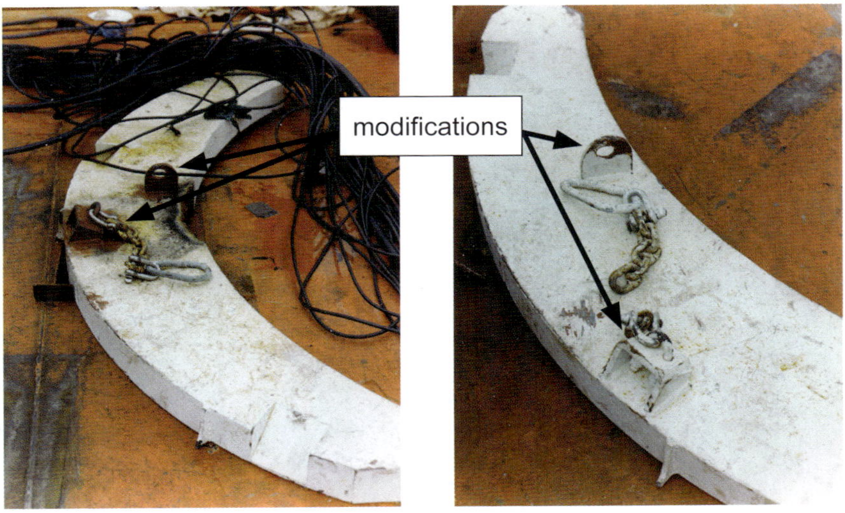

Drop weights showing the modifications.

In defense of this action, Supervisor Frank Montafia says it was argued that the diver could exit the bell during an emergency, cut those lashings, and then climb back inside to release the weights. That rationale is deceptively attractive but it falls completely apart in a worst-case scenario if the bell becomes marooned on the seabed with its bottom hatch in the mud, a situation entirely possible in this case because the *Wildrake* bell had not been fitted with a stage. Møllerodden had been directed to deliver only a "sitting bell."[36] Nothing more. And with no clump weight or underframe for the bell to sit on, it was obvious to the crew that if something should go wrong and the bell did land up on the seabed with its bottom door in the mud, the divers would be trapped inside and unable to exit to cut those lashings.

That would be a very unwelcome scenario, but it was not a hypothetical risk. Eight months earlier, an accident in the Beryl Field (the *Canopus* accident) produced exactly that set of circumstances with fatal results. The two divers, Tony Prangley and Michael Ward, became stranded on the seabed inside the *Canopus* bell without life-support services because all connections to the mothership, including the bell umbilical, the main lift wire, and the clump weight system had been ripped away during the accident. Moreover, Prangley and Ward could not save themselves by releasing the drop weight from inside because the weight had been secured on the outside of the bell with a set of "secondary locking pins." Had the bell been fitted with a stage, their chances for survival would have been greatly enhanced.

By nature, divers live and breathe attempting to anticipate problems that might arise. They play worst-case scenarios in their minds trying to foresee every contingency under which something might go wrong. The whole act of planning a dive, for example, *is an exercise in foresight*. They know that it is never a single problem, but a whole constellation of unattended problems that combine to produce a cascade of failures that leads to a catastrophic event. They are keenly aware of this because they carry around the memory of previous fatalities to remind them of the consequences of not taking care of business.

That the *Canopus* scenario was possible, and had recently occurred, should have acted as a deterrent to Infabco's supervisory personnel not to lash the drop weights particularly on a bell with no underframe. They knew it was safer to dive the bell with a stage because they had a discussion about fabricating one.[37] Yet, in spite of the fact that they understood the safety reasons for using a stage, and the magnitude of the danger if one were absent, diving operations carried on without one.

It was clear that the *Wildrake* diving system was seriously flawed. The davit was capable of launching the bell over the side, but unable to bring along the safety element of its clump weight; the bell would not snub into the cow-catcher; the swivel was too long; the moonpool too big; the drop weight system was considered unsafe; there was no bell stage; the aeration system did not work. All of these weaknesses—without the discovery process of prior field testing—put a well-respected superintendent, a man who "always had an answer to a problem," in the untenable position of permitting safety to be degraded, ultimately slipping past the threshold of what was considered reasonable into the realm of the illegal.

When the drop weights were tampered with, the entire diving system lost its certification then and there. "It's like your car insurance," says Commander Warner of the Department of Energy, "if you alter the engine, you negate the insurance."[38]

The UK diving regulations specifically states that "all diving plant and equipment used in connection with diving operations" must be examined and tested *after* it has been altered or repaired, and *before* it is put back into service.[39] Approval of the alteration can only be granted by an outside Certifying Authority such as the DNV, not the interested diving company.

In addition, the diving bell must have an approved secondary means for returning it to the surface. When the DNV sanctioned the *Wildrake* bell on the third of July, they authorized the drop weights as satisfying this requirement. The clump weight system was not tested and therefore not authorized as the secondary recovery system because the DNV inspector recognized that the bell could be used without it.[40] On page two of the DNV certificate, it was stated that if essential alterations were made to the diving system without prior DNV approval, the certification became invalid.[41] Lashing drop weights with nylon rope was an essential alteration because the drop weight system constituted the only authorized secondary means for recovering the bell.[42] Thus, the *Wildrake* bell was not in compliance with the regulations.[43]

As the employer of divers, Infabco Diving Services had statutory as well as common law duties to furnish a diving system that was safe and in good working order whether Møllerodden provided a defective system or not. If the drop weight system was in such questionable condition that there was no alternative but to violate the safety regulations by disabling it, then clearly the bell should not have gone into the water.

Beleaguered by problems, Nadeau was nearly put in the position of having to put his foot down and tell Wilhelmsen, "I refuse to dive." But there were other options available to him. Møllerodden could have been asked to prove their release mechanisms were properly engineered and safe. A bell stage could have been fabricated when the clump weight system was abandoned. When the swivel was discarded, a better one could have been procured from a rigging store. The same foresight used to predict the "possible" failure of the release mechanisms should have been able to recognize that no stage, lashed drop weights, and a jury-rigged connection to the bell formed a deadly and illegal combination.

That Nadeau would try to keep the job rolling by modifying the system is understandable; divers routinely make minor changes with the intention of enhancing safety. But it is difficult to see how these alterations advanced that goal. As isolated events, each change may have seemed inconsequential, perhaps even harmless. But in time, their true import would come into sharper focus. Seventeen years later, Dave Lane would candidly remark that "they were so prepared to cut corners just to get the job done."[44] John Joly would conclude that the tangle of problems resulted from the fundamental conflict between the lack of time to prepare the system and Infabco's zeal to get the *Wildrake* on contract.[45] Gary Legere and Tim Newson would recall that there was "a rush" to get the vessel to sea. The rush, the lack of time, the illegal modifications—these were all symptoms of yet another, more insidious dynamic at work: that safety was being sacrificed to avoid upsetting BNOC's work schedule.

Nowhere is the pressure to get the job done more keenly felt than in the oil industry. This fact was memorialized in the Department of Energy's very first Diving Safety Memorandum:

> There is no doubt that, at all times, everybody involved in offshore exploration and exploitation for minerals are under considerable pressure due to the vast amount of money invested. However, this situation must not be allowed to put unnecessary pressure on diving supervisors to continue operations when they consider that conditions are unsuitable.[46]

Like a ripple in a pond, job pressure begins at the top of the food chain, in an oil company boardroom, and by the time it falls upon the shoulders of a diving superintendent it can reach the proportions of a tsunami. Wielding colossal and absolute power, it is the oil company's schedules and bar charts that overlay all others, and delays in their programs created by subcontractors are not accepted very graciously. Delays impact negatively on reputations, on government expectations for self-sufficiency, and on profits. And no lowly diving superintendent wants to be caught in the crossfire holding up the job.

The quest for speed was unrelenting. And the sense of urgency—the need for an "over-riding objective" to secure the most thorough and rapid exploitation of the Continental Shelf was coming home to roost on ships like the *Wildrake*. Moneylenders had to be paid back and delays in schedules invariably translated into increased development costs,[47] generating pressure on contractors to take shortcuts in safety in order to meet economic goals.[48]

Even before the SALM contract was signed, Infabco had spent more than three weeks mobilizing the *Wildrake* at its own expense.[49] Revenues would not begin to flow until services were rendered. Had Nadeau shut down diving operations to allow Møllerodden's engineers to effect repairs, the *Wildrake* would have immediately gone off-hire. What would happen next would depend largely upon the relationship between BNOC and Infabco. If BNOC officials were supportive and the *Wildrake's* problems minor, then they would have probably cut Infabco some slack and said, "Go fix the diving system and bring it back," with no damage to the contract. But if the problems were significant, taking a lot of dollars and many weeks to fix, then the financial consequences for Infabco would have been more serious. Former diving company owner Ric Wharton says that in these early days the oil companies were "ruthless" about terminating contracts if their schedules began to suffer.

"Their ability to run people off the contract is the starting point for negotiations," Wharton says. "They can terminate it whenever they like."[50]

And that "ability to run people off the contract" was clearly spelled out in the "Conditions of Hire" in the telex BNOC sent to Infabco inviting Masterson to bid on the re-installation of the SALM. The first condition read: "Period of hire would be a minimum of seven (7) days and thereafter *on a day to day basis terminable at any time.*" (Italics added.) And condition four emphasized the need to perform: "The contractor's performance during the period of hire *is of the essence of the contract...*"[51] (Italics added.)

British Petroleum ran Comex off the *Thor* in the Forties Field in 1975; BNOC eliminated Sub Sea from the Thistle Alpha platform in 1978. The diving industry was a very cold-blooded cut-throat business. Møllerodden's system had a host of systemic problems and there was no *a priori* reason to believe that these "flat tires" would not lead to a similar fate for Infabco.

Faced with that unpleasant threat, Wharton says contractors tend to cover up and keep their problems to themselves.[52] This leaves the superintendent in the unenviable position of trying to maintain a high degree of safety while trying to keep the coffers filled for his company. In a perfect world, he has 20/20 vision and sees the distinction between these conflicting roles and answers to the one with the higher priority. And when he is brought to the threshold of choosing between right and wrong, he is sure to make the right decision and face down client pressure because he is convinced that his first duty is not to the client, not to the company for which he works, but to the diver whose life is depending upon his good judgment. He also knows the limits of modifying a diving system, and that, regardless of the legal aspects, there is no moral justification for sending a man to the bottom of the sea with lashed drop weights.

Strangely, there are those in the diving industry with hearts of blue steel who defend the practice of cutting corners. They've adopted the corporate view that the needs of the job justify the sacrifice of safety. Their attitude is that divers are highly paid and should expect and accept whatever degree of danger the company chooses to impose upon them, whether legal or illegal. Not surprisingly, this perspective typically emanates from the rough-and-tumble environment of a well-heated office, not from within a cold bell at 500 feet. And every time a diver is killed, these "office divers" trot out pathetic excuses to explain why the fault lies somewhere other than the policies they create and enforce. But that's all they are, just excuses in the face of tragedy and after the tombstones have been laid.

The reality is, oil companies push hard to stay on schedule. They try to cajole, threaten, and even play one superintendent's weaknesses off another to achieve their goals. But the superintendent can turn around and tell the client that the conditions are unsafe to dive, that they might even kill the diver, and 99.9 percent of the time the client will back off because he doesn't want to be the cause of another man's death.

For the superintendent who does his job well, the real test of character is not how fast he can get the job done by circumventing the law—anyone can do that—but whether he protects his diver when the job heats up and the pressure is on. Is he able to recognize a catalogue of defects as a red flag waving in his face? Does he have the integrity to turn to the client and say, "We have to stop work to make repairs."? Because if he can't do that, then he has no business sitting in the driver's seat. He doesn't belong there because he won't be able to prevent his job from becoming a death trap, and when the crisis comes, he won't have the courage to make the right decisions to get his diver back alive.

Years later, several Infabco divers would question why Nadeau didn't insist on having elements of the diving system fixed right away. One former superintendent would comment that, while Nadeau was "a top-notch sort of guy," he may have gotten "too big for his britches" during this period of time. Still another diver would remark ruefully that Nadeau was "just a superintendent of the times."

He was also a man who had trains to keep running. The swivel that was too long, the unmanageable clump weight, the absence of a bell stage, the lashed drop weights—these were all safety-net issues that were dealt with expeditiously with no charge to anyone and no delay to BNOC's schedule. The job could keep rolling, the vessel could remain on contract. Redesign of a drop weight system would cost money, a new swivel would have to be ordered, a bell stage fabricated, the moonpool re-engineered. This all takes time, and the choice of turning the *Wildrake* bow around and heading for Stavanger to take that time was not an option that John Nadeau chose.

Chapter Ten

Hobson's Choice

It was one of those deals where—this is the way we are going—and it was either make the best of it, or leave the job.

—— John Joly, Infabco diver[1]

On the day Mark Hansen began decompressing the six *Wildrake* divers, Richard and Skip were beginning their long journey back to Scotland. It was July 19 and both men had been asked to join the *Wildrake* in Norway to complete the re-installation of the Thistle SALM.

In West Springfield, Skip had been home for only 25 days, barely enough time to forget about the North Sea, but long enough to see Rick and Sheila's new baby, Rebecca, now just two months old. He was so proud of his sister and so overwhelmed with joy that it took him the better part of a week to stop grinning.

At his sister's house, he collected his mail and found good news in a month-old letter from the law firm of Hale and Hamlin.[2] The property he wanted to buy in Maine was free and clear of all legal encumbrances, so Skip mailed a $12,000 check and instructions to the firm to finalize the purchase of the land. At a family celebration on the Fourth of July weekend in Maine, he took Sheila's husband up the coast to show off the property.

"We're going to do fish farming," he told Rick as they walked the ground.

Rick took one look at the marshy coastline and said, "What? How do you plan to fish? Victor, what are you going to..."

But Skip wouldn't listen to Rick's doubting words. "Don't worry about that," he shot back. After nearly four seasons of living on the end of a thin rubber hose, he knew the challenges he would face in the future would pale in comparison to those he had already met on the bottom of the sea. They drank a couple quarts of Budweiser and then Skip talked about the new boat Infabco had on contract. "I can't wait to get out there," he said.[3]

Several weeks later, he was driven to Bradley Airport by family friend Harry Tranghese for his flight to Scotland. They sat down in the lounge and had a drink and Skip mentioned that when he got back, his land sale in Maine would probably be finalized, and then he took a wad of cash out of his pocket and passed it to Harry. "I don't need this over there," he said. "Give it to my sister."[4] And then he was gone.

That same day, at Los Angeles International Airport, Richard was in an inconsolable mood. Marisa was now 14 months old and he was leaving her again. The North Sea was the answer to his financial worries, but leaving Jeanne and Marisa for another six-week tour was unbearably painful for him. Just before boarding the plane he pulled out his diary and began writing:

> LAX again. Broke down and sobbed last night. Flight to London in ten minutes and I don't feel good about leaving my fine two loves...I'm really broken hearted. How I love them...[5]

Two days later, Richard joined Skip in Aberdeen, and with fellow Infabco divers Lacey Lahren and Dave Mitchell, took an early morning flight to the Norwegian port city of Bergen. There, they transferred to a small single engine seaplane waiting at the dockside. Their destination was the *Wildrake* lying in a remote fjord about 26 miles south.

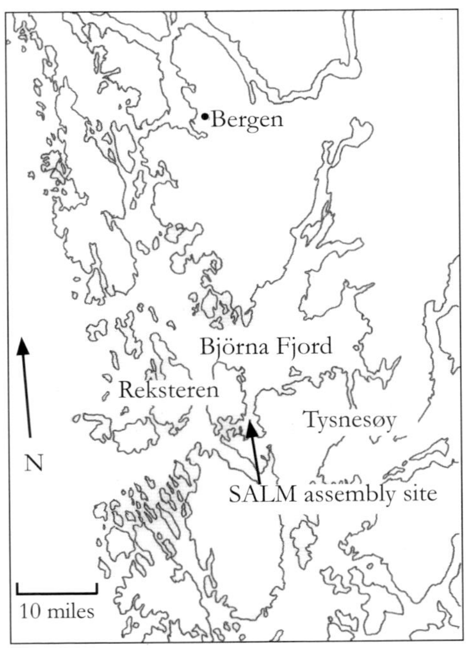

The Cessna 206 took off, quickly gained altitude, and began flying under a blanket of silver clouds and over one of nature's most spectacular sculptures, the fjords of Norway. For tens of thousands of years, the Nordic coastline had been etched and carved by glacial ice to produce a maze of seaways and inlets that must have inspired the Vikings to look beyond their horizons.

From the plane, the four divers watched as the picture-postcard coastline of Rauna Fjord meandered in and out like a puzzle piece for fifteen miles until it emptied into the much larger Björna Fjord. Here and there, small olive-green islands capped with verdant forests of birch and ash popped up like mushrooms from the blue-gray sea. Ten minutes later the pilot began his glide path down a narrow channel of water which opened

Courtesy of Marcus Taylor Ltd.

The *Wildrake* in Björna Fjord.

into a protected body of water trapped between the islands of Reksteren and Tysnesøy.

As the pilot maneuvered to land, sea gulls danced on air streams high above the matted surface of the water. There were no villages, no roads, no industry, no evidence that man had ever been here, only a rumpled landscape of forested mountains spilling down to the water's edge. The serenity was breathtaking, and were it not for the brightly painted ship floating in the water below, time would have been lost in this almost prehistoric setting.

The *Wildrake* with two small coastal tug boats and the refurbished riser were floating in the water below. The buoy was still in Amsterdam undergoing repairs.[6]

At 1100 hours, the four divers stepped aboard the *Wildrake*.[7] John Nadeau was still in charge of diving operations. Richard and Skip received their cabin assignments and were put on the dayshift under Supervisor Ted Roche, while Lahren and Mitchell were assigned to the nightshift under Supervisor Frank Montafia.[8]

Courtesy of Steve Lobban

The *Wildrake* bell.

After changing into their work gear, Richard and Skip walked around the ship to familiarize themselves with the diving system. They saw the davit hovering over the bell like a giant praying mantis. By the standards of Sub Sea and Ocean Systems, it was enormous. The shiny orange bell perched within its white protective framework looked like a pumpkin inside a bird cage. A series of emergency gas bottles clamped to the framework stood like Roman sentries, while molded blocks of syntactic foam, used for buoyancy compensation, crowned the steel capsule. The bell had three 14-inch observation ports, not the silly six-inch peep holes of earlier models. A small blue plaque fixed to the outside said Møllerodden, Norway.

Strangely, for all the money spent on the system, very little seemed to have gone into its most crucial life-preserving component—the bell umbilical. It was handmade and taped together, like those Skip had seen divers build in Sub Sea's parking lot—veritable stone age technology—not something he expected to find on a system billed as state-of-the-art.[9]

Richard and Skip went below deck where Mark Hansen was on duty in the sat van. A bank of depth gauges across the control panel indicated the divers were approaching 228 feet.[10] It would be another three and one half days of decompression before the hatches were opened.

With the SALM project now entering the second phase of the contract, Nadeau selected Richard and Skip as the first team to go into sat, Lacey Lahren and Dave Mitchell as the second team, and Roger Palmer and Dave Boyd as the third team. All six divers were highly experienced, had worked on loading buoys before or in the Thistle Field for years.

Overall, the objective was to reconnect the buoy to the riser. Of the four methods BNOC considered, it was ultimately decided to join the two sections in the vertical, rather than the horizontal orientation.[11] Once the two components were upended and aligned end to end, the male coupler piece on the bottom of the buoy would be drawn into the yoke on top of the riser by means of a "haul-down-wire" jack system. Then, the connecting pin would be forced through the U-joint. In theory, the operation was much like inserting a bolt through a couple of holes, except in this case the bolt was 14 inches in diameter, weighed over a ton, and was machined with only a 1.06 millimeter clearance through the hole in the coupler piece.[12]

BNOC's man in charge of the re-installation was Chuck Alexander, the same engineer who rescued the buoy seven months earlier when it became detached from the riser. Also on board was Infabco owner, Brian Masterson, who joined the *Wildrake* the day before when the ship stopped over in Bergen.[13] He had decided to personally oversee the remainder of the project and serve as his company's liaison with BNOC.[14] Later, people would wonder why Masterson would take on a duty normally performed by the diving superintendent. He could have stayed in touch with the operation via telex. But Masterson wanted to be at the heart of the operation. After winning the contract to re-install the world's largest tanker loading buoy, he now had the golden opportunity to showcase a brand-new vessel and demonstrate that his company had the expertise and manpower to handle big construction jobs. A successful outcome would likely generate additional contracts. It might even cause BNOC to reconsider its present relationship with arch rival, Wharton & Williams, who currently ran the *Stena Welder* as the in-field utility diving vessel. For Masterson, usurping Wharton & Williams's position at BNOC's throne was part of the game of conquering the competition, and he'd come all the way to Norway to make sure that everything would proceed according to plan.

The next day, July 22, the riser was upended. Dressed in a red checked shirt and wearing a blue hard hat with a radio strapped to his belt, Nadeau initiated the operation by completely flooding the tail section of the riser. This created a small list in the riser which was then towed by a small tug to a safe distance from the *Wildrake* where a bypass valve was opened to allow water to begin filling an adjoining compartment.[15] As the end of the riser gradually picked up weight, the upper portion slowly began to lift out of the water. Twenty minutes later, Björna Fjord looked as though it had a giant straw sticking out it. On the fantail of the *Wildrake*, Nadeau, Alexander, Masterson, along with SBM engineers Pastourel and Davidson, congratulated each other and posed for a round of photographs.

Courtesy of Marcus Taylor Ltd.

Upending the Riser.

Courtesy of Marcus Taylor Ltd.

From left to right: SBM engineers Pastrourel and Davidson, Diving Superintendent John Nadeau, BNOC Project Engineer Chuck Alexander, Infabco owner Brian Masterson.

Courtesy of Lacey Lahren

Courtesy of Marcus Taylor Ltd.

Nadeau and Skip rehearsing the coupler pin assembly.

Over the next several days, the riser was moored against the side of the *Wildrake* and inspected for damage. Nadeau and Skip ran through a dry rehearsal for installing the pin.

Just after 10 o'clock on the evening of the 24th, Tim Newson, John Brook, Dave Lane, Butch Wright, Dave Rhodes, and John Joly emerged from sat gaunt-faced and exhausted.[16] It had taken them more than five days to reach the surface. And it was then that Lacey Lahren began to hear criticisms of the system he was about to dive, that there were a lot of "bugs" in it, and that the *Wildrake* "hadn't undergone sea trials" to put it to the test.[17] Those criticisms caused him and Skip to take a closer look at the bell. Almost two years later, at a public inquiry, Lahren explained what happened next.

"One of the things that Skip and I particularly discussed, that bothered us, was the condition of the umbilical, which was similar to a type we had used before, but it was missing one important safety factor."[18]

That safety factor was a "strain relief," an essential piece of gear that should have been mounted near the top of the bell to prevent the life-support lines from being ripped from their terminal connections in the event that they took the weight of the bell. Normally a clamp or braided wire sleeve (called a Chinese Finger) would have been fitted to relieve any undue strain, but Møllerodden did not supply one and no one from Infabco rectified the situation before diving operations began on the 7th of July. Diver John Joly says the lack of an umbilical strain relief bothered the divers, but because "there was enough pressure on to get things going," it was left unresolved. As a result, for 17 bell runs, topside merely reminded the lockout diver to make sure that the umbilical had enough slack in it before he swam from the bell to the job site.[19]

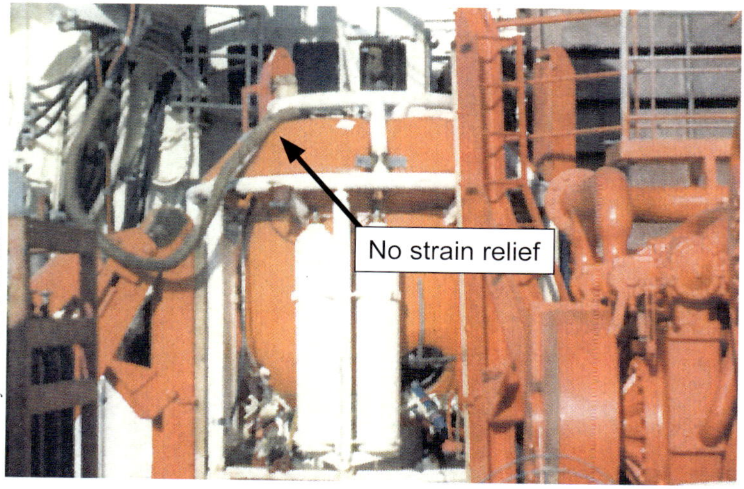

Courtesy of Steve Lobban

But Skip and Lacey were not happy with the arrangement and were granted permission to fabricate and install a clamp, which they did over the next several days.[20]

No doubt, in talking with the other divers, Richard and Skip heard about the deficiencies of the moonpool and the reasons for abandoning the clump weight. No doubt, they saw the shackles on top of the bell and the jury-rigged drop weight system, a modification that Lahren would later characterize in open court as being "a ridiculous procedure" and "a fairly dangerous practice."[21] And, according to Lahren, all of the divers, "to a man," objected to not having a stage on the bell. "We wanted a way of being able to get out of the bell with the umbilical free, the lift wire free," he recalls.[22]

But as so often happens in the diving industry, divers tend to voice their grievances obliquely, amongst themselves, rather than directly at the one person who has the power and control over the situation. In time, friends, relatives, and colleagues would wonder why Richard and Skip—two very capable and experienced divers—would subject themselves to the added risk of diving in a bell with no stage and lashed drop weights. Why didn't they, or someone during the first sat, object or refuse to dive if they were dissatisfied with the system?

At the inquiry, Lahren tried to shed light on this seemingly incomprehensible behavior when he was asked if he had expressed his criticisms to anyone in authority before he used the bell.

"I believe I made comments in passing to people in authority," Lahren replied, "but I did not make a specific point to anyone in particular—largely for the reason that this is not something that is customarily done. It is obvious in that I realized what the hazards were and subjected myself to them."[23]

Regarding his dissatisfaction with the drop weight system, the examiner asked, "Did you do anything to have the drop weight system improved, or at least to have the criticisms brought to the attention of someone who could deal with such a problem?"

"No, it is more or less the situation—if I were taking a flight from Aberdeen to London, I would not go up and discuss the procedure with the pilot, although I am riding in his plane."

"You would be committing your life to the bell and its safety devices?"

"Yes, and I do when I get into an aeroplane."[24]

Later, Lahren would come back to the issue of trust, and try to explain more fully how deep the bond must be between the diver and the people who are managing his safety.

"Knowing that you have mechanical faults in the system," he told the Court, "you rely very heavily on your supervisory people to take care of you. The man in the saturation control van is your mother and father, he feeds you, he flushes your toilet, he gives you shower water. The man who sends you into the water is like a God to you. You are dependent on him to bring you back....You must have faith in people who are taking care of you."[25]

And Lahren had complete faith—almost blind faith—in Nadeau because Nadeau was an iconic figure, one of the "gods on water." And supreme beings have a way of generating a kind of persuasive inertia among their crew, even to the point where the diver is willing to accept—rightly or wrongly—their judgment to dive a system with mechanical faults.

Still, the core issue—what prevents an experienced diver like Lahren from complaining about the system directly to his superintendent—had not been fully explained in his reply. And so the question was raised again by a different examiner.

"Why did you not voice your complaints to authority, this is what I have difficulty in understanding. There were people in authority there?"

"Yes."

"There was a diving superintendent there?"

"Yes."

"You had your serious doubts about a vessel, in which not only would your own life be entrusted, but the lives of a number of your colleagues?"

"Yes."

"Why did you not voice your complaints or fears?"

"Fear is an integral part of diving. It is something that you harness and control—in some instances to your own detriment. It is like a fear of flying."[26]

"I appreciate that, I can readily appreciate that even someone accustomed to it may find the idea of diving to these depths with such, perhaps, a flimsy lifeline, frightening. For these reasons, I would have expected people in your position to be most concerned that before a diving vessel went down that it was in as safe a condition as possible?"

"People in my position now are—people in my position then, basically, are not, because of economic necessity."

"Economic necessity, by which I take it you mean the need to earn large sums of money?"

"To make a living," Lahren replied, countering the barb. And then he extended his answer by exposing one of the harsh realities of the diving industry. "The situation is such that if I were to go on the system and complain about it, somebody else would take my place."[27]

And when pressed for a third time as to why he didn't bypass Nadeau and take his complaints directly to company owner Brian Masterson, Lahren focused his answer more sharply on the real threat that permeates the offshore industry.

"The diver wouldn't go and say 'I don't like your system, I won't dive it,' unless you want to be unemployed."[28]

That tactic—if you don't dive, then I'll replace you or run you off the job—was skillfully used throughout the industry to compel divers to do things they otherwise wouldn't do. In the 1600s, Thomas Hobson owned a livery stable in Cambridge, England, and as the story goes, he used to give his customers only one choice when renting them a horse: the one closest to the door. If the customer indicated he wanted something else, Hobson would tell him to go find another stable when he knew there was none within a day's journey. More than 300 years later, Richard and Skip were now faced with Infabco's version of Hobson's choice: either dive or leave the job.

Offshore, there were no police, no regulatory officials, no moral authority standing in the way of a superintendent who established an unsafe practice as standard operating procedure. He was the linchpin of the operation and he ran his deck the way he saw fit. He decided who went into sat, who supervised the job, and who got sent to the beach. He was responsible for keeping the safety net in place. For all practical purposes, he was the law. The client, the supervisors, the divers—they all relied upon his good judgment to do the right thing. Whether that superintendent adhered to conventional law depended entirely upon his moral code. And on board the *Wildrake* in July 1979, the threat of losing one's job had a sobering and foreboding presence. As John Joly recalls, no one was happy with the bell, but "it was one of those deals where—this is the way we are going—and it was either make the best of it, or leave the job."[29]

In a culture that operates under quasi-military autocratic rules, the old adage—the nail that sticks up gets pounded down—is never more fittingly applied. A diver can speak up concerning safety issues, but never too loudly. He can state his objections, even quote regulations if he dares to, but he does so at his peril. He may be on good terms with his superintendent, but that relationship suddenly takes a shift in direction when the sharp edges of safety are being ground smooth—a diver's replacement value never more firmly stated, than when his choices have dwindled to "either make the best of it, or leave the job."

Richard and Skip didn't have to ask themselves—What chance do we have in persuading a powerful, well-respected superintendent to halt diving operations and effect repairs?—because they knew implicitly the answer was, no chance. Just the power gap between their respective positions was sufficient to avoid risking a confrontation, particularly when it was evident that the superintendent had crossed the Rubicon and demonstrated that safety was taking a back seat to the client's schedule.

When Richard and Skip arrived on board the *Wildrake*, the attitude towards safety had already been established. The system had been modified. Seventeen bell runs had been performed, and the supervisors, who were legally bound not to dive a system with so many obvious defects, were following along to protect their jobs. For Richard and Skip to step forward and complain about the standard Nadeau had set would have been a job-terminating move. As Lahren correctly stated in court, the diver was not going to voice his objections to anyone in authority, not unless he *wanted* to be unemployed. This was 1979, and the same ruthlessness that oil companies unflinchingly employed to run diving contractors off the job was being applied even more capriciously to divers.

So divers kept quiet, effectively forced to compromise their values, if they expected to keep their jobs. They formed no unified front of protest because they knew they'd be picked off one by one, sent to the beach, and blacklisted as "troublemakers." Unable to protest, they harnessed their fears and did their best to eliminate whatever problems fell within their realm of expertise such as fabricating strain reliefs for umbilicals, while at the same time expecting their superintendent to have the courage to call in engineers to restore confidence in such things as drop weights and swivels, and to build bell stages when they abandoned clump weight systems.

Later, divers who knew Richard and Skip would speculate that, when they saw the lashed weights, they just assumed it was okay. But nothing of the sort happened. Before bell operations began in the fjord, Lacey and Skip talked about cutting the ropes off the drop weights, but decided against it because, as Lahren later testified, it would have been an "unauthorized alteration" and "contrary" to the way they were being supervised.[30] In plainer words, they knew there would be repercussions directed at them if they did so.

Did Richard and Skip bear any responsibility for diving in a bell with lashed weights and no stage? Yes, of course. All divers assume certain risks every time they enter into saturation or make a surface dive, and they take those risks seriously because they know that if something goes wrong it's their necks on the chopping block. But what Richard and Skip did not anticipate, after flying half way around the world, was to be placed onto the horns of a dilemma where they had to balance their jobs against an unreasonable risk. They didn't expect the superintendent to pass the safety buck onto their shoulders. Nor did they expect to see their supervisors backing away from their responsibilities. Their work was dangerous enough without having to second-guess whether the people above them were taking their own jobs seriously.

It used to be in Britain that if a person had knowledge of the danger, then that was enough to say he had "assumed the risk," and therefore could not hold his employer at fault for whatever injuries he might sustain as a consequence.[31] But in the late nineteenth century British courts began to find that having knowledge of the risk was not enough to provide the employer with a complete defense to an action of negligence. In other words, just having knowledge of the risk didn't necessarily mean that the person voluntarily agreed to it. It was held that the person not only had to consent to the risk, but he must also "assume the risk freely and voluntarily."[32]

In *The Law of Torts*, Harry Street points out that the courts came to recognize that "economic pressures" can weigh against voluntary conduct as indicated by L.J. Scott in *Bowater v. Rowley Regis Corporation:*

> "For the purpose of the rule...a man cannot be said to be truly 'willing' unless he is in a position to choose freely, and freedom of choice predicates, not only full knowledge of the circumstances on which the exercise of choice is conditioned, so that he may be able to choose wisely, *but the absence from his mind of any feeling of constraint* so that nothing shall interfere with the freedom of his will."[33] (Italics added)

Clearly Richard and Skip were under pressure to remain employed. And it would not be an overstatement to say that they must have felt a sense of duty and obligation to stay on the job as part of Nadeau's hand-picked sat team. And yet, the people who were there ostensibly to look out for their safety, were not returning that duty in kind. When Richard and Skip were on leave in the United States, no one called them up and said, "Oh, by the way, we've disabled the drop weight system, jury-rigged the bell wire connection, tossed the clump weight aside, and, no, we're not going to build a bell stage even though we know we should." Instead, the people at Infabco let them fly to Norway to be presented with Hobson's choice.

While Richard and Skip had a duty to protect themselves, Infabco had a much higher duty to ensure their safety because it had the power to dismiss them and control the conditions under which they dived. Further, it had the legal duty to comply with the diving regulations and to do whatever it could to eliminate the dangers of diving in a bell with no stage and lashed drop weights. But far from fulfilling those duties, the company was committing fraud by representing the *Wildrake* as a safe system, when in fact it was not.

At 0600 on July 27, the material barge *Happy Rider* steamed into the glassy waters of Björna Fjord. Two white massive cranes towered above her deck like goal posts on a football field; and lying across her deck was the newly refurbished 324 metric ton yellow-tipped buoy.[34]

After the *Happy Rider* anchored a safe distance from the *Wildrake*, the buoy was given a final inspection by a team of engineers, then raised from its stanchions and placed into the fjord. By eight o'clock that evening it had been towed across to the *Wildrake* and secured to the starboard aft railing with heavy mooring lines.[35] The two sections of the SALM were now lying at right angles to each other. At this stage the plan was to upend the buoy, vertically align it over the top of the riser, then use the haul-down wire jack system to draw the two pieces of the SALM together and slip the U-joint pin through the connection.

Courtesy of Marcus Taylor Ltd.

The *Happy Rider*

Over the next 24 hours, the riser was attached to the end of the *Wildrake* crane hook and flooded until it had a 5-ton negative load. The haul-down wire, coming through the top of the riser, was then shackled to the bottom of the buoy and the crane wire slacked off in order to transfer the full weight of the riser to the end of the buoy via the haul-down wire. After the crane was disconnected from the riser,[36] a series of compartments within the buoy were stage flooded. Shortly after midnight, on the morning of the 29th, the top of the buoy stood out of the water like a lighthouse in the cold night air.

Courtesy of Marcus Taylor Ltd.

Later that morning, at 1100, Richard and Skip entered the *Wildrake* bell dressed in their hot water suits and began blowing themselves down into sat. This was bell dive No. 18, and they were being given the first shot to earn their bragging rights by reassembling the world's largest tanker loading buoy.

As the needle on the internal depth gauge climbed, Richard looked around the interior. It was considerably bigger than an Ocean Systems bell,[a] and everything was new and fresh and hadn't been "diseased" yet. Dozens of shiny valves controlled gas and hot water flow; gauges read pressure and depth, and electrical cables supplied power and communications. A small analyzer detected oxygen

a 43 percent bigger.

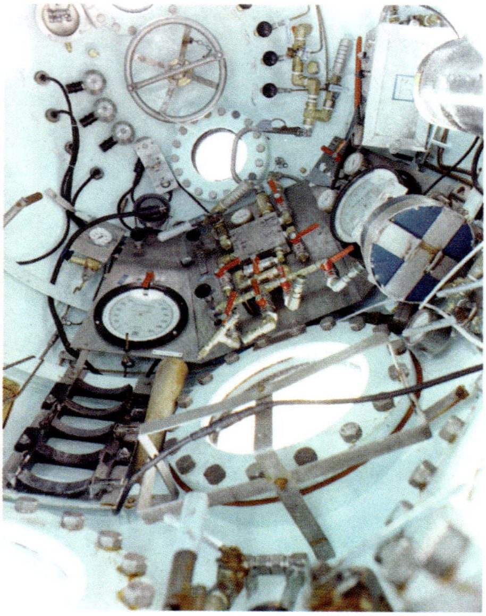

Courtesy of Marcus Taylor Ltd.

Inside the *Wildrake* bell.

levels, while a fan inside the CO_2 scrubber whirled. On opposite sides of the bell, two racks of dive umbilicals were neatly coiled on steel horns.

At 140 feet, Richard closed the blow down valve as Skip dressed in to dive. A voice over the bell speaker announced the beginning of the launch sequence. As hydraulic motors whined, the bell was raised off the deck and paused to allow the deck crew to close the bottom door. Then, it was brought into the cow-catcher and rotated clockwise over the port side of the ship. At 1130 hours the main winch wire began spooling off the davit drum and Richard and Skip were quietly lowered into Björna Fjord on their way to capture a little glory for themselves.

Twenty minutes later Skip dropped out of the bottom of the bell into open water and glanced around to get his bearings. At 160 feet, visibility was excellent with the water color ranging from shades of emerald green above, to a dark indigo below. And then his eyes locked onto the SALM 70 feet away. Its massive shadowy presence rose out of the abyss with silver-colored ingots of sacrificial anodes dotting the colossal structure like a zipper mark.

He swam to the riser then looked back to check the slack in his umbilical. Through the liquid haze, the once brightly painted orange bell was now just a gray cutout against a slate-blue backdrop, its protective bumper frame glowing like a white-hot force field with the bell lights shooting rays of light into the universal sea. Nothing distinguished its size or its relationship to the surrounding water. There were no fish swimming by, no plant life, no ocean floor to mark its place in time or space. The capsule was hanging on the end of a thin wire like a Christmas ornament. If the bell suddenly plunged to the seabed 531 feet below, Skip knew he would be dragged to the bottom along with it. It was thoughts like this he quickly put out of his mind.

During the first few hours of his dive, he made adjustments in the rigging to align the riser to the buoy. He inspected the path of the wire coming from the haul-down jack to the bottom of the buoy. He checked the hydraulic hose from the surface to the jack. And then he ordered topside to begin stroking the jack, and ever so slowly, the distance between the buoy and the riser began to close.

By 1600 the hole in the buoy coupler piece was roughly in-line with the egg-shaped holes of the riser yoke.[37] Skip was ready to stab the pin. He released the retaining straps on the pin tray and adjusted its pitch by turning an alignment wheel. Moving to the other side of the riser he operated a small hand jack whose wire led through the coupler piece to the end of the pin resting in the tray. By pumping the jack, the pin would be drawn through the hole to make the SALM complete, but only if everything was lined up, and as things turned out, there were problems with the alignment of the two objects nearly the length of two football fields.

For the rest of Skip's dive and for all of Richard's, there were more adjustments, but the pin would not slip through.[38] After thirteen hours, Richard and Skip were brought back on board and the bell locked onto the saturation complex. Later that night, Richard made an entry in his diary:

> July 29th - 11:00 am Skip G. & R in water - blew down to 140' in bell. I had 5 1/2 hrs in water. Out at 12:00 mid. 13 hour bell run to attempt to stab pin. No joy city. Stand up chambers.

Three hours after Richard and Skip returned, Lacey Lahren and Dave Mitchell were committed to sat. At 0342, bell dive No. 19 began. For the next six hours, Lahren and Mitchell struggled with the troublesome pin, and eventually, after using a variety of tools which Lahren says didn't work, they ended up rigging the ship's crane to the pin through a series of snatch blocks and giving it "a hell of a jerk" to snap it into place.[39] At 1400, the bell was pulled and Lahren and Mitchell joined Richard and Skip in chamber one. Eight hours later, Roger Palmer and Dave Boyd logged an eleven-hour bell run to add the rest of the U-joint components to keep the massive pin locked in place. By midnight of the 30th, the SALM was a single unit again.[40]

Earlier that day Mark Hansen's relief arrived by seaplane. When the door of the aircraft opened, a blonde, bespectacled, six-foot-one American stepped from the cabin carrying his gear. His name was Dave Olson. As a life-support technician, Olson was well-respected. According to Mark Hansen, "Oly" was someone who stood up and told it like it was, never minced words, and never stabbed anyone in the back. "He's not a vicious kind of guy." Hansen says. "He just knows what he believes in and he'll tell you right to your face—doesn't hold anything back. Great guy to have on your side."[41]

After working on the *Gulnare* for the better part of a year, Olson felt good about the people he worked with. Most of the divers and supervisors were ex-Sub Sea while the gas crew running the saturation chambers had been hired away from Taylor Diving and Salvage, and they were some of the best in the business. "It was like a crew of top-notch people," he says reflecting back.

People who came onto the job and didn't perform were weeded out and sent to shore, never to return. As a result, the divers on the *Gulnare* jelled into a tight crew where everyone knew each other's skills. Olson was used to seeing familiar faces, people he could trust. But, when he arrived on board the *Wildrake* that day, the first thing he noticed was a lot of unfamiliar faces. Of the two dozen divers, three quarters of them had only been with the company a matter of weeks, [42] and this made Olson feel uneasy.

The next day, July 31, the SALM was moored to the side of the *Wildrake*. The buoy had been dewatered and the riser ballasted to her tow-out draft.[43] For the next several days, the three teams of divers were sent down on bell dives 21 through 24 to install the jumper hoses to link the riser/buoy flow lines around the U-joint.[44] By Thursday, August 2, the hoses had been installed. The job was running smoothly and the second phase of the program was on schedule.

Throughout that week, there had been more crew changes; people coming and going in dribs and drabs. Steve Tegt, Dave Rhodes, and Butch Wright were no longer on board. John Joly and Brian Hoye were due to leave the next day. Thirty-three year old Dave Lane was moved into the position of deck foreman on dayshift, while John Brook took over the same duties on nightshift. Dayshift Supervisor Ted Roche was relieved by newcomer Don Cox and nightshift Supervisor Frank Montafia was replaced by Brian Reid, who had been present during the mobilization of the *Wildrake* in Norway, but had left the ship during the preparation of the SALM base to spend several weeks at his home in southern England.

And then John Nadeau's relief arrived.

As soon as Peter Holmes stepped on board, Dave Lane recognized him immediately from his days in the Persian Gulf. It had been more than five years since he'd crossed paths with Holmes, and Lane was gob-smacked by how high he had advanced up the chain of command.

"As an organizer, he was quite good, but he wasn't a strong diver. He didn't really have that much of a diving background. So, when he came on board as a superintendent I was totally flabbergasted."[45]

Although Holmes had not held the post of superintendent on any job for more than a year,[46] he'd been recommended to superintend the SALM installation by Craig Roberts after the two men met during the outfitting of the *Wildrake*. In Aberdeen, Roberts's recommendation was passed to Infabco Project Manager Bill Hemmings who reviewed Holmes's CV which indicated he had "extensive experience" as a diving superintendent and a unique knowledge of the *Wildrake* diving system. To Hemmings, Holmes "fitted the bill," but curiously, it was not until the day before Holmes arrived on board the *Wildrake*, that Hemmings decided to hire him.[47]

A month had passed since Holmes had walked the deck of the *Wildrake*, and once on board he soon realized that the diving system had been altered. But if Holmes had any inclination to object to its condition, he would have faced a tough, uphill battle—one that would have been, in all fairness, unrealistic to expect a newly-hired superintendent to fight. Nadeau had conducted twenty-four successful bell runs; Holmes had conducted none. Nadeau had been a business partner of Brian Masterson's—a shareholder of Infabco; Holmes was an unknown quantity.

BNOC was waiting. A $6.4 million SALM was strapped to the side of the *Wildrake* and all eyes were focused on the man who had just arrived. To turn things around, Holmes would have had to face Brian Masterson and argue a safety case using words like: needs modifications, repairs, and further testing—words that tend to glaze eyeballs, chafe against deadlines, and offer only the promise of additional expense.

Caught in this dilemma, Holmes saw no reason to agitate his new employer, and so he left things the way they were, thereby reinforcing the age-old theme: "We've been working fine up to now, so what's the problem?" But in reality, they were playing Russian roulette with a pistol held to the diver's head.

For thirteen days, while the *Wildrake* bobbed on the waters of Björna Fjord, the deficiencies of Møllerodden's ill-conceived diving system were left unresolved. Like dominos set to fall, Nadeau's stopgap measures had effectively become permanent fixtures. And now, virtually the entire crew had been replaced including its top leadership. Whatever team continuity existed over the last few weeks simply evaporated. Men were literally walking around asking themselves, *Who's this guy? Where did he come from?* Years later, Mark Hansen would reflect on the mixture of events that transpired that week and remark, "Nothing had jelled yet, everything was still up in the air—the people, the divers, the supervisors, the boat—everything was just all up in the air."[48]

Over the next several days (August 3-4), the sat complex was blown down to 240 feet. Bell dives 25 and 26 established a transponder and two inclinometers on the SALM.[49] Richard and Skip performed a short dive (No. 27) at midnight on the third, and at 1000 hours on the morning of August 4, Lahren and Mitchell made the last bell run (No. 28) in Björna Fjord to ballast the riser to a tow out draft of 468 feet. The SALM was then secured against the starboard side of the *Wildrake*. The job was heating up now; the push was on to get the SALM out of the fjord.

After Project Engineer Chuck Alexander received a favorable weather forecast, the *Wildrake* raised her anchors and began shepherding the SALM through the narrow stretch of water separating the islands of Reksteren and Tysnesøy. The trip was painstakingly slow and took longer than Alexander anticipated because the channel was only 508 feet wide and extreme care had to be taken not to stray off course and ground the SALM against the underwater ravine. By 1300 they entered the main body of Björna Fjord. Six hours later, the SALM was handed over to the oceangoing tug *Starmi* to complete the final passage to Thistle.[50] In the meantime, the *Wildrake* began sailing for Bergen to pick up supplies.

With phase two of the SALM project completed, Richard began writing a letter to his best friend, Doug Campbell:

4 August 1979

Doug -

Richard here—at 240 feet in the saturation chamber, bored after 6 days in here, looking at possibly 20 more or so. I hope all is well with you guys.

I've been playing with timetable strategies, money juggling and barn plans, as much as space will allow with 5 other guys. Hard to keep gravy off my graph paper. We just spent 10 days in a small, deep fjord [south] of Bergen, before saturating. It's been so nice, I even went swimming—diving off the bow about 80 feet off the water—even got some rays on deck—lots of work—we just finished assembling the top section of a huge device they call a SALM (single anchor leg mooring) to be towed out and installed out in the BNOC Thistle Field. It's just a glorified, swivel jointed, tanker filling riser that has been rebuilt.

We've been doing 12-15 hour bell runs, getting a little ragged, dealing with simian engineering. Some very heavy rigging in the water. The buoy for this turkey weighs 340 tons. The boat is Norwegian, brand new and 20 million worth of elaborate diving vessel—it's a diving and fire fighting boat—the sat system is below decks; woe to those of us that like to see sun out the ports. And, I can stand up in all the chambers. That's a first.

God, I don't know how I'm going to keep this up. Lots of qualms these days. This business of being all-of-a-sudden locked into commitments has been hard to get used to, but my better judgment sez it's the thing to do at this point in time. The reality of choosing the available jobs and staying gone from Jeanne and Marisa is really heartbreaking. God, how I love that little girl. You wait, she will tend me out in the kelp in a few years, whilst I dive for groceries.

I hope I make enough headway before the winter rains. After this trip, I'll be able to pay the generator off, get it installed, and have enough to build a bridge and barn almost. David Kayes, installation cohort, tells me that they can have the tower up and generator installed in 3 weeks...I want to be tucked into the trees up there by Christmas and have fire crackling in the rain, watching the creek swell up and roar at the doorstep. Impossible by then I guess, but maybe we can ride the winter out in a trailer—next to the creek near the barn site would be perfect—can not be seen and is private—ah, yes, but for now, I'm in here, squeaking, on the way through Bergen at this moment, and we will be at 540 feet by this time tomorrow. Be glad you didn't become a deep-sea diver. I hope the job goes fast. Even at $500 a day, I still want out as soon as possible.

And then Richard set his pen down, deciding to finish the letter later.

As the *Wildrake* sailed towards Bergen, a voice suddenly came over the horn speaker. It was Nadeau, informing the six divers that he would not be sailing with them to Thistle. "I've got a baby due...," Lahren recalls him saying, "and no matter what I'm doing, that's where I'll be."[51]

At 2130, the *Wildrake* steamed into Bergen.[52]

In sat, even though a diver is entombed in steel, he still feels the pulse of the job. Everything he hears and sees as he passes from chamber to bell to water is recorded somewhere in his mind, played back, then compared to what he thinks should be happening to him. Tiny nuances—an off-handed remark over the comms, a hint of frustration in someone's voice, delays in the launch sequence, and unexpected equipment breakdowns—speaks volumes to the men inside the chamber. And in chamber one, Lacey Lahren was feeling a shift in the chemistry of the job.

"We were left with supervisors that we knew nothing about, that we didn't know their histories, hadn't worked with them previously. We didn't have much voice in the matter. At that point in time, the only consolation you had was $530 a day—comfort zone kind of evaporated at that point."[53]

Tender Steve Lobban also sensed the change in mood. "Just not a good feeling—Peter Holmes—nobody knew where he had come from, who he was, what he did....And there was always something wrong with that boat."[54]

Just after midnight, as dark clouds gathered in the heavens, Captain Hedin cleared the island chain protecting Bergen's approach and pointed the bow of the *Wildrake* into the gray North Sea to begin the long journey to Thistle.

Chapter Eleven

Prelude to Disaster

They had us diving when they shouldn't have. I just made them pull us out of the water. It was just too much. If it would have been anybody else—if it would have been anybody with less experience—they probably would have left them down there.

—— Roger Palmer, Infabco diver[1]

At 1430 on the fifth day of August 1979, the *Wildrake* sailed into the Thistle Field.[2] On the bridge, Captain Hedin adjusted the ship's heading and took up a position directly over the SALM base about a mile from the northeast corner of Thistle Alpha. The window-lined control room was separated into a fore and aft bridge with a small communications room at its center. An internal staircase led down to B deck where the captain, the diving superintendent, and the chief engineer were quartered. Mounted over the stairway, the ship's clinometer, a liquid-filled piece of tubing with a bubble trapped inside, indicated the degree to which the *Wildrake* was rolling on Thistle waters. Across the aft bridge, a sweeping console controlled propulsion, the ship's fire fighting cannons, and the Kongsberg dynamic positioning (DP) system. On one of the display monitors, a symbol glowed brightly, registering the distant echoes of a transponder attached to the SALM base more than 500 feet below.[3]

Although dynamic positioning is widely used in the commercial maritime industry, it was initially developed for the military in the mid-1960s. In the diving industry its primary purpose is to keep the vessel in a fixed position without the need to lay anchors which is an expensive and time consuming process. In essence, DP is to a diving ship what an autopilot system is to an airplane.

The *Wildrake*, like all vessels, was subject to the environmental forces of wind, wave and tidal movements. In order to quantify these forces and make them available to the DP computer, the ship was equipped with a series of position sensors that constantly fed data into the computer. There were anemometers to measure wind speed and direction; gyro compasses to determine heading, and vertical reference systems to measure the angle of roll.

But these sensors only gathered data about the movements of the ship and the external forces acting upon it. What the computer also needed, in order to direct the *Wildrake's* bow and stern thrusters to drive the vessel back on location, was information about its position and heading relative to a fixed object such as the SALM base or a nearby platform. That information was provided by several Position Referencing Systems working in tandem with each other. One very accurate system was an antenna unit called the Artemis Position Fixing System which was mounted on the *Wildrake*'s mast. This represented the "mobile station." The corresponding tracking antenna or "fixed station" was located on the corner of Thistle Alpha. During operation these two antennae locked on to each other to establish a microwave link which indicated the vessel's distance and bearing relative to the platform.

The *Wildrake* also had a Hydro Acoustic Reference System which consisted of deploying one or more transponders on the seabed to broadcast an intermittent stream of acoustic signals. Even the *Wildrake* bell had a transponder attached to it for this purpose. The signals emanating from the various transponders were then picked up by listening devices (transducers) mounted on the hull of the ship. On the aft bridge, the transponders, as well as the *Wildrake*'s position, were interpreted by the DP system and displayed as symbols (X, Y, etc.) on the computer screen marking their location relative to each other. A third reference system, Taut Wire, had not been installed.

Besides obviating the need to anchor the ship, dynamic positioning can play a decisive role during emergency situations because of its quick setup time. But DP comes with risks to the diver. Problems with software, under-powered thrusters, and unexpected surges in wind speed and changes in wind direction have caused systems to fail. And when the ship's DP suddenly goes haywire, the diver can expect to find himself being dragged along the seabed like a lure on the end of a fishing line. This can be an extremely dangerous situation if there are seabed obstructions around. Depending upon the depth of the dive, water conditions, and the type of mask the diver is using, DP thrusters can be deafening and interfere with the diver's ability to hear his supervisor clearly.

Without the use of DP, Captain Hedin could manually direct the ship's engines and thrusters by using a joystick on a portable console called the Poscon (Position Control), but then technically the *Wildrake* would have been "underway." And while the practice of "live-boating" was employed in other parts of the world, it was considered too dangerous for North Sea operations and was prohibited by law.[4]

Courtesy of Marcus Taylor Ltd.

The bell handling controls.

When the *Wildrake* arrived in the Thistle Field, sea conditions were marginal and problems developed in setting up the DP system due to the amount of swell.[5] Thirteen hours later the ship was holding station, and at 0330 on the morning of the sixth, Supervisor Brian Reid stood in the corner of the bell van deep in concentration. As he stared out the window onto the back deck of the *Wildrake,* his hands gripped the bell handling controls. Thirty feet away the bell hung from the cow-catcher like a giant water bubble against a darkened sky. Standing nearby under the glare of the davit floodlight, Reid's deck crew waited on his next move.

Slowly, the thirty-six-year-old supervisor drew a long black lever labeled "Slewing" towards his body. Instantly, the sound of hydraulic fluid began rushing through a valve as the davit rotated towards the side of the ship. When its payload was clear of the railing, Reid relaxed his grip and the bell stopped in midair. Taking hold of two levers marked "Main Winch" and "Umbilical Wheel," he nudged them forward. The bell dropped out of the cow-catcher and within a few seconds it was on its journey into the dark churning waters of the North Sea. It was bell dive No. 29, and Roger Palmer and Dave Boyd were in the bell.

As it descended, Reid watched the main umbilical snake out of its holding basket like a giant cobra. It rose 20 feet into the air, looped over the top of the umbilical wheel, then down into the water. At the fifty-foot mark Reid stopped the bell and slewed the main lift wire in towards the ship's railing. One of the deck crew picked up a small bracelet of chain with a snap-hook on it, threw a double hitch around the wire, then snapped the hook onto a Dee-ring lashed to the bell umbilical. With a signal from the deck crew, Reid pushed his levers forward and the bell continued its descent. The attachment of this "chain-stop" was to prevent the umbilical from developing a dangerous bight of slack in it. Every fifty feet another Dee-ring appeared and another chain-stop was snapped into place. In this way, the two life-support lines became married to each other as they entered the water.

After twenty minutes, Palmer and Boyd felt the bell reach its final destination with a sharp tug. The depth gauge next to the gas panel read 485 feet. In the bell van, Reid pressed a lever on the Helle radio and informed his divers that they could go for the door. Palmer dressed in, exited the bell and located the SALM base. Standing on the seabed, he gave Reid orders to move the *Wildrake.* He wanted the bell above, and to one side of, the base to illuminate the work site with the bell lights.

In dive control, Reid relayed the distances he was given to the bridge. The captain inputted a new position into the DP computer and within a few moments Palmer saw the glowing orb bounce across the jet black sea and come to a stop where he predicted it should. He told Reid he was satisfied with the bell's location. Reid turned and pressed a power button on the wall to shut down the davit's noisy hydraulic motors; then he informed his deck foreman, John Brook, that the bell was in position. Immediately, one of the deck crew took a piece of polypropylene rope and secured the bell umbilical to the corner of its holding basket. Without this "tying-off-line," the shiny new umbilical had a tendency to slip over the umbilical wheel and fall into the sea.[6]

In the water, Palmer immediately felt the disparity between diving in a flat-calm fjord and the sea he now faced. The bell was rising and slamming down on the end of its wire like a steel pile driver. Most of the equipment he needed would be sent down to him from the ship by crane, but some of it was tied off to the side of the pitching bell which Palmer wasn't thrilled about. "That was the other reason why we wanted a stage," he says reflecting back, "because we didn't have any place to put anything."[7]

Over the next several hours he mounted a haul-down jack on top of the SALM base, fed the haul-down wire through the base yoke, then set up the retraction wire for pulling the coupling pin through the U-joint.[8] The procedure to reconnect the SALM to its base was essentially the same that was used in the fjord for connecting the buoy to the riser.

On the seabed, there were no currents and visibility was crystal clear. At one point, Palmer crossed paths with a slew of conger eels slithering in the dark around the base. *Dave Mitchell is not going to be pleased*, he thought. Palmer had worked with the Canadian before and knew he absolutely hated the sight of these creatures.

Meanwhile, on the surface, weather conditions continued to deteriorate into the rudiments of a storm. Wind speeds of 25 to 30 knots began whipping across ten-foot seas stirring up white horses.[9] Five hundred feet below, the only indication Palmer had that the weather was worsening was the way the bell was now swinging and "snatching" on the end of its lift wire.

Boyd must be getting the piss knocked out of him, he thought.[10]

In the bell, Dave Boyd sat watching gas gauges while flexing his lower jaw. Every time the *Wildrake* fell into a trough, the water level rose in the trunking, compressing the gas bubble he was sitting in. This caused his paper-thin ear drums to flex painfully towards his brain. And when the ship climbed a swell, they expanded back the other way. Minute by minute he yawned and wiggled his jaw to ease the pressure. The davit was equipped with a heave compensation system to counteract the motion of the ship. But the *Wildrake* had sailed on contract before it was commissioned.[11] Had it been in operation at this time, the bell would have remained virtually motionless, and the shackles on top of the bell would not have been wearing against each other.

At 0600, Supervisor Don Cox came on shift and took over for Reid. Several hours later, Palmer began to complain about his hot water supply. There didn't seem to be enough flow coming down the hose. On the surface, Cox verified that everything was okay; the machines were pumping, no alarms were sounding. But Palmer knew he was losing hot water so he went back to the bell, which was now surging up and down like a bucking bronco.

"I got beat up—getting smacked by the bell—trying to get in," he recalls. And once inside, he decided that continuing the dive was sheer folly. "That bell was just getting knocked all over—snatching," he says. Over the speaker he told Cox to get them out of the water, and by 0900 they were back on board the ship and locked onto the sat system.[12]

Later that morning, deck foreman Dave Lane discovered the reason for Palmer's hot-water complaint. Within the bell umbilical, the hot water supply hose had split. Normally this hose is made of a thick, elastic material—what divers call "Gates hose." But when the *Wildrake* bell umbilical was put together, someone had used Synflex hose instead, a plastic-like material normally used for transferring gas.[13] Synflex was not used in dive umbilicals because it had a tendency to kink and fracture, and after 29 bell runs (18 in the North Sea, 11 in the fjord), it had done just that.

Taking advantage of the storm to make repairs, Lane directed his deck crew to begin duct taping a new Gates hot water hose to the *outside* of the bell umbilical.[14]

One deck below, in chamber one, Richard was having doubts about the job. The recent wholesale crew change, and the fact that Palmer called off his dive (rather than topside), left an uneasy feeling of worry in his mind. He picked up his unfinished letter to Doug Campbell and began writing with feelings of excitement and uncertainty:

> 6 August - We are at a living depth now of 460 feet—position for the job. The boat doesn't take a swell like we'd been told, and there is moan from below somewhere. They say it's the surge baffling system. The first team got back 3 hours ago with stories to tell. The motion compensator doesn't work so well,[a] and the bell has been pitching up and down and sideways—and the one we've been waiting for: 7-8 foot conger eels on the SALM base plate. Yes, it's starting to get exciting—you'll probably hear some sea stories after this one. We are the next team down, but for now we are waiting on weather. There's a Force 10 coming so it should be a few days. I must be a crazy daredevil to do this crap—I'm all excited now.
>
> I'm almost ready to pack my stuff and get out of here (quit) but, I won't.

And then Richard talked about Hollister Ranch. At the end of the letter he made a final note:

> God, I want to be at the ranch this winter. I just about deserve it after all these trips. Well, write soon if you can partner.
>
> Love to Patti,
>
> Richard

---

a Richard is incorrect here. The motion compensation system wasn't working at all.

Brent Field, twenty miles south

While the *Wildrake* had stopped work in the Thistle Field, twenty miles south in the Brent Field, the dynamically operated *Uncle John* was preparing to make its first dive of the day, bell run No. 260. It was 11:45 in the morning, and the *Uncle John* was moving into position alongside the Brent Bravo platform. A 12-inch subsea crossover valve had to be removed.[15]

Courtesy of Marcus Taylor Ltd.

The *Uncle John.*

Under contract to Shell Oil, the *Uncle John* was far more stable than the *Wildrake*, and for the next six hours and forty-four minutes it remained on DP. Then, as sea conditions grew worse, the world's most capable diving platform was forced to pull her bell and retreat about a mile east of Bravo to wait out the storm. Later that day, Comex Construction Superintendent Don Higham recorded the event in his diary:

> Dive aborted due to adverse weather conditions. To[o] dangerous to keep barge so close to platform in conditions like this. Wind SE to S. Max 38 knt.[16]

Gulnare, Thistle Field

A mile from the *Wildrake*, bell operations on the *Gulnare* had been terminated as well. But not by the weather. Eleven days earlier, the rig had been hit by a catering strike which began in the Forties Field, then spread like a plague northward through Claymore, Piper, and Heather, before descending on Thistle.[17]

The catering staff on the *Gulnare* handled cooking, bed making, and laundry duties. Without those essential services, operations were brought to a grinding halt, forcing the oil companies to beach large blocks of personnel until the wage dispute was settled. In the coming weeks, British Petroleum and Shell Oil would airlift more than 1,400 workers from Forties and Brent back to Aberdeen.[18] As for BNOC, it immediately ordered Superintendent Tom Belcher to suspend operations and decompress the divers he had in sat. For Infabco diver Rod Pelzel, the strike couldn't have come at a worse time.

Pelzel was a US Navy veteran, an avid Southern California surfer and an ex-tropical fish chaser. In 1977, he graduated Santa Barbara City College and moved to the Gulf of Mexico to begin his diving career. Eight months later, he became another "escapee" when he realized the company he was working for seemed to be setting up roadblocks to his only ambition, which was to get in the water and do what he'd been trained to do. During all the months he worked for McDermott International, he never made a single dive.

If there was anything in Pelzel's life that served to propel him towards the North Sea, it was this period of disenchantment with the Gulf. Looking back, he remembers thinking to himself, *"This ain't happening. It's not going quick enough for me; I want to do better, I want to go faster, I want to get in the water and make the money. I mean, already I can see that I'm not going to be doing this the rest of my life. So, I want to get in, I want to make the dough, and get out."*

On the advice of some classmates, he flew to Aberdeen in May 1978 and quickly signed on with Sub Sea International. But after a week of shop work, he quit to join Infabco. In Pelzel's words, Sub Sea looked as though it was going to be "another one of those bullshit things, like in the Gulf"—using divers as cheap labor with little opportunity to get wet. Infabco, on the other hand, was fresh and new, and it offered better wages and a lot of upward mobility.

In April 1979, eleven months after joining the company, Pelzel went into saturation for the first time with Lacey Lahren. But after four bell runs and ten days of sat, he had to come out when Lahren got sick. Three months later, on Saturday, July 21, he was put into sat again, this time with a diver by the name of Gage Perth. Pelzel was looking forward to a long sat and seeing a fat paycheck at the end of the job.

But five days later, while, again, on his fourth bell run, Thistle was hit with the catering strike. It was the 26th, and he and Gage had just been pulled out of the water when news of the labor action (and its financial implications) was delivered over the horn speaker. Pelzel couldn't believe his bad luck. That night he wrote about it in his diary:

> July 26 - Was awaken at 3:00 for bell dive #4. Finally got down at 5:30. I locked out and welded anodes to first Lockheed[a] on bottom. Scared shitless, but knew I must drop down inside. Went pretty smooth except surface support really fucked-up in job preparation. [Lockheed] Chamber 2 had absolutely zero visibility. Wire rope everywhere. Lock out a second time for 4 1/2 hours to tack 1 anode on the son-of-a-bitch. Got the job done.
>
> Finally back at [the] system. We were told that OCS [Offshore Catering Services] is going on strike. 37 rigs in the North Sea shut down. We begin to decompress from 440' immediately, two hours after leaving bottom. WE'RE COMING UP AND I CAN'T FUCKING BELIEVE IT.

Six days later, on the morning of August 1, Pelzel emerged from sat and was put on the nightshift under Supervisor Tom Almon.[19] Also on board were Chris Roper (Skip's partner during his first sat on the *Galnare)*, Bob Gemmell (Skip's classmate from dive school), and Field Superintendent Craig Roberts.

---

a The "Lockheed" was a chamber built around the junction of two pipelines.

Pelzel's work now shifted from the exhilarating to the tedious: taking inventory of inspection gear, working on air tuggers (winches), and chipping paint in the bell.[20]

On August 6, as the *Wildrake* bobbed at sea waiting for the weather to clear, Pelzel called his wife, Sue. Hearing her voice, however, only dampened his spirits and made him feel homesick. He still had almost two weeks left on his tour—or at least he thought so at the time. What Pelzel didn't know was that he was just hours away from witnessing what he would later describe as one of the most "preposterous" events of North Sea history.[21]

Chapter Twelve

Bell Dive No. 30

I think everybody has the right to say, 'I don't feel right, and I'm getting out.' There's all kinds of pressures you'll put on yourself, too: 'What's my bell partner going to say? What are the other guys going to say? Am I wimping out here? Am I feeling right?'... That's the decision that the diver has to make, and would it be more accepted now in 1996 than it would be in 1979?—I don't feel right, I don't think this is a safe operation—absolutely. You'd be listened to a lot more now because safety is a whole different scenario.

—— Ted Roche, Infabco Supervisor[1]

Tuesday: August 7, 1979

Like a bombardier over enemy territory, the crane operator on Thistle Alpha stared down at the payload swinging over his target several hundred feet below. With skilled practice he pushed a lever forward and the shipping container slammed down with a loud crash onto the back deck of the red and white ship from Göteborg. Almost immediately the pennant wire jerked tight and the container lurched off the deck a few feet before it skidded to its final resting place. A half second later the headache ball and rigging lines collapsed against the steel box with a thunderous clap. From the aft bridge, Captain Allan Mellor watched as one of the ship's crew unshackled the lifting bridle, allowing the steel ball to drift away like a balloon; and then Mellor put his vessel in motion against a fitful sea, glad to be away from the noisy citadel towering above.

Originally a supply vessel, the *Stena Welder* had been pressed into diving service when Wharton & Williams hired her to support its contract with British National Oil Corporation. Unlike the *Wildrake,* the *Welder* was not a DP vessel, but had to run anchors to conduct diving operations. Its diving system, made by Aqua-logistics, was old and problematic. One former supervisor characterized it as "a real hodgepodge" of secondhand equipment. The main chamber and entry lock were welded to a launching skid, while a smaller mismatched chamber, was not. Every time the supervisor called a dive, the crew had to disconnect the smaller chamber and retract it out of the way to skid the main chamber and entry lock outboard to lower the bell over the side. Even Captain Mellor shuddered every time the bell went into the sea. "The launching system we had on the *Welder* was terrible," he says. "It was bloody useless; it was ancient."

Courtesy of Don Cox

The *Stena Welder* diving system.

It was also a constant headache for the people who had to keep it functioning.[2] In recent weeks the crew was having problems with the communications system which would suddenly break up during a dive. Technicians suspected the failure was caused by sea water getting past a bad seal inside an oil-filled junction box mounted on the outside of the bell where the communications and power lines terminated.[3]

To cure the problem, dayshift Supervisor Mike Allen instructed his crew to begin stripping down the bell for a much needed face lift. The main umbilical had recently been replaced,[4] and now Allen wanted his crew to begin changing out the divers' tethers with the ones that had arrived in the shipping container that morning. It was hoped that these repairs, along with work on the junction box, would solve the nagging problems they were having. In the company log, Allen made a note of the work in progress:

> started derigging diver's umbilicals from [the] bell and installing new umbilicals.
>
> electricians to complete re-wiring of diving umbilicals as first job in the morning.[5]

On board the Wildrake

It had been almost four days since Richard's last dive and he was growing restless and worried. His world had collapsed to a cold steel envelope 19 feet long by 7 feet in diameter with bunks slung from the ceiling by bits of chain as in a prison cell. He'd been inside now for nine days and the atmosphere had mutated into a warm smelly stew of catsup, body odor, and the acrid smell of disinfectant. The food on the *Wildrake* was terrible and Richard's ear drums had turned into a soggy mess in the tropical environment because Infabco had no ear drops to give him.

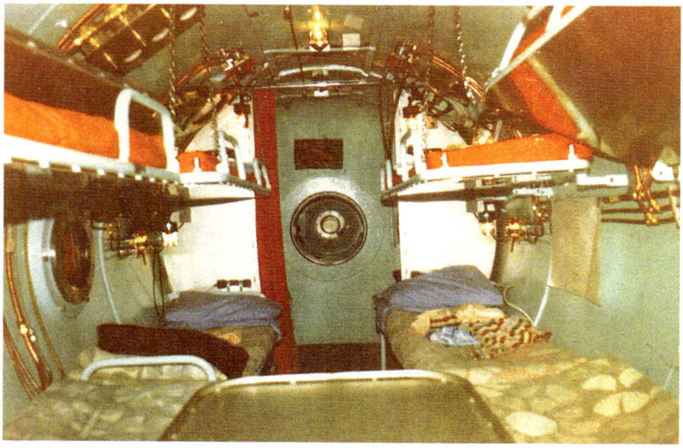

Courtesy of Gary Legere

Inside Chamber One, 1979.

Across the chamber, a long blast of hissing gas over a whirling scrubber suddenly brushed his face, the injection of oxygen testimony that topside judged the atmosphere was getting too lean.

Richard lay in his rumpled bunk thinking about how he could make it all work with the family, the ranch, and his life. It was difficult keeping his mind on the job; images of Santa Barbara, Jeanne, and Marisa kept flashing in his mind. Oh how he missed his family. He should be home for them, especially to hug Marisa and be her Daddy. Caught in an emotional trap, Richard didn't know what to do. No matter how much he tried he couldn't seem to stretch the walls of his chamber to include the people he loved, nor suspend his feelings and confine his thoughts to work. Not anymore. Being a father and a husband had changed everything. At thirty-two, Richard was not the same man that arrived in Scotland in the autumn of 1973. He was older, wiser, and more reverent of his profession. The excitement and glamour had been washed away by the cold and the danger. He was a mercenary in a foreign land, and the price of the job was the persistent ache he felt in his heart for Jeanne and Marisa. It was a harsh reality he found difficult to reconcile, and at times, too much to bear.

"I'm almost ready to pack my stuff and get out of here (quit)," he'd written to Doug Campbell.[6]

Richard had lost faith in his keepers. Since his arrival in Norway he'd become more and more disillusioned with them, and every time he crawled inside the bell he had to convince himself that this was just another dive. But the uncertainty about his safety had a looming and foreboding presence, and it constantly invaded his peace. The blind faith he relied upon to dive had now withered to a sense of dread and raw fatalism. Richard wanted out, but he desperately needed the money.

Since July 21, he had been writing each day in his diary—short notations of what he'd done, what he'd seen, and what his moods were. He tried to keep his spirits up, but after realizing that the conditions on board the *Wildrake* were not what he'd expected—that he was working for people he didn't know, didn't trust, and for a company that didn't seem to care—he reached over and picked up the blue spiral-bound notebook that served as his diary and began to pour his heart out, describing an industry that he had once loved and respected and now despised. He wrote long heart-wrenching passages of fear, despair, and love as though he were a condemned man writing his last thoughts on earth. Richard was scared, and he wrote it all down, for all time.

These were old battles that had haunted him before. And he knew that asking to come out of sat was his only option. "...but, I won't," he'd said to Doug. Richard knew he had to sell a portion of his soul to stay employed and he didn't like it. He also knew he couldn't confide his thoughts to anyone but his diary. Basic freedoms and rights didn't exist for divers who questioned their company's safety policies. Richard had learned not to complain in sat because he could never be sure how the complaint would be received on the other end of the radio. He didn't complain because he knew all too well that the squeaky wheel didn't get the oil, it got decompressed.

Sitting in that chamber, both Richard and Skip understood what they were up against. Left with terrible doubts, all they could do was hope that topside was taking care of business. All they could do was interrogate themselves with nagging uncertainty: *Am I comfortable with this? Am I confident that these guys are doing the right thing? Do I know enough to refuse this dive?*

In the bell van two decks above, Supervisor Don Cox turned to his left and picked up the wall phone and dialed the sat van informing Dave Olson that the weather was coming down. He wanted one of the divers from the next team to go into the bell to begin internal checks.

At thirty-two, Cox was articulate, college educated, and a consummate professional. He'd grown up in the hills of North West Durham near the steel town of Consett, and as a boy had been fascinated by a popular BBC1 program called Hans and Lotty Haas, about diving and the world beneath the sea.

That fascination intensified when he learned to use scuba in college and later how to buy submerged wrecks lying off the coast of England. In 1973, the same year Richard landed in Aberdeen, he transitioned from amateur diver to professional with the help of a friend working for Comex; and for the next four years, Cox worked his way through the gas fields off the coast of Yarmouth, crewed on various SEDCO and ODECO rigs, and to gain more hands-on experience, left the North Sea for the oil fields in the Persian Gulf, where there wasn't the extremes of weather. Instead of one bell run every three or four days in the North Sea, Cox was diving nearly every day.

In 1977, he was home on leave in England, when Wharton & Williams (2W) offered him a job on the *Stena Welder* in the Thistle Field. He accepted the job, and, because of his experience, was quickly promoted to sat supervisor. But in

July 1979 he quit 2W for Infabco because, after two years on the *Welder*, he'd become frustrated, in part, with the diving system. "I just couldn't relish the fact of spending another full season on the *Welder* with all the inherent problems and the mismatch."

Getting hired at Infabco proved to be easier than expected. Because of Masterson's expansion plans, the company was desperate for divers and supervisors. And the newly hired project manager for the *Wildrake*, Bill Hemmings, was a former Comex Service Manager who knew Cox from his days with that company. Hearing that Don was unhappy with 2W, Hemmings hired him immediately and told him to report to Infabco's Aberdeen office, where he was appointed to the position of *Wildrake* diving supervisor on the tenth of July. Shortly thereafter, he was flown out to the vessel in the Thistle Field to learn Møllerodden's system from nightshift Supervisor Frank Montafia.

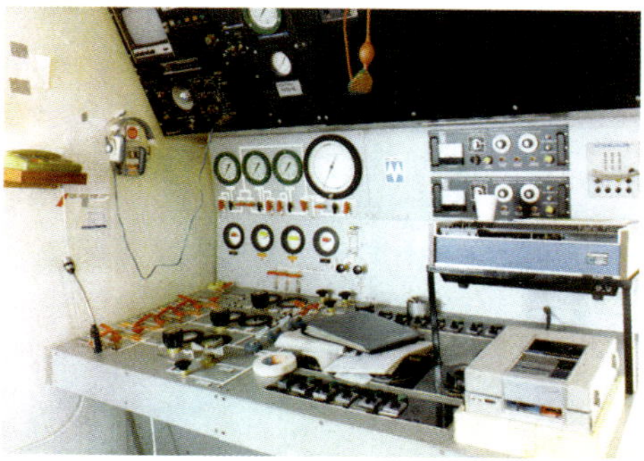

Courtesy of Marcus Taylor Ltd.

Supervisor's gas panel.

Now on board the *Wildrake*, Cox pulled one of Infabco's Dive Sheets out and wrote in the upper right hand corner Dive No. 30, and below that, 7 August 1979. He listed the job location, the client, members of the crew, and the equipment they were using. There were three large regulators on the gas panel in front of him. Cox noted their upstream pressures, analyzed the oxygen content in each mixture, and entered the information on the form. He checked the temperature of the hot water supply and made certain bell electrical power was on-line and the video camera was functioning. Twenty-nine hours had passed since Palmer's aborted dive, and as the *Wildrake* was coming back into position over the SALM base, Cox glanced out the portside window. All he saw was a stream of long gray swells, eight to ten feet high, running down from the north.

When Skip received the message from Olson, he tied a bandanna around his head and crawled through the two-foot diameter hatchway that led to the entrance lock. As he closed the hatch behind him he noticed Richard still writing in his diary.

In the entrance lock, the ceiling door was now hanging down like a tin can lid. Skip put one foot on the toilet, and with the other, got a purchase on the door and began climbing the rungs of the five-foot-long passageway. In the cool bell atmosphere the scrubber motor was humming as water droplets slowly crept down the walls. Skip took a spray bottle filled with a mild disinfectant and cleaned the hats, then began organizing the gear when a voice echoed over the bell speaker, "Let me know when you're ready to start bell checks."

"Roger," Skip said.

At 1415, Skip was ready and Cox began calling out the names of valves on his check list.[7]

"Gas sample line."

"Open," Skip replied.

"Diver 1 Kludge."

"Open."

"Diver 2 Kludge."

"Open."

"Internal depth."

"Open."

One by one, the valves were called off and their status verified. They ran through the equipment list and then Skip flow-checked the hats and gave a comms check on each, "One, two, three, four, five, how do you read me topside?"

"Loud and clear," Don replied.

After 25 minutes, bell checks were complete and Skip was told the bridge was still trying to set up on DP. Just before he climbed back down to wait in the chamber, he looked into the camera lens and said, "Gee, Don, you're not bad for a limey," which put a grin on the Englishman's face.[8]

Outside on deck, tender Ron Francisco was performing external bell checks under the supervision of deck foreman Dave Lane.[9] They were checking valve positions, on board gas supply, and loading the bell with tools. There were 17 items on the checklist, and items number one and two were to ensure that the shackles on top of the bell were secure, and that the bell transponder was fixed to the bell frame.[10]

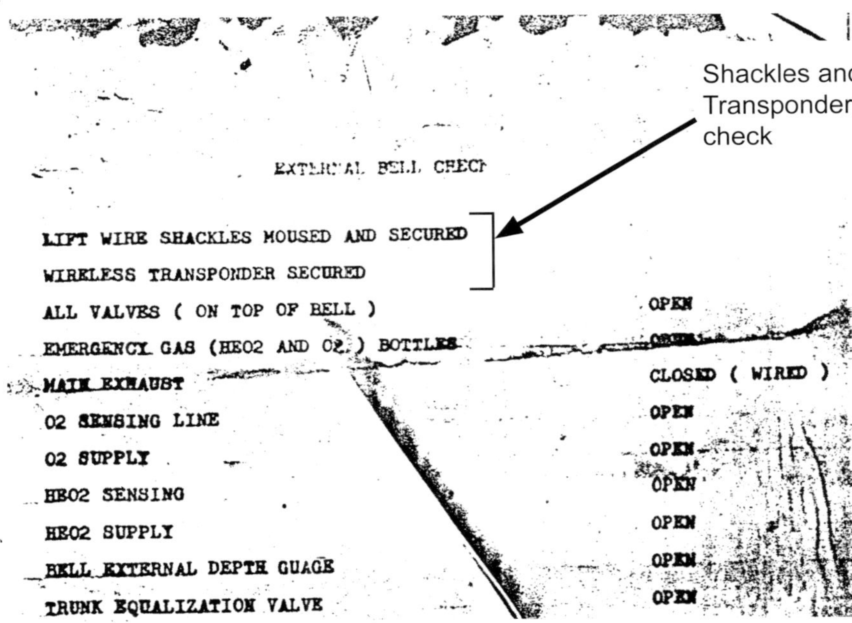
EXTERNAL BELL CHECK

LIFT WIRE SHACKLES MOUSED AND SECURED	
WIRELESS TRANSPONDER SECURED	
ALL VALVES (ON TOP OF BELL)	OPEN
EMERGENCY GAS (HEO2 AND O2) BOTTLES	OPEN
MAIN EXHAUST	CLOSED (WIRED)
O2 SENSING LINE	OPEN
O2 SUPPLY	OPEN
HEO2 SENSING	OPEN
HEO2 SUPPLY	OPEN
BELL EXTERNAL DEPTH GUAGE	OPEN
TRUNK EQUALIZATION VALVE	OPEN

Courtesy of Sheila Borsari

Wildrake external bell check list.

When Brian Reid came on shift three and a half hours later, problems with the DP system were still unresolved. It was the Artemis system that was not functioning properly.[11] Cox gave Reid an update on the present situation and told him that Richard and Skip were standing by. By 1830 the DP system still wasn't working so the decision was made to run anchors and begin diving.[12] At 1945 the *Jon Viking* arrived alongside to begin laying the first of four anchors.[13] Two and a half hours later they were set and tensioned and the *Wildrake* lay bobbing at the crossroads to four outstretched cables like a giant "daddy longlegs."

Down in chamber one, the atmosphere was still and hushed. Skip was dressed in his ragged orange woolly bears and Richard in his white polo sweater and red thermal trousers.[14] All they had to do was pull on their hot water suits and go. They'd been waiting on the "hot seat" for more than seven hours with a thick serious feeling of apprehension. They tried to relax; sleep was impossible. The only thing they could do was lie in their bunks and try to keep their heads clear and visualize the job. Somehow that always seemed to help.

The clock dragged on for another hour. The team was slipping back into the diving mode, some of the other guys had already put their heads down.

And then the calm was broken.

"Okay guys. They want you in the bell."

Rust wavering
Lovely fenders and grills
on the wind
splayed foxtail
summer sun ground
God help us.

—— Richard Walker

Twenty minutes before eleven o'clock on the night of August 7, the proud father from Santa Barbara and the Italian rebel from Massachusetts donned their hot water suits and climbed the entrance lock tunnel into the bell. They unwrapped a fresh soda sorb canister and snapped it to the top of the scrubber motor. They lowered the heavy internal door, stood on the hatch, and blew the bell down another ten feet. After notifying Reid they had a seal, they heard the muffled sound of escaping gas; Gary Legere had turned a valve, venting the tunnel and isolating them from the chamber system.

In cramped quarters, Richard and Skip made themselves comfortable while footsteps scuffled outside. They listened to the whine of hydraulic motors and deck plates reverberating as the bell clamp retracted below them, unlocking the capsule from the docking station. Suddenly they felt the bell jostle, then float into the cow-catcher as though they were riding a giant Ferris Wheel. They heard the bottom door being closed; seconds later their ride swung clockwise over the side. Through a porthole Richard and Skip saw the dive station awash in harsh light, and then the floor of the bell seemed to collapse beneath them when they broke free of the davit on their journey into the abyss below.

The first wave rolled along the hull of the *Wildrake,* picking the seven-ton bell up on a hill of water and filling the side ports with a swirling mass of foamy bubbles. Almost instantly, the wave crest vanished, plunging the bell into the trough and leaving the lift wire and shackles singing under the mechanical strain.

As Brian Reid launched the bell, Superintendent Peter Holmes looked on.[15] Also in the bell van were Ron Wilson, BNOC's Quality Assurance Inspector, and Jim Burgess, BNOC's diving consultant. Wilson wanted to see a video of the yoke on top of the base.

As the bell entered the water Burgess began recording the events in a small red log book.

2251 - Bell in the water, going down (Skip Guiel and R. Walker).

And then, before a chain stop was put on the umbilical, Reid unexpectedly brought the bell back up on deck and married it to the system.

2254 - Bell in cow-catcher. Being brought back onboard - seas too rough.

2255 - Bell on deck. Bell married back on system.

According to Holmes, the bell was brought out of the water in order to change the heading of the ship to give a more calmer sea on the leeward side of the vessel.[16] But that rationale lacks a certain degree of logic to it. The bell was already in the water, through the interface, and, as the log stated, "going down." Why bring it back up and subject it to another run through the splash zone to change the ship's heading? The position and heading of the *Wildrake* could be changed later as it normally would be when the diver calls for a move.

Holmes's explanation also conflicts with his supervisor's. According to Brian Reid, the bell was brought back up because someone reported to him that the video camera had come loose when the bell hit the water.[17] But John Brook, who was running the deck, gives a different story.

Brook had been standing at the ship's railing when the bell entered the water and noticed that something had indeed come loose in the turbulence. But it was not the camera, it was the bell transponder. From the dive shack, Reid couldn't see this because his view was obstructed by a 13-foot drop from the railing to the waterline. When Brook informed him what had happened, Reid brought the bell back on board. The deck foreman immediately set about to re-lash the transponder to the bell frame, a job that would have taken a few minutes to complete, but was told instead to cut it off.[18]

Remembering that fateful decision, Brook says, "There was a good deal of pressure at the time from BNOC and the powers that be to get the bell in the water—get the dive underway, etcetera, etcetera—which was a general attitude towards everything. And I don't think I'm speaking out of turn when I say this, that that was the attitude. And for a client to see the bell go in the water and then to see the bell come out again—questions would be asked....Having informed the dive control that the transponder was loose, I was told to remove it and get the bell back in the water as quickly as possible, and I wasn't there to ask questions."[19]

Down in chamber one, Lacey Lahren had turned in shortly before Richard and Skip's departure and when he heard the dull thud of the bell landing back on the entrance lock, he thought to himself, *"Oh shit, its our turn."*[20] Not having kept track of time, Lahren sat up wondering what was going to happen next. The bell had been down for a short while and the question on his mind was, "*Were they going to send Richard and Skip back down or were they going to send a fresh team down?*" And then, just as he was about to ask for a coffee, the bell winch suddenly fired up again. Listening intently, he heard the sounds that confirmed the bell was being recommitted. So he went back to sleep wondering whether the dive would be aborted, or whether Richard and Skip would do their eight- to ten-hour bell run.

After five minutes on deck, the bell was launched over the side. As it descended, the deck crew began applying the chain-stops to the umbilical.

2300 - Bell unmarried from system. Will try again.

Inside, Richard and Skip watched the bubbling water rise in the ports. During the descent they called out the depth to topside. Fifteen minutes later, they came to a stop at 485 feet and Reid notified them that they could go for the door.

Wasting no time, the two divers worked in harmony together. Skip turned a small valve on the inside hatch and equalized the pressure between the trunking and the bell. They raised the door and Richard sat down. He pulled a neckdam, dubbed the "toilet seat," over his head and donned his emergency bailout bottle. He clipped his dive hose to a stainless steel Dee-ring on the left side of his harness and leaned forward while Skip connected his bailout regulator to his bottle and opened the supply valve. Richard applied a small amount of baby shampoo to the inside of the face plate of his diving helmet to keep it from fogging; and then he pulled his dive gloves on. Skip taped them to his suit and opened the valve to Richard's breathing gas on the manifold. Richard checked the flow to his hat, and taking a deep breath, put the helmet on and secured it in place. The piercing smell of panacide, the disinfectant Skip used to clean the hat, filled his nostrils. Small droplets went down his throat on his first breath. It tasted like antiseptic urine.

Richard performed a quick comms check with Reid while Skip reached over and turned a couple of valves diverting hot water to his partner's umbilical. Richard connected the end of the flowing hose to the hip connection on his suit, and as the water began circulating, he kicked the dogs on the bottom door open and slipped his fins on. When Richard was ready, Skip reached over and slowly cracked the blow down valve to press the bell from 470 feet to 485. Within a minute, the hissing gas had forced the bottom door to drop. At twenty minutes before midnight, Richard Walker glanced into the eyes of his partner and then quietly disappeared into the dark waters below.

With snapping champagne bubbles swirling about him, he quickly darted clear of the surging bell and announced his presence, "Diver in the water."

The illuminated bell danced on its steel wire in a sea of darkness like a spider on the end of its silk thread. Richard couldn't see the bottom from within the glowing aura. He pulled the bottom door back by its lanyard and secured it to the bell frame. He ordered topside to "slack the diver as he goes" and then he released his grip on the bounding capsule.

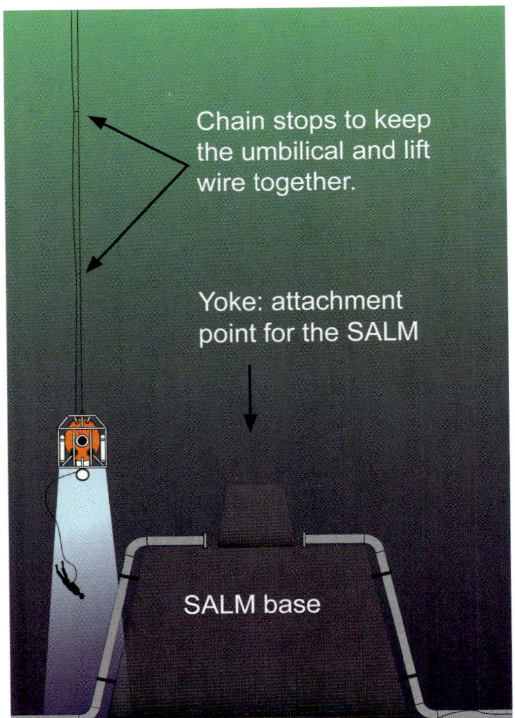

Like a feather drifting through timeless space, he watched in awe as a suspension of miniature sea life leaped about him under artificial light. With a wave of his arm he conducted a phosphorescent orchestra as his umbilical slithered back and forth like a sea serpent—thrusting itself down and then back up again—growing in length as he continued his fall. His eyes were fixed upon the abyss below, anticipating an image of another world, and within seconds, like a developing photograph, the surrounding blackness became a moonscape. Landing softly on his feet in a cloud of silt, Richard looked up to see a crystal spaceship hovering inside a halo of blue violet haze—surreal and beautiful. Off in the distance, dark shadowy images flitted in and out of the monochromatic gloom, and then, against the black curtain, he saw the faint outline of his destiny. Slowly, he turned towards the dark object and plodded off, umbilical trailing after.

2348 - Diver located the base.

At the foot of the SALM base Richard called for topside to take a depth reading. Resting in 526 feet of water the base was enormous, a colossal carom piece with six sides gently sloping towards a plateau about 25 feet off the seabed.

The corners were labeled with large white numerals. The yoke that Brook had inspected three weeks earlier stood at its center and towered another ten feet. Coming out of the jet, two 16-inch pipelines approached the base from opposite directions. These were the ballast and crude lines that came from the platform over a mile away.

Wanting the bell closer to the job, Richard ordered Reid to move the ship ten feet to port.[21] Three minutes later, he had topside raise the bell five feet to keep it clear of the yoke, and then he went back to the bell to retrieve the video camera.

2353 - Pick bell up 5'. Bell surging up/down 10'

At the dive station, the deck crew was notified that Walker had the bell in position; and just as they had with previous runs, the bell umbilical was secured to its basket by the tying-off-line.[22]

Gripping the surging bell tightly, Richard unraveled a rack of thick orange video cable and took the camera over to the base. In the near weightless environment, he crawled up the side effortlessly and began the survey.

In the bell van, Burgess and Wilson watched as Richard directed the camera at different parts of the yoke, but after four minutes, Richard stopped the survey when he noticed the bell surging dangerously close to the SALM base. Even on anchors, the *Wildrake* was not rock solid. As the ship's Offshore Manager would later recall, "There was a long swell, and the boat was wobbling sidewards."[23]

At midnight, Richard began to reposition the bell. The first move was 15 feet forward, and then the bell was raised another five feet.[24] Still unsatisfied with its location, he ordered the *Wildrake* 30 feet to stern, then down on the bell five feet. Ten minutes after midnight on the morning of the 8th, he continued the survey.

Several miles away the *Stena Welder* was trolling lazily back and forth when Mike Allen was relieved by Supervisor John Perrett. Among Perrett's divers were Eddy Frank and Phil Kasey-Smith. As the shift began, Eddy could see that dayshift had already taken most of the gear out of the bell, and that technicians were working on the environmental control units in the chambers. And then he was approached by Superintendent Jim Hurney who said, "Tonight, okay Ed, you're familiar with all this sort of shit. Test it out and put the umbilicals back in the bell."[25] And then the twenty-six-year-old American went off to bed. He was sick with fever and suspected he had a liver infection.[26]

0025 - Lowering cover plates and down-line to diver.

Inside the Wildrake bell, Skip wiggled his lower jaw and shifted his weight on the metal seat. He glanced over at the gas panel. The needle on the pressure gauge reading Richard's gas flow was in its normal position. There had been no problems with the hot water supply, and from time to time, at topside's request, he paid out and took up slack in Richard's umbilical.

At 0025, Richard laid the video camera down and established a down-line on the base for receiving equipment from the surface. The bell was outboard the ballast line near corner number four. In dive control, Ron Wilson decided he had seen enough and went to bed leaving Jim Burgess to oversee the rest of the operation.[27] Over the next hour and half, items for the pin assembly—cover plates, swivel pin washers, and a 5-ton hydraulic jack—were lowered down to Richard in two lots. But the items were landing on the seabed, far from the base, and Richard found himself wasting a lot of time and energy dragging them to the job site.

0155 - Diver has washer and 5 ton Jack. Dragging them over to base plate.

To reduce his workload, he decided to reposition the ship to have the next load land closer to, or directly on top of the base. At 0200, he began a succession of four moves in a clockwise direction. The first move was 15 feet to port. Standing on the seabed, Richard watched the bell "pogo stick" to its new location. When Reid reported they were "all stop" on the ship, he asked them to move another 15 feet in the same direction. At 0215 he ordered the ship 20 feet forward, and then five minutes later, in a final move, an additional 15 feet.[28]

0220 - Move vessel 15' ahead.

During these last two moves, the bell had crossed over the 16-inch ballast line and was now hovering on the platform side of the SALM base between corners six and one. At this point, Richard had been in the water for over two and a half hours and decided to secure the video camera on the bell.[29]

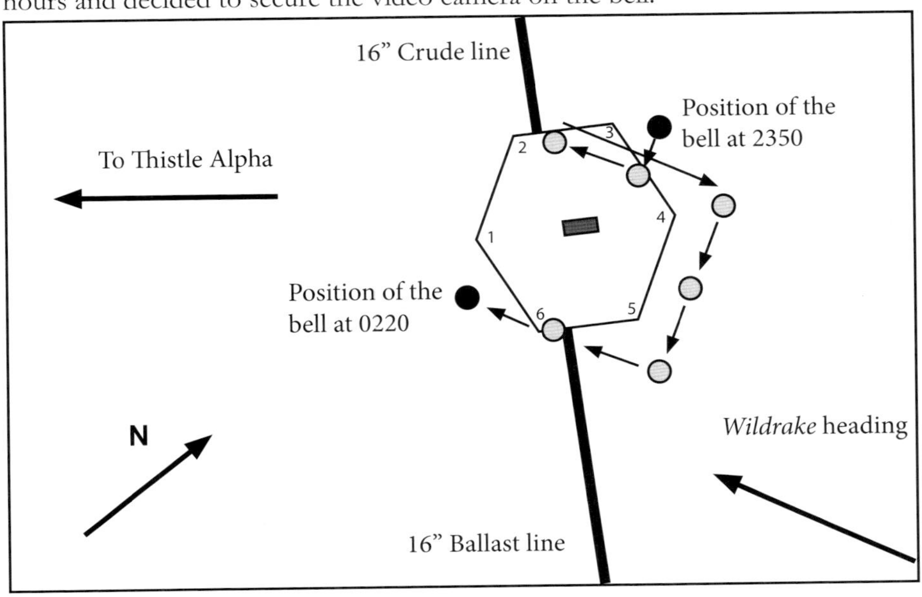

Richard moving the bell in a clockwise direction to avoid the yoke.

In his cabin, Kurt Baukau was lying in bed, unable to sleep. He kept hearing a sound and feeling a vibration in the ship that unnerved him. *Boom*, and then after a long pause, another *Boom*. Baukau was Wilhelmsen's Offshore Manager and had once been a diver with Ocean Systems. He'd also seen his share of accidents. In 1974, he was on board the *Drill Master* when Pier Skipness and Robert Smyth accidentally surfaced in their bell from 320 feet.[30] After giving up diving he landed the job as Wilhelmsen's representative and had been with the *Wildrake* from its inception.

Boom. There it was again. And then Baukau realized what was keeping him awake. It was the bell wire banging in the sheave on the end of that "Mickey Mouse bell launching system."[31]

After remaining on duty until quarter-past two, Superintendent Peter Holmes reported to Brian Masterson on the progress of Richard's dive, then decided to get a cup of coffee before retiring to his cabin.[32]

Three decks below, near the dive station, diver Tim Newson was inside the ship's machine shop working on a project for Chuck Alexander. After coming out of sat on July 24, he planned on going home to Cornwall, but at the last moment agreed to stay on to work with Alexander.[33]

One deck above, John Brook was standing in the bell van with Brian Reid monitoring the progress of Richard's dive when he heard something out of place.[34] There was a loud bang on deck which sounded like a small explosion.[35] At first it was difficult to attribute the sound to any specific event because it was mixed in with the general vessel noise, but Reid heard it too, and while looking out the window he saw the bell wire suddenly jump on the end of the davit.[36] It was a sound that both men would never forget for the rest of their lives.

Eighty-eight fathoms below, the bell rocked to one side flinging loose gear across the interior and hurling Skip out of his seat. Shaken, and with freezing water rising in the trunking, he had no time to think about what had just happened. He reached over and cracked the blow down valve. After a moment the water stopped its ascent and then, slowly, it began to recede.

In the water, Richard was making his way back to the bell with the camera when he looked up through the watery gloom to see the bell strangely cocked and hanging from the umbilical closer to the seabed.[37] Suddenly, everything came to a streaking halt in an odd, surrealistic way, as though time had split leaving him standing inside a polythene bubble peering from another dimension at a world gone awry. The surging bell, the sparkling sequins of sediment, the wavy movement of water—everything magically stood still for an instant in syncopated time—the trigger to set it in motion again, the blink of his eye.

Under the glow of flood lights, Richard instantly recognized the danger before him, and somewhere within his soul, an inner voice took control, calmed his breathing, and spoke to him on the most fundamental level. There was no training for the kind of reality he was facing at that moment and there was no time to be afraid. His eyes flickered, and with total presence of mind, he aimed the camera at his nightmare and said, "Topside, we have a problem, the lift wire is parted."[a]

Even before he saw the images flashing over the video monitor, Brian Reid suspected what had happened. But it was only when he turned and faced the picture Richard was sending him, that his suspicions turned to sickening reality. The bell was hanging at a 45 degree angle on its life-support umbilical,[38] and looking at the gauge on the console, he understood that it had also dropped in depth. Fearful that Richard would be trapped outside the bell if it came to rest on the seabed Reid ordered him to get back inside and get a seal.[39] Seconds later, the video monitor went black, and according to Reid, all communications with the bell were lost at this time.[40]

Richard had already ditched the camera when he shouted his response. Frantically he drove his body across the sea floor with a single goal in mind—*get inside the bell.* Without the support of the main lift wire, the bell was slipping dangerously close to the seabed on its chain-stops. After timing his entry, he and Skip closed the inside door and blew the bell down sealing themselves inside. The bottom door was still tied back; there had been no time to close it.

Just as Reid was ordering Richard back to the bell, John Brook had exited the bell van for the dive station. As he hurried down the metal staircase and approached the umbilical basket, he heard a hissing sound.[41] Looking up, he saw the bell being supported by its umbilical, and that the strain of the load was on the tying-off-line.[42] What's more, the polypropylene rope had garroted the bell umbilical and fractured the hot water line taped to the outside of the umbilical. Like blood gushing from a broken artery, hot water was spilling onto the deck instead of running down to the bell.

In the bell van, Reid dialed Peter Holmes's cabin but got no answer.[43] At the same time, he continued trying to reach the bell over the radio with no success and thought the communication cable must have broken.[44] Suddenly a regulator on his gas panel moaned as the gauge monitoring internal bell depth climbed to 560 feet. From this, Reid understood that Richard was safely back inside the bell.[45]

Down in the machine shop, Tim Newson was busy working on his project when Brook hurried in and shouted, "Tim can you give us a hand on deck

---

a In court, Reid used the word "parted," but BNOC engineer Ron Wilson testified he was "certain" Reid understood that the wire had "separated" from the bell, and that the "shackles failed" as Richard first noted with the video camera before losing communications. *Source*: Walker/Guiel FAI p. 673-675, 698.

because we've got a problem with the bell."[46] At the dive station, Newson saw the umbilical under "severe strain" with hot water spurting from it.[47] When one of the tenders asked him if he should shut down the hot water machine, he told him, "No, leave it running. Go and call Peter Holmes and get the other shift out." And then he ran up the staircase to dive control to ask Reid what was going on.[48]

Reid informed Newson that both divers were back in the bell with a seal and that he was unable to reach Holmes in his cabin.[49] According to Newson, Reid ordered him to "Go and get Peter Holmes." When Newson said he'd already sent someone to do this, Reid shot back, "*Well, go and double check!*"[50]

As Newson left dive control, the bell was still hanging from its umbilical and no attempt had been made to recover it.[51] The journeyman diver climbed the two flights of stairs to the superintendent's cabin and found him awake with a towel wrapped around his waist.[52] When he explained the emergency, Holmes told him he'd be right out.

Meanwhile, Reid had phoned the sat van and asked Gary Legere to come up and establish emergency communications to the bell.[53] After consulting with Reid, Legere began setting up the through-water wireless unit which involved lowering a tomato-sized hydrophone on the end of a wire cable over the side of the ship. If he could get it close to the bell, then Reid might be able to speak to Richard and Skip. But under the best of circumstances these early systems didn't work very well: thermoclines, distance to the bell, and noise affected the quality of communications.

At about this time Ron Wilson stepped into the bell van and was given the impression by Reid that the lifting cable had "separated" from the bell.[54] Wilson had been sleeping in his cabin when Burgess called him and said he thought that, as the only BNOC employee on board the vessel, he ought to be there, now that there had been an accident.[55] When Wilson arrived, Reid was directing someone to go and wake up other members of the crew, people whom he thought should be present because of the accident.[56] Shortly thereafter, Legere left to alert Don Cox and Dave Olson.[57]

When Holmes arrived in the bell van, he stood and listened while his supervisor explained what had happened. The bell was no longer attached to the main lift wire; the two divers were inside and reasonably safe; they had no comms; the deck crew was rigging up the through-water telephone, and Richard and Skip's only heat source was bleeding all over the back deck.[58] It was every superintendent's nightmare, and on that night its dark shadow would put Peter Holmes's leadership abilities to the test.

Nearly nine hours earlier, Supervisor Don Cox had gone off shift, had a meal, watched a video, and retired to bed, when the next thing he knew, Gary Legere was shaking him saying, "We've lost the bell. Get up, we've lost the bell." And then the tall American vanished out the door. All over the ship people were being awakened with those very words, "We've lost the bell. We've lost the bell."[59] And, like Tim Newson, when they arrived at the dive station they found the main umbilical under a terrific strain.

To the crew, it was obvious that, in order to repair the ruptured hot water hose, they first had to take the weight of the bell off the tying-off line. According to Brook, "a more substantial piece of line" was tied around the umbilical and secured to the side of the ship. But when the tying-off line was cut, the weight of the bell was immediately transferred onto the new line. "We basically transferred one problem straight onto another," Brook says.[60]

While the deck crew dealt with the umbilical, Don Cox arrived in the bell van in the midst of a discussion between Masterson and Holmes who were considering whether a rescue ship should be called at that time. After a few moments, and without telling either man what he intended to do, Cox exited the bell van for the bridge in order to radio the nearby *Stena Welder*.[61] But before he left, he says Masterson and Holmes had decided not to declare an emergency just yet, but to "evaluate what was available."[62]

Having worked on board the *Welder* for two years, Cox knew the crew and the capabilities of the ship. And he knew that if the *Welder* had divers in sat, they would most likely be at the same depth as Richard and Skip. But when he radioed the ship to inquire about its operational status, he discovered the *Welder* did not have anyone in saturation at that time and that the system was down for maintenance. Although Cox reported to the *Welder* that the *Wildrake* had a "bad problem," he did not reveal any specific details about the accident, nor did he tell them they had an emergency.[63] As he later testified at the inquiry, "The initial conversation was just to let them know that we had a problem and that we may require assistance."[64]

By the time Cox returned to the bell van, Peter Holmes had managed to lay his hands on the umbilical winch manual and was discussing an alternate means for recovering the bell with Reid and Masterson.[65] They were also reviewing the availability of rescue ships if one were to be needed. They knew the *Stena Welder* was close at hand and under contract to BNOC. And they knew that the diving vessels *Seaway Swan*, *Uncle John*, and the *Talisman* were working twenty miles away in the Brent Field because Masterson had given Shell Oil prior notice that the *Starmi* would be towing the SALM into the area.[66]

At about this stage of the conversation, Cox turned to Holmes and informed him that he had been in touch with the *Stena Welder* and that it was not operational. This bit of news, however, did not go over well with the company owner. According to Cox, when Masterson heard what he'd done, he "threw a wobbler."[67] Today, when he looks back on the possible reasons for that behavior, he points out that a diving company owner wouldn't want anyone, especially his competitors, to know what had just happened. Accidents can produce negative repercussions, both economic as well as legal. By law diving contractors were required to report accidents, whether fatal or not, to the Department of Energy. And no contractor wanted the government nosing around asking a lot of questions about practices that might be considered unsafe and against regulations. In the North Sea, the prevailing attitude was: if an emergency could be handled without officialdom becoming the wiser, so much the better.

So Holmes, Reid, Masterson, and Cox stood in dive control and considered their options. They had a bell down suspended by its umbilical in 500 feet of water and no means to recover it using the main winch. If the clump weight system had been in service, they could have raised the bell back to the surface using the guide wire winches. But, because the clump weight had been abandoned, Holmes and Reid were now forced to advance what they believed was the *Wildrake's* true secondary means of recovery—the bell umbilical.[68]

Both men subscribed to this view[69]—Reid even going so far as to later testify that using the umbilical was the textbook method of secondary recovery[70]—although neither man had ever witnessed a test which certified the umbilical for this purpose.[71] Reid supported the idea because someone from Møllerodden told him that the umbilical winch was the secondary means of recovery. He'd even been shown a brochure full of colored pictures which stated as much.[72] Indeed, the umbilical winch manual that Holmes had produced stated on page two:

> Emergency:
>
> In case of wire breakage, the diving bell may be lifted to the water level with the umbilical wheel.
>
> - Shut the emergency valve and tighten the force button. This will give enough pulling force to lift the bell in water.
>
> - It may be necessary to increase the tension wheel force, to avoid slipp between the tyre and the umbilical [*sic*].[73]

And mounted on the bell handling console just below one of the wheel valves was a plaque which clearly read:

EMERGENCY

SHUT TO GET

EXTRA PULLING

POWER[a]

By law, every supervisor engaged in diving operations is required to read, understand, and follow his company's emergency procedures when a bell has lost its main lift wire.[74] Whether Reid and Holmes consulted Infabco's diving manual during this period is not known, but it is clear that both men believed the only

---

a Despite this evidence, Møllerodden would later claim that the bell umbilical was "not designed" to be used as a means of raising the bell. *Source*: Civil Case Closed Record, November 9, 1983, p. 35, 54.

practical way for saving Richard and Skip was to follow an obscure document written by a small manufacturing company in Norway which had nothing to do with diving, but everything to do with hydraulically operated power wheels.

As Reid would later claim, the recovery procedure was "quite straightforward," and there was no reason why they shouldn't try it,[75] notwithstanding the fact that Richard and Skip's only means of staying alive was the umbilical.

One other factor might have influenced these early considerations. Reid had been employed by Sub Sea International, and within that company it was always rumored that, in an emergency, the umbilical could be used to recover the bell.

"Everyone knew that Sub Sea bells were 500 pounds negative," one former Sub Sea supervisor recalls. "Well that was something that we were all fed, but all of us thought about it at the time, is that really true?"

There was, however, another option available to Holmes. He could set the bell on the seabed, repair the hot water hose, call in another diving company to effect a rescue, and then wait. As long as hot water was supplying heat to Richard and Skip, they would be safe for the time being. They had a reserve supply of mixed gas and pure oxygen. And they had enough soda sorb to scrub the bell atmosphere of carbon dioxide for perhaps another eight to twelve hours. After that, CO_2 poisoning would become a real threat. But for now it was their lack of hot water that concerned everyone.

As the discussion in the bell van progressed, Cox says everyone expressed their opinion of what should be done, and how it should be done, but ultimately Holmes had to make the decision because he was the superintendent and that's what he was there for. And he had that manual in his hand. "Holmes had looked at the manual," Cox says looking back, "and Holmes had assured us—all of us—and he had convinced Masterson...that the manual said the bell could be recovered on the main umbilical."[76]

But Holmes also knew that if the lift failed, his only recourse would be to call for outside help. Cox had already let the cat out of the bag by notifying the *Welder* they had a problem. And, despite the fact that Cox had informed Holmes and Masterson that the *Welder* was not operational, the decision was made to bring her alongside in case they needed assistance.[77] According to Masterson, the *Welder* was selected because it was the closest vessel to the accident, and she had more experience in the Thistle Field than any other vessel.[78] To gain its release from its normal work duties would be a straightforward matter, he would say later. "All we had to do was to ring the Offshore Installation Manager on the platform."[79]

But in order to secure the release of the other vessels in the Brent system, Masterson thought he would have to go through the Shell marine department for permission which would have taken longer.[80] And when Holmes asked him if he should ring the *Welder*, he replied, "Yes, I think we have got to get a rescue vessel in."[81] And with that, the plan to lift the bell using the umbilical wheel was temporarily suspended while Holmes and Masterson went to the radio room to contact the *Welder*.[82] On the bridge, Holmes called the platform and received immediate permission from the Installation Manager to have the *Welder* come alongside and give assistance.[83] He knew it was not a DP vessel and that its efforts to conduct a rescue operation very close to the *Wildrake* could be frustrated by the *Wildrake*'s anchor lines, but, as he would later explain, it was the only vessel in consideration.[84]

So Holmes contacted the *Welder* and talked to the marine crew, but he never asked to speak to Superintendent Hurney; he did not receive a report on their state of readiness, nor did he inquire.[85] When asked at the inquiry if he thought it was important to know whether the *Welder* was in a position to dive right away or not, Holmes replied, "Yes, but I was still getting ready to make an attempt to lift with the umbilical."[86]

> 0257 - Wildrake reports diving emergency. Stena Welder directed to assist.
>
> —— Thistle Radio Log

Chapter Thirteen

The Beginning of the End

The most dangerous thing in the sea is the people who are supposed to be helping you.

—— John Joly, Infabco diver[1]

On board the *Stena Welder*, diver Eddy Frank was transfusing his body with cups of bad freeze-dried coffee as he plodded through another cold night of boring maintenance work. Repair of the *Welder's* dive system was in full swing. The environmental control units had been stripped out of the chambers,[2] and at Eddy's feet the bell umbilicals were strung across the deck, ready to be fitted with new terminal connections. Inside the two man bell, the hot water inlet valve and the main gas supply gauge had been replaced by the previous shift.[3] The interior was small, very small; but strangely, without the hoses coiled on their horns and the scrubber unit, Eddy thought it almost seemed big. Along the circumference of the bell, copper tubing snaked from valve to valve as black electrical cables draped down from the ceiling like licorice whips. There was still a lot of work to do.

Suddenly, people began bustling and running about the deck. *"There's something up here,"* Eddy thought to himself.[4] And then someone explained that the *Wildrake* was having problems and that they might be asked to make an emergency dive. At the same time Superintendent Hurney appeared and began giving orders to the crew. The life-support technicians were told to prepare the chambers for blow down; deck hands were instructed to check available rigging gear, and then Hurney turned to Kasey and Eddy and told them to get the bell ready. They would be the first team down if it came to that.

These were "strapping young lads," Hurney says reflecting back. Big and strong, and the best he had.[5] And then he hurried off to call his office in Aberdeen.

When Don Cox contacted the *Stena Welder* on an open airwave earlier that morning, the effect of that broadcast was like the sound of drums beating across a silent jungle. Within minutes of the transmission, word of the accident quickly spread to Infabco's diving office on the *Gulnare*. Superintendent Tom Belcher was sitting at his desk completing that day's paperwork when a voice on his radio suddenly interrupted his concentration. It was the Thistle Alpha radio officer asking if he was aware of the *Wildrake* problem. Belcher said no, sensing an uneasiness in the man's voice, almost as though he felt uncomfortable possessing information no one else knew about. The voice on the radio asked if Belcher wanted to speak to someone on the *Wildrake*. Belcher said he'd get back to him after he spoke to his boss, Field Superintendent Craig Roberts.[6]

When Roberts subsequently contacted the *Wildrake*, he offered his assistance, but the conversation ended without it being accepted.[7] A short time later, however, Masterson called back and requested him to come to the *Wildrake* and bring help.[8] Roberts then selected nightshift Supervisor Tom Almon and divers Randy Gunter, Chris Roper, and Rod Pelzel to accompany him to the *Wildrake*.[9]

While Captain Mellor was steaming the *Stena Welder* towards the accident site, a high-speed Zodiac was sent to pick up the *Gulnare* divers. At approximately 0320, Roberts and his team climbed down a ladder system on one of the rig's massive legs and jumped into the Zodiac. Overhead, a thick blanket of clouds had filled the night sky. There was just over a mile of open water to cross, and as the boat faded into the night, Belcher looked northward and saw the *Wildrake* glimmering like a cold star. Suddenly, he felt hollow inside. Deep within his soul, he realized that somewhere on the bottom of this unfriendly sea, two of his friends' lives were in peril and that he might never see them again. Memories of Richard came flooding back to him, of their days at City College, of the softball games in the park, and how they swam and hiked on weekends in the foothills of the Santa Ynez mountains. Tom knew Richard was a man who dearly loved life—as Skip did—and now they were down there fighting for their lives. For a brief moment he closed his eyes and prayed for them and hoped he would never have to think of this day ever again.

The bell was only suspended by the umbilical. Now I formulized that the divers in the bell were aware of how the umbilical was attached to the bell, and the connection was rather flimsy. Therefore, I would have thought that the divers would be rather anxious not to sever those lines to the bell. Without their gas, hot water, they would be in a very serious condition.

—— Roy Giles
Department of Energy Senior Diving Inspector[10]

Like the creak of an old rocking chair, the sound of the bell surging up and down had an almost rhythmic, hypnotic effect. Every time it surged downward, Richard and Skip hoped that the small loop of chain securing the umbilical to the bell frame would remain in tact. Both men knew it was not made to carry the weight of the bell, and if the mooring parted, the individual supply hoses would break free from their penetrators. It was a prospect neither man could erase from their thoughts.

Since the accident, the bell had cooled somewhat. Hot water supply had been erratic—one minute it was there, then suddenly it disappeared, and now it was flowing again through the small heater mounted on the wall. They had no comms, but the scrubber was still working. The wireless radio had remained silent. Richard and Skip had tried every configuration: switching to high transit mode, then low, dialing up the volume, then down, but there were no voices offering hope.

No lift wire.

Richard had showed them on the video what the problem was. *Did they see it? Did they know?* He exhaled heavily and shook his head. He talked with Skip about the measures topside should take to recover them, but what topside would actually do was anyone's guess.

Set us on the seabed and call for help!

Richard and Skip didn't even want to think about the consequences if they were to lose hot water permanently. Helium was not like air. It tries to bring everything around it to its own temperature as quickly as possible. And the greater the density of the gas, the faster this process takes. Studies have shown that at depth "the rate of heat loss with helium can be as great as the heat loss which occurs on immersion in water."[11] Instead of days, survival could be calculated in terms of hours.

Within a year, the Norwegians would conduct a series of experiments and discover that once heat had been cut off from a bell, even one that is insulated, the temperature of the atmosphere approaches the temperature of the sea in approximately 150 minutes.[12] The average temperature of the North Sea where Richard and Skip were diving was 4 degrees centigrade.[13] Helium also reduces the insulative properties of clothing such as woolly bears.[14] Heat loss can occur through a variety of sources: conduction, respiration, radiation, convection, and even through urination brought on by the cold.

Lost in thought, Richard and Skip barely noticed the initial shudder when it came. At first it was subtle, blending in with the regular pattern of tugs on the umbilical, and then it happened again. Skip glanced up and Richard's distant expression dropped off his face. There was another jerk on the bell, and for a moment it was difficult to tell what was actually happening, but one thing was certain—they were moving; they just didn't know which way. The two divers looked at each other and waited—their faces frozen in anticipation, their ears listening carefully to every sound, their minds trying to interpret every motion. Then, there was a new sensation, as though they were floating...

They're hauling us up on the umbilical. No, No, No....

Skip immediately got on the radio, "Put us on the bottom; put us on the bottom." And then suddenly everything stopped moving, as in the eye of a storm.

> *I questioned immediately why in the fuck would you possibly lift the goddamn bell with an umbilical? If you could maintain the umbilical, you could keep these guys going for a long time. Why would you ever try and lift the bell by the umbilical? I don't care what they write. And when you looked at that umbilical it was not a very big umbilical. It just seemed insane to me—when you had a stabilized situation that you could deal with for awhile—bring in somebody else or drop a surface diver to get you sorted out and get you hooked back up and then bring back the bell. You certainly never risk the lifeline.*
>
> —— Dave Olson
> *Wildrake* Life-Support Technician[15]

In the bell van, tender Nigel Winterburn was holding a small tape recorder near the radio as he tried anxiously to re-establish contact with Richard and Skip. The entire conversation leading up to the accident had been recorded through the hardwire connection to the bell. But that link was now broken and Peter Holmes knew he would have no record of any future conversations. So he had gone to his cabin and retrieved a small recorder for Winterburn to use; but the only sound pouring out of the speaker was raw static.[16]

Having requested the *Stena Welder* to come alongside, Holmes ordered Reid to proceed with the lift.[17] Reid radioed Brook on deck to standby for a bell recovery using the umbilical wheel. Then he fired up the hydraulic motors, closed the emergency valve on the bell handling console, and adjusted the lifting torque of the umbilical wheel.[18] From the window of the bell van, he could only partially see the wheel 30 feet away on the other side of the davit arm.[19] When the lift began, everything seemed to be all right—the bell actually started to come up as expected—but within a few seconds Brook was shouting into the radio, "All stop! All stop!"[20]

The umbilical had jammed in the sheave.

For the past half hour, while the crew had focused on gathering tools and materials to repair the ruptured hot water hose, little attention was paid to what was happening 20 feet above their heads. With each motion of the ship, the service lines of the umbilical had been taking a beating on top of the wheel. To Brook, the wheel looked like a "Morris Minor tire" sandwiched between two flat circular plates, and like most tires this one was slightly rounded on the sides.[21] The whole thing rotated by hydraulic power. It's construction had been subcontracted to a company called Karmøy Mekaniske Verksted AS in Norway that specialized in high pressure hydraulic equipment. According to the specification, the tire was supposed to be "rubber filled."[22] But unbeknownst to anyone, it was in fact empty at the time.[23] And because the umbilical was under tremendous strain, the tire had flattened and the duct tape bundling the service lines together had split, allowing various cables and hoses to migrate along the surface of the tire. Some of the smaller cables had actually lodged themselves in-between the tire and one of the side plates. Reid was unaware of this and when he began turning the power block, instead of the cables rolling off the back of the wheel as expected, they became entangled in the winding gear of the power block like in a meat grinder. Once again, hot water began cascading onto the deck.

Brook explained to Reid what had happened and suggested he reverse the winch. The weight of the bell should pull the jammed cables and hoses free.[24] But when they tried this all they succeeded in doing was wedge the cables in deeper. At this point, Tim Newson says there was a real fear that the umbilical might break in half and be lost over the side, so the crew lashed it to the railing with a second line.[25] Then Newson and diver Ron Arscott took a crow bar and climbed the davit pedestal and walked to the end of the boom.[26] Like a gangplank, it was dangerously narrow and slippery, and while Newson leaned precariously over the wheel, Arscott held his legs to keep him from falling into the sea. As Newson wedged the crowbar under the umbilical, Reid used the hydraulics to rock the wheel back and forth to try and ease the umbilical cables out, but they were jammed solid.[27] As steaming hot water continued to shower the deck below, it was clear that the mashed umbilical was stuck fast and that rotating the power block had made matters worse. Now, the bell could be neither raised, nor lowered.[28]

Everyone at this point—Cox, Masterson, Reid, Holmes, BNOC personnel—was offering a suggestion here and there, but nothing was working. Standing nearby, Brook was beginning to detect a general feeling of frustration and disbelief.[29] No one could understand why a system that was supposedly designed to retrieve the bell had failed to do its job.

"It is getting to be a very fraught situation now," Brook recalls, "because nothing is going right."

And what happened next demonstrated truly how desperate the decision-makers were becoming.

While the lift wire was no longer attached to the bell, the umbilical still was. Moreover, the two lines were married to each other by the chain-stops. And it was thought that, by hoisting in the lift wire, the chain-stops would transfer the load to the umbilical and thereby raise the bell to the surface.[30] No one was certain whether a dozen loops of garden-style chain could take the weight, but they decided to try it. As the main winch began to wind in the heavy bell cable, Brook, Newson, and Cox witnessed the chain-stops being ripped from their securing points. Cox says the attempt "damaged the umbilical."[31] And damage the umbilical it did. Newson would later testify that, "parts of the umbilical were breaking off."[32]

For Superintendent Holmes, each new decision seemed to unravel one vexing problem after another. It was like pulling on a stray thread that refused to break, but instead triggered another series of blunders nudging Richard and Skip ever closer to the edge of a horrifying precipice. The bell was stuck ten feet off the bottom;[33] the davit was out of commission, and the umbilical was inextricably snarled in the power block. Without communications, Holmes had no way of knowing the condition of Richard and Skip. The natural inclination was to get them out of the water as quickly as possible. And with conditions growing worse, stabilizing the situation by maintaining hot water to the bell and turning the rescue over to the approaching *Stena Welder* would have been a prudent course of action. He could even delay his next move and seek the advice of Craig Roberts who was due to arrive at any moment. But Holmes was impetuous and didn't want to wait. Instead, he decided to gamble Richard and Skip's lives on one last available resource—the ship's 30-ton Brattvik crane.

Commenting year's later on Holmes's decision, Brook says, "I think people had come to terms with the fact that whatever had happened up in that sheave was permanent."[34]

According to Newson, Brook, Cox, and Kurt Baukau, the plan was to lash the umbilical to the crane hook, raise the load, stop off the umbilical to the gunwale, lower the hook back down, take another bight of umbilical, raise the load and repeat this series of snatches—snatch, secure, snatch, secure—to get the bell close enough to the surface for a diver to attach a new lift wire. It was an unprecedented move, and several divers, including Newson, had serious doubts about the plan and warned against it.[35]

Since the early days of bell diving, one of the most fundamental precepts has been that the main umbilical should be protected at all costs because it is the only thing that separates a diver from eternity. Everyone knew the *Wildrake* umbilical was the most basic of bell umbilicals; it was handmade; it had no external reinforcement; it was already falling apart, and no one could say for certain that it had ceased to function at this point. If it broke in half now it would be an understatement to say it would further endanger Richard and Skip's lives and complicate a rescue mission.

Holmes, however, argued that the core of the umbilical was equipped with a Kevlar strength member designed to bear the weight of the bell. Newson was aware of this but expressed no confidence in the plan. "I mean, it was only sort of half-inch diameter stuff that was supposedly made out these new materials," he recalls.[36] And Reid would later testify that, during the outfitting of the *Wildrake*, doubts had been expressed about the umbilical and its ability to raise the bell. "We didn't think it was man enough for the job it was supposed to do."[37]

To reassure himself, Holmes retrieved a copy of Møllerodden's drawings which detailed the components of the umbilical, and saw the number "10" with the letter "t" next to it. Thinking this meant the strength member could withstand a 10-ton load, he decided to confirm this directly with Møllerodden and promptly left the dive station to call Norway.[38]

As time progressed, the bell van saw a flurry of bodies coming and going. Decisions were hastily formulated and directions given to various parts of the ship—the sat van, the bridge, the deck. Winterburn remained on the through-water comms while Reid was preoccupied with the emergency and monitoring gas flow through the dive panel. In a corner of the cramped room, BNOC engineer Ron Wilson had been so engrossed by the unfolding events that it suddenly dawned on him neither Reid nor Burgess had been keeping a log since the start of the accident. Burgess had been making entries up until that point, averaging one notation every seven minutes, but for nearly an hour, the log had remained silent.[39] Wilson suggested to his diving consultant that he continue the log by recording everything he saw and heard, a measure that would ultimately produce the only complete record of the accident and the rescue.[40]

Thinking back, Burgess estimated the time of the wire separation and bundled all the events since the accident under his first entry:

> 0245 - Bell cable either parted or came loose from bell. Bell has a seal, and comms. Both divers inside. Depth 510'—on way up bell umbilical jumped off tire, and wedged between tire and side plate.[a]

Meanwhile, Holmes had returned from the bridge with news that the strength member was indeed designed to withstand a 10-ton load. But before he set his new plan into motion, BNOC requested that Captain Hedin reposition the *Wildrake* to relocate the bell away from the SALM base.[41] When the move was completed, the bell was still suspended off the seabed, but now it was well away from the base and in clear territory.

> 0325 - Capt. of Wildrake moved 20 to 30 yrd to port from base.

---

a This log entry would cause considerable confusion at the subsequent inquiry. See Appendix for explanation.

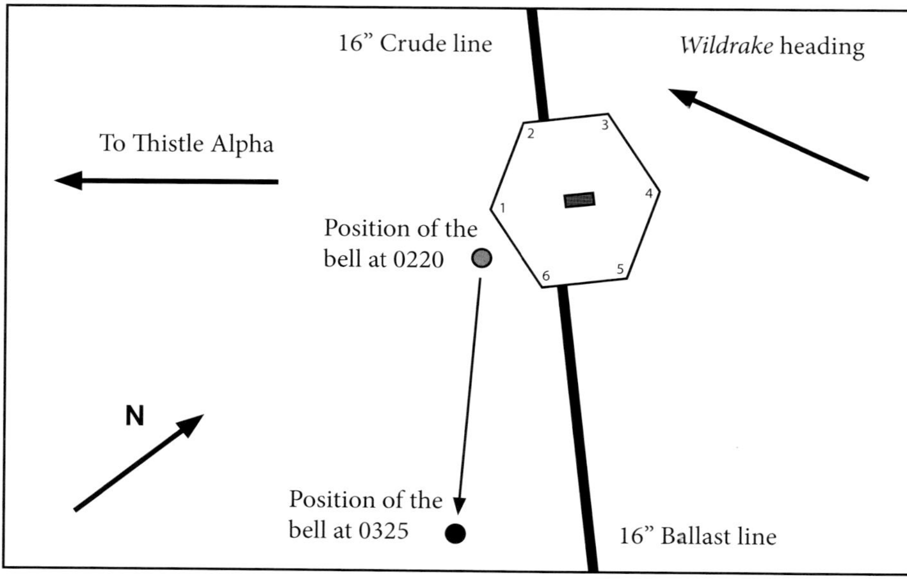

Minutes later, the *Stena Welder* closed the distance to the accident scene, taking up a position "head to toe" with the *Wildrake*, 200 yards off her portside.[42] While efforts were under way to put the rescue bell into service, Captain Mellor received a radio transmission informing him that the *Wildrake* crew "were going to attempt to lift the bell on the umbilical."[43]

No one out there was saying no.

—— Gary Legere
Infabco Life-Support Technician[44]

On board the *Wildrake,* collective anxiety was painting the deck black. To John Brook, it seemed "every man and his dog" was now involved in the operation.[45] People were dashing about, "shouting and arguing,"[46] as they gathered rigging gear in preparation for the dangerous and unthinkable task that lay before them. German born Kurt Baukau stood nearby and thought all the activity lacked direction: "There was no man who had his head on his shoulders. They were all running around like plucked chickens."[47]

Plucked chickens or not, the *Wildrake* divers were not a cohesive, efficient work force, nor were they familiar with each other's talents, skills and habits. And faced with orders to commit sacrilege in a culture where orders are to be obeyed, not questioned, they could hardly rely upon each other for moral support if one of them were to take the quantum leap and cry out, "STOP!"

It was a moonless evening, and a light rain had begun to fall. Some of the crew were donning bright orange slicker suits against the cold drizzle. Opposite the dive station, on the starboard aft deck, perched high within the crane tower, a faceless man sat waiting in a translucent cab with his hands wrapped around the controls of a machine powerful enough to raise thirty *Wildrake* bells from the seabed. The boom was in position, the crane hook had been lowered, and the lives of Richard and Skip dangled by a fragile 500-foot rubber thread—a thread they were about to jerk on.

> 0336 - Deck crew hooking up nylon sling around umbilical to clear from power sheave (used a 1 1/2" nylon rope).[48]

Out of the crowd someone produced a nylon rope sling. And despite the earlier example of the tying-off line, and the fact that it was *only the Kevlar strength member* that had a 10-ton rating, the loop of one end of the sling was taken around the entire umbilical and passed through the loop of the other end, and dropped like a noose over the crane hook.[49] And then a man with a radio turned and gazed up at the crane driver. As hope-filled eyes interlocked momentarily in the night, thoughts of doubt resonated against fear. With Holmes looking on,[50] the man pressed the talk button and fired the final command, "Up easy on the load."

Down in the sat van, Steve Lobban had been assisting Gary Legere and Dave Olson when dive control called: "Everybody that's available get up on deck."[51] Less than a minute later, he broke through the doorway nearest the dive station to see several men on top of the davit and a solemn-faced crowd gathered at the portside railing. A choker hitch had been fashioned around the umbilical; the crane hook was elevating, and from the sheer gravity of the situation, the young tender sensed something profound was about to happen. Within a few seconds, the rope sling ran out of slack and began to strangle the rubber lifeline. Like the sound of leather twisting, gas lines stretched to their elastic limit, and then, one by one, they began to burst with a shocking bang in the crisp cold air.[52]

Still, the lift continued. And then more hoses snapped and popped like firecrackers. Lobban saw hot water "flying all over the place,"[53] and then suddenly the lift stopped. When Holmes asked why, Reid said they had contact with Richard and Skip. The message was very faint, but a voice could be heard over the static, "Put the bell on the seabed. Put the bell on the seabed."[54]

The problem in this situation is that they are cooling, not only because of loss of the hot water but they are also breathing cold gas...If you cool people centrally, in other words, you cool the core without cooling the skin, then you may become cold and develop symptoms or signs of hypothermia without actually feeling particularly cold. Now, these guys were not in that position because they had lost hot water, and they had no heated gas. So, they're losing heat both through the skin and through the central—through the core.

—— Dr. Stephen J. Watt
Department of Environmental and Occupational Medicine
University of Aberdeen[55]

It was the death of the small lamp, casting them into darkness, that finally confirmed Richard and Skip's worst fears. They now understood with brutal certainty that topside had tried, and failed, to raise them by the umbilical. Whatever remaining confidence they held in their crew must have been crushed when the light went out. Were they angry? Did they shout obscenities? Probably. No one can say for sure. But whatever emotions they unleashed would have been quickly diverted by a sudden feeling of fullness in their heads. The bell was loosing gas. Somewhere along the umbilical were broken gas lines. They flipped a switch and put the bell on emergency battery power, and dialed up the Variac. A dim yellow light began to glow and the scrubber motor starting whirling again. They closed valves to re-seal the bell, then took an inventory. Out of the 17 service lines only two survived the trauma—main gas and topside internal depth. Bell power and hot water were gone.

They were losing heat rapidly. And they knew that if topside didn't restore their heat source, the steel ball they were in would very soon turn into a freezer. Skip looked at Richard's face, trying to gauge his condition. It was cruel irony that something as necessary as breathing was draining calories of heat from his body. Worse yet, Richard was wet which would make his plunge into hypothermia more accelerated.

They unwrapped their Strentex survival suits which under the circumstances must have seemed like another head-shaking joke to them. They were just thin foil blankets made for weekend campers. They were not practical for the environment Richard and Skip were in. They had been designed for spaceflight to reflect infra-red heat back to an astronaut's body. In Richard and Skip's case, most of their heat was being lost through respiration and convection,[56] something which only special respirators and insulation could have helped. Indeed, several years earlier, three scientists at the Royal Air Force Institute of Aviation Medicine demonstrated that the material used in space blankets given to Richard and Skip, offered "no better insulation" than ordinary polythene bags.[57]

Skip looked at the silent bell heater. It was only coiled copper tubing stuffed in a tin can supplied by hot water through a rubber hose. *High fucking technology! The billions spent on subsea operations and still there was no backup heating system for a diving bell. Where was the support from industry when it came time to protect the hard working diver?*

Skip knew the answer to that one. If contractors weren't forced to raise safety standards, then why would they spend a dollar to do so? He opened a metal box containing the carbon dioxide analyzer and pulled out a small, straw-shaped glass Drager tube. He nipped both ends off, stuck one end into the mouth of a hand-held pump, compressed the bellows a number of times, and then read the blue dye trail like a thermometer. The CO_2 content was close to the limit, but he and Richard decided to take the soda sorb to the extreme, then use their last canister. After that they would have to go on...

Suddenly they had that weird feeling of floating again. The bell, still hanging at a 45 degree angle, rocked from side to side. There was a thud, and then another, followed by an even greater impact, and then the capsule tilted to an upright position. They were on the bottom.

Huge drops of water were collecting on the skin of the bell, and Skip could see that Richard was beginning to shiver.

> *Who are these people? Where did they come from? Nobody knew them. Nobody seemed to know too much about them other than the fact that they really didn't know what the fuck they were doing.*
>
> —— Rod Pelzel, Infabco diver[58]

Peter Holmes and Brian Reid stood around the small radio box listening to Winterburn repeat: "Can you hear us? Can you hear us?[59] Reid didn't think the messages were getting through, but the command from Richard and Skip was clear: they wanted topside to stop what they were doing and put the bell on the bottom.

Suddenly the door of the control room opened and Craig Roberts, Tom Almon, Randy Gunter, and Chris Roper stepped inside.[60] Questions and explanations were exchanged, and in the highly charged atmosphere, tempers flared. Viewed as noisy and disruptive, Roper and Gunter were promptly ejected from the crowded room.[61] Roper says the attitude expressed to them was: "Go find something useful to do."[62] They were given no direction nor delegation, so they joined Pelzel on deck who was mystified by all the unfamiliar faces around him and the fact that Infabco had sprouted another job over night.[63]

Roper walked over to the davit where long 15-foot braids of mangled umbilical hung down like strands of spaghetti. Water was cascading onto the deck while people tried to repair the hose. The veteran diver was immediately struck by how shiny and small the damaged lifeline was. It did not have an external braid to it and it did not convey to him any positive feelings about its integrity. "It was a pretty rinky little chunk of hose," Roper remembers.[64]

In the bell van, it was now woefully apparent to the decision makers that the rescue could no longer be confined to an in-house operation, but that Infabco would have to enlist the services of Wharton & Williams to save Richard and Skip.

Exposed to the elements on the bridge-wing of the *Stena Welder*, Captain Mellor stared across the gray gap separating the two ships. In his right hand he held the control stick of the Poscon, manually directing his ship's engines and thrusters to stay on location. He had been maintaining a safe distance of about two football lengths when he received a message over the VHF radio.[65]

"Okay Captain, we want you to do a recovery for us."

The time was 0345.[66]

At this point, Masterson decided to send Don Cox to the *Welder* to act as his personal emissary because of his prior working relationship with the *Welder's* crew.[67] Ten minutes later, Cox boarded a Zodiac and as he was being ferried across to Mellor's ship, he saw the power lines of the torn umbilical arcing in the water.[68]

By this time the *Wildrake* crew had managed to free the umbilical from the power block by using the crane and a thicker rope wrapped below the damaged section. A strain was taken and Roper says the umbilical was "physically manhandled" out of the sheave. The bell was then lowered to the seabed.

Efforts to mend the broken umbilical were undertaken, but most of the crew considered it to be so badly damaged they had little faith that any of the services were getting down to the bell.[69] Randy Gunter and Tom Almon would later co-author a statement to the police in which they declared that only the main gas supply and topside's ability to determine internal and external bell depth had survived.[70]

As Captain Mellor began sidestepping the *Stena Welder* closer to the *Wildrake*, Jim Burgess recorded the alarming consequence of a reckless plan:

> 0400 - Most of bell's umbilical hose pulled apart—Bell internal press. is 560' Outside depth is 522'

Chapter Fourteen

Who's in Charge?

I know that shit happens, but when you're getting the pay for being the number one guy in the field, you'd better earn it.

—— Dave Olson
Infabco Life-Support Technician[1]

When Randy Gunter and Chris Roper were first notified about the accident, they assumed their assistance in the campaign to save Richard and Skip would be welcomed with open arms. But what actually occurred in the coming hours would leave both men wondering for the rest of their lives, why they had been called to the *Wildrake* in the first place.[2]

After being expelled from dive control, Gunter and Roper went below to the sat van thinking they might learn what was being planned, who was in charge, and in what way they could be of help.[3] There, they found Dave Olson and Gary Legere feeling helpless and in the dark.[4]

Gunter was an outspoken individual. He had an intense "go-for-it" style that defined the word gung ho. He was talented, "could fix anything," and wasn't afraid of getting into someone's face if he thought they were making a mistake. Roper and Olson came from the same mold, and the three headstrong divers saw no reason why they should be sitting around twiddling their thumbs while two people—their friends—were slowly freezing to death on the bottom of the sea. There was no information coming out of the bell van, and if there was a plan of action being formulated, no part of it was being made known to them. This was unacceptable to them, and as Gunter would say later, he and Roper "deemed it necessary" to do whatever they could to save the lives of Richard and Skip, even if it meant jumping a surface diver down 500 feet to reload the crane wire on the bell.[5] And this was precisely what they planned to do. They understood the terrible risks involved, but emotions were running high and Roper thought it seemed like "the logical thing to do—do a surface dive."[6]

That logic appealed to Olson, too, because he knew that an emergency surface dive to the bottom had been done before.[7]

It was August 28, 1973, and the pipelay barge *L.B. Meaders* was constructing the 32-inch pipeline from the Forties Field to landfall at Cruden Bay when something went wrong with Taylor diver Paul Havlena's diving gear. Havlena and his partner had been lowered in a bell to repair the pipeline after it buckled in a recent storm. The helmet Havlena was wearing was part of a prototype gas reclamation system called the Compressor/Depressor (CD) system.[8] It was designed to cycle his exhaust gas back to the surface through a hose, scrub it free of carbon-dioxide, fortify it with oxygen, then deliver the sanitized gas back to him.

Mounted on the rear of Havlena's helmet was a back pressure regulator which prevented the recovery system from applying too much suction on his lungs. Unfortunately, the regulator malfunctioned during the dive and a vacuum was drawn on the helmet which ultimately caused Havlena's lungs to collapse.

When topside relayed the emergency to the bell, Havlena's bell partner tried, but failed, to pull the unconscious diver back inside. The decision was then made to send a surface diver down to give assistance. George Cundiff dressed in, jumped into the water and descended 320 feet into the darkness. Ninety seconds later he reached the bell and helped push Havlena's lifeless body back inside.[9]

After the accident, company divers began referring to the CD system as the "Certain Death" system and it quickly fell out of favor.[10] But over the years, the story of Cundiff's heroic rescue attempt was never forgotten by a young rookie saturation technician by the name of Dave Olson.

Now, six years later, Olson, Gunter, Roper and others, began assembling a makeshift dive station on the starboard side of the ship. A portable gas panel was brought out, a radio located, and short lengths of hose (whips) were made up to connect the panel to the ship's gas supply outlets. The crew knew the bottom depth, but they didn't know if the surface diver's umbilical was long enough to reach it, and when they flaked the huge bundle out on deck and measured it, they discovered that it was too short.[11]

Roper and Gunter considered retrieving a longer umbilical from the *Gulnare,* but time was critical, so they decided instead to connect two surface tethers together. Once the splice was made, Olson pressure tested the hose and a dive hat was put on-line. Finally, there was a poignant moment in which everyone in the group paused and looked at each other. The question on their minds was: *Well, who's going?*[12]

When the decision was finally resolved, the light of dawn was reflecting off silver clouds as Dave Lane walked out on deck dressed in for the rescue mission. To some of the crew, jumping a diver 500 feet was insane. The most obvious question was: if Lane was unable to raise the bell, where was his safe haven?

Insanity or not, the thirty-three-year-old Englishman felt he had to do something to take control of the emotions going on inside of him, even though he knew his odds for survival were slim. His only chance of coming out alive was to join Richard and Skip in the bell, because, by the time he got them off the seabed, he would be in serious decompression territory. Still, Lane was determined and had he suddenly changed his mind, someone else would have picked up the mask and gone. It was madness, and they all knew it, but they also knew Richard and Skip would have done it for them.

At 0415, the regulator on Reid's gas panel began moaning again, evidence that Richard and Skip were flushing the bell to keep the CO_2 level down.[13] To give them a better chance for survival, Olson and Legere changed the main gas supply from 3.5% HeO_2 to 6%.[14] Meanwhile, Roberts had positioned himself on the radio box hoping to hear from his two friends. But that hope soon faded as the *Stena Welder* drew closer with the sound of her thrusters roaring in the water like the engines of a 747.

For nearly an hour Captain Mellor had been edging his ship towards the *Wildrake*, trying to stay within the narrow confines of her outstretched anchor wires. His objective was to avoid a collision long enough to allow the rescue diver to find the *Wildrake* bell and get a new line on it. All of his ship's available means of power and mobility—her main engines, thrusters, and rudders—were linked to the small joystick in his hand. His only positional aid was a series of cross references he took by azimuth bearing on various platforms silhouetted against the breaking dawn.[15] Without dynamic positioning, Mellor knew this was not the most efficient method of maintaining position, but it was all he could do.

At 0429, Kasey and Eddy were dressed in and ready to begin bell checks.[16] And then, just as the crew was lowering the clump weight, the drive chain on the winch broke.[17] Technicians frantically set about to repair it.

At 0440, as Supervisor Mike Allen commenced bell checks with Eddy, Don Cox received a radio communication from Brian Masterson. Up until this point, the two supervisors assumed that, as the rescuers, they would take charge of the operation and organize a plan to save Richard and Skip. But that assumption was quickly dispelled by Masterson when he radioed Cox telling him what the plan would be.[18]

> 0440 - Instructions from Wildrake to Stena - put guide wire on top of our bell then send 30 t crane wire down to hook up to bell pad-eye.

During an emergency recovery, it is customary for the superintendent requiring assistance to speak to the superintendent of the rescue vessel in order to formulate a coordinated plan to save the lives of the stranded divers.[19] Among other things, available resources, diving system capabilities and rescue strategy have to be clarified. But, as precious minutes ticked by, neither Holmes nor Roberts spoke to Superintendent Hurney on board the *Stena Welder*.[20] It was as though a communications blackout had descended upon the two ships, and as a result, the object of saving Richard and Skip fell to a singular plan that excluded the judgment, experience, and knowledge of its would-be rescuers.

According to Masterson, Holmes conferred with Reid on a plan of action then carried the proposal to him where they discussed it and ultimately decided that the *Wildrake* would be the one to raise the stricken bell, not the *Stena Welder*.[21] Curiously, no other suggestion was made, not even to follow standard procedure.[22]

Normally, when a diver is ordered to recover an object from the seabed, the bell is lowered and the diver locks out to make a circle search. Once he locates the object, he gives directions to topside to re-position the crane hook directly over the load for a vertical lift. He then connects the lifting strop (sling) to the load, stands out of the way, and orders the lift to proceed. When the object clears the seabed by roughly 20 feet, the diver orders topside to stop the lift, and then he returns to the bell. The bell is recovered to the surface and topside recovers the object. An ordinary sat diver can make an open water recovery of this type in under two hours. The procedure is so routine that thousands of recoveries are made every year without incident or problem.

The *Wildrake* was unable to make the recovery. Therefore the only logical course of action was to allow Wharton & Williams to perform the hookup using standard procedure. With their davit, they could bring the stricken bell to the 60-foot level, then hand it off to the *Wildrake* crane to be mated to their system.

But Masterson and Holmes were insisting on the completely unorthodox approach of using their crane to make the lift. This was a crucial mistake.

Everyone understood the *Stena Welder* had to be over the accident site in order to get its bell near the stricken bell. This required Captain Hedin to move the *Wildrake* off location which meant the *Wildrake* crane hook would not come down near the stranded bell, but too far away for the rescue diver to reach it.[a]

Holmes's solution to this problem was to employ a two-stage technique widely used to lower gear when diving operations are not conducted directly overhead. First, the rescue diver would attach a guide wire from the *Wildrake* to the stricken bell. This wire would serve as a "shot line" for the heavier crane cable. In theory, the crane hook would slide down the guide wire close to the *Wildrake* bell where the rescue diver would connect the hook to the attachment point on top of the bell with a wire sling. The plan sounded plausible. But it had serious flaws to it, and anyone with ten years experience, as Holmes had, should have foreseen the problems that would develop from attempting to get a hookup from an angle.

The crane wire was two inches in diameter and the steel headache ball above the hook weighed at least 600 pounds. Just the extra rigging involved for the operation would delay getting Richard and Skip out of the water, and they didn't have the luxury of time. It was an ill-conceived plan, and Holmes should have

---

a The umbilicals in the rescue bell were only about 100 feet long, information that would have been helpful to Holmes had he conferred with Hurney.

known better not to divert from a method with a proven track record in a time of crisis. Not only was he complicating matters for the rescue diver, but he was gambling the plan's success on several crucial factors: that the *Stena Welder's* erratic diving system was up to the challenge and that Captain Mellor would be able to hold his ship on station without the benefit of dynamic positioning.

On the portside bridge-wing, Captain Mellor continued to struggle to keep his ship from colliding with the *Wildrake*. The slightest movement of the joystick sent the 1,135-ton vessel lurching in the rolling sea. At 0406, he'd run across one of the *Wildrake*'s anchor wires and it caused him great anxiety.[23] When the incident happened again at 0442, Mellor reluctantly advised the *Wildrake* that he was going to ask the anchor handling vessel, *Jon Viking*, to come in and lay starboard anchors to hold his ship a safe distance away from the *Wildrake*.[24]

At 0458 the *Jon Viking* radioed that they would begin running anchors in thirty minutes after retrieving the anchor buoys from the *Gulnare* where they had been stored.[25] Mellor couldn't wait for this. At 0501, he sent a message to the *Wildrake* captain suggesting, if he would slack off on his portside anchor cables, he would bring his ship in on Poscon to make a quick dive since they were now ready to launch the rescue bell.[26]

What Mellor was proposing was extremely dangerous for Kasey and Eddy, and he personally warned the two divers that he would be holding his ship on engines alone.[27] He knew the SALM base, with its tangle of pipe work and steel projections, was a dangerous hazard that could easily snag a diver's umbilical. And he knew that any maneuver on his part could cause his ship to run off station, shred the umbilical, and probably kill the diver.

Mellor also had the weather to worry about. If it suddenly turned against him, he'd have to abort the operation and abandon Richard and Skip to their fate. He didn't like these odds, so he turned to the BNOC representative on the bridge and advised him that he should immediately bring in a backup vessel, and he specifically requested the *Uncle John* knowing it was in the Brent Field. The BNOC representative passed this request to Masterson, but according to Mellor, "Mr. Masterson said it would not be necessary at this time."[28a]

Meanwhile, in the *Wildrake* bell van, Brian Reid recorded another vent by Richard and Skip at 0455.

As the yellow glare of an arctic sun broke the gray horizon, the portside anchor wires of the *Wildrake* were slacked off. Without tension on the lines, the ship began to slowly yaw back and forth like a compass needle, slipping towards the starboard direction. The deck crew paid out 60 feet of umbilical into the water[29] and Captain Hedin managed to put the *Wildrake* on dynamic positioning.[30]

---

a At the inquiry, Mr. Masterson could not recall whether the suggestion was made to him by the *Stena Welder*. *Source*: Walker/Guiel FAI p. 598.

By this time Roper and Gunter had completed final arrangements for the surface dive to the bottom. But when Roberts learned about it, he quickly vetoed the plan explaining that the risk was too high. They would wait on the *Stena Welder*. Rod Pelzel remembers a frustrated and angry Roper returning with the bad news: "Oh fuck, they don't want to do it. They shut it down; we're not doing it."[31]

That frustration was understandable, but Roberts saw a 500-foot surface dive, on the end of a spliced rubber hose, in the middle of an unpredictable sea, as a suicide mission.[32] Two lives were already in jeopardy, and he did not know the condition of Richard and Skip. Assuming that Lane had no problem reaching the seabed, it was getting him back alive that worried him. If Lane lost hot water, or comms, or if the weather blew into a storm—the odds against his survival would be overwhelming. His only recourse would be to get inside the bell. And even if he did manage to raise it off the bottom, there was no assurance that Richard and Skip would be conscious and not lying on top of the inside door. At that point, Lane might as well kneel down in the mud and hold his own funeral service. Roberts knew this and made the unpopular, but necessary, decision not to risk another diver's life, and in doing so he almost certainly saved Dave Lane that day.

Holmes was not a leader of men.

—— Chuck Alexander
SALM Project Engineer[33]

With the surface dive shut down, Gunter and Roper shifted their energies to deck operations already in progress under the direction of Peter Holmes. By this time, Roper was beginning to sense a rift developing between the largely British *Wildrake* crew and the small contingent of American divers that had come over from the *Gulnare*.

"We weren't welcomed with open arms and—'Come on over, this is the game plan, and this is where you guys can help,'" Roper recalls. "There wasn't open hostilities, but there was not the team effort that one would expect."[34] Indeed, Holmes would later testify that, with the exception of Roberts, he regarded the *Gulnare* crew as unhelpful interlopers.[35]

Neither Roper nor Gunter had ever met Holmes before, and astonishingly, even at this stage, Gunter was still not aware that Holmes was the *Wildrake* superintendent. Looking back, Roper says it was Holmes's demeanor that caused the confusion, that he was out on deck "doing things that he would have been told to go out to do."[36] As a result, Gunter was operating under the erroneous impression that Holmes was the deck foreman, not the superintendent.[37] And when he was shown the rigging that Holmes had made up for the rescue diver, alarm bells suddenly went off in his head.

Near the end of the 30-ton crane wire was a massive steel headache ball—a 42-inch-long, barrel-shaped swivel, followed by five feet of chain with a large hook on the end. Because the headache ball was substantial and heavy, and because the crane hook could not be connected directly into the pad-eye on top of the bell, Holmes had taken lengths of half-inch wire and fastened the ends back upon themselves with bulldog clips[a] to make "temporary eyes."[38] This was something a diver might do under dire circumstances if no other slings were available. But Gunter knew that the *Wildrake* had a rigging locker and that fashioning slings from raw wire was a waste of time and effort, notwithstanding that it was too dangerous to be pulling a 7-ton bell out of the water with two live divers inside. He suggested they use heavier, manufactured slings with permanent eyes, which was the normal practice.

But Holmes rejected Gunter's idea and stood his ground. Consequently an argument broke out with Holmes becoming "very upset" with Gunter.[39] When Gunter indicated he was going to get a stronger set of slings out of the ship's rigging locker, he was told "only the *Wildrake* personnel could open the chain locker themselves."[40]

Eventually Holmes relented under the combined protestations of Roberts, Almon, and Roper, with Tim Newson eventually going below deck to retrieve two larger slings (1 inch and 1.5 inch) which were then shackled end to end and dropped onto the crane hook.[41]

By 0510, repairs on the *Stena Welder* clump weight winch had been completed and the weight was in the water.[42] As the *Welder* waited on the *Wildrake* crew to send over the guide wire, Tom Almon began discovering one bad rigging practice after another which had to be corrected, thereby further delaying the launching of the rescue bell. The operation Almon was most concerned with involved "fair leading" the guide wire from a winch mounted on the davit through one or more snatch blocks to redirect its travel between the two ships—the kind of basic rigging familiar to a first year diver, but performed by Holmes in such a way that caused Gunter and Almon to judge the superintendent as having "a total lack of understanding of rigging practice."[43]

At 0534 the end of the guide wire was sent across in the Zodiac to the *Stena Welder* to be taken down with the rescue bell. At the same time, the *Wildrake* crew began slacking off another 60 feet of bell wire and umbilical as the *Wildrake* moved to starboard for a second time to allow the *Stena Welder* access over the accident site.[44]

In dive control Reid listened to the moan of the gas regulator on the console. The clock on the wall read 0543. On a piece of paper he scribbled the time and recorded that Richard and Skip were flushing the bell.[45]

---

a Bulldog clips, or Crosby clips, reduce the safe working load of a wire by 20 percent.

When divers breathe cold oxy-helium at depth the loss of heat from the respiratory tract becomes colossal.

—— Evan L. Lloyd
Hypothermia and Cold Stress, p. 111

In the dim light Skip stopped the vent and turned his eyes towards the gleaming porthole. It looked like a molten pool of obsidian. For the hundredth time he gazed out into the black universe and wanted to swear this wasn't happening to him. But the static from the radio confirmed it, the cold in his bones confirmed it, and worst of all, the look on Richard's face confirmed it. It was Skip's brick wall and he couldn't get past it.

He hated the cold—always had—and now the thought he might die from it momentarily transported his mind to dark frightful places where it hadn't visited since childhood. If he had to die, he would have wanted it to be fast and painless, but this—this was going to be death in slow motion. He closed his eyes and shook off the depressing thought, half expecting that when he opened them, he would be free of his tomb.

God, get me through this.

They were in cold stress now, shivering, feeling miserable, their hands numb and stiff. Breathing cold, very dry, helium felt like breathing razor blades. Their core temperatures were being rigorously defended. Eventually their arms and legs would be sacrificed for vital organs. They gripped their space blankets tight and huddled together, but nothing could stop the heat loss from respiration. Not only were their bodies trying to warm the inspired gas, they were trying to humidify it as well. As a result, they were becoming dehydrated. They had fresh water to drink, but cold depresses thirst.

Three hours had gone by since the accident and Richard and Skip knew they had to deal with the drop weights before they would be unable to. Hypothermia would alter their thinking and behavior. They were already feeling clumsy. Soon, they would feel lethargic and sleepy. To stay alive, they would have to fight the urge to close their eyes. They speculated on what topside was doing, but the radio at the top of the bell gave no clues and sleep was pressing down upon their eyelids.

Huddled together under space blankets, they knew they were in serious trouble. Trapped in a bell on the bottom of the sea was all the horror they had feared. They were running out of time. The policy for releasing drop weights varied between companies. In some, if rescuers had not made contact within four hours, then the divers should drop their weights.

What was topside doing?

Richard and Skip decided to wait a little longer.

I don't recall anybody saying, 'You take over.'

—— Craig Roberts
Infabco Field Superintendent[46]

By now Randy Gunter had the impression that there had been a total meltdown of authority. Vital information seemed to be known at one end of the vessel but not at the other. Conflicting orders were issued which caused the crew to fragment and work at cross purposes to itself. This ruined the rhythm and timing of an operation whose main business now was to get Richard and Skip out of the water alive as quickly as possible.[47] As Gunter would later say at the court of inquiry, there were times in which the left hand didn't know what the right hand was doing, and with commands coming from multiple directions, the crew began asking itself, "Who am I supposed to listen to? Which way am I supposed to do this?"[48]

This confusion over who to listen to depended upon one's vantage point. Dave Lane saw people taking orders from whomever "shouted the loudest."[49] Almon and Gunter said it was not clear to either of them "who was in charge of the operation."[50] Brian Reid thought Craig Roberts was in charge because "he was the most senior one there."[51] Kurt Baukau, like Almon and Gunter, saw no one taking charge during these initial hours because neither Holmes nor Roberts stepped forward to direct the rescue mission.[52]

When Holmes was asked at the inquiry, "Who was controlling the whole operation, was there anybody?" he responded, "Yes, it was being done on deck by myself and by Masterson."[53] This view coincides with how Chris Roper saw the situation, but with the added distinction that Masterson was the "ultimate decider."[54] Indeed, as the events of that day would unfold, Holmes would constantly defer to the company owner.

Meanwhile, Roberts, with a reputation in the industry for being "probably the best that you could get," effectively hung back and stayed out of the way, playing only a nominal role in the rescue. Some of the older hands who knew him well were mystified by this uncharacteristic behavior and would later declare, with great reticence, that he "should have been assuming more of the leadership responsibility," and that he clearly "dropped the ball."

But Roberts did not drop the ball, he simply didn't want to run with it—an attitude still present nearly two years later when he gave testimony at the inquiry. Asked whether he thought it was a good idea that one person should take overall charge of a rescue, he gave the astonishing answer, "One person should not do it."

"One person should not do it?" the examiner asked.

"No."[55]

He knew he was the senior diver on the scene. And he knew that others regarded him as such. When Tom Belcher was notified of the emergency, he didn't take control of the situation; he hurried off to inform Roberts, because Roberts was the field superintendent. And when Masterson sought help, he asked for Roberts because Craig was his top expert. Even Holmes regarded Roberts as the "overall superintendent" for Infabco.[56]

Willing to accept the rewards and benefits of field superintendent, Roberts, however, was reluctant to confirm his title and position at the inquiry. It took two attorneys on four separate attempts to get him to finally admit that he was "Superintendent on the *Gulnare*."[57] The attorneys were persistent about this issue because it was their job to uncover any defects in the system that produced the accident. That included the management of the rescue. They needed to know who organized the rescue and who gave orders. Logically, it had to be either Holmes or Roberts. But Roberts kept equivocating and downplaying his role during cross-examination, saying he didn't know what his position was and that he was in the Thistle Field merely "to evaluate the crews and the diving systems."[58]

When asked what part he actually played in the rescue, the response he gave was, "I did a few things; I spent a lot of time just speaking on the through-water comms, trying to get through."[59]

Understandably, when he arrived on the scene, he found as bad a situation as a superintendent could find. Two divers were trapped in a bell on the seabed with a life-support umbilical trashed by an unsuccessful recovery attempt. Clearly Roberts saw that mistakes had been made and that the operation was bordering on chaos. But instead of taking overall charge and declaring, "Okay, this is my responsibility; this is what we're going to do," he took the rather diffident approach of allowing major decisions to be derived from a collection of "joint inputs."[60] Indeed, when asked who acted as the brains of the operation, he said the brains were "more of a combination of everyone."[61]

Roberts could have announced at any point he was calling the shots, but he never did. It might have looked bad for Holmes to be overridden by a visiting superintendent; it might have created some friction. But Holmes had been with the company for a mere six days and presumably Roberts was there in the first place to bail him out of his predicament.

The situation at hand demanded that one man square his shoulders and take the reins. The lives of Richard and Skip depended upon it. But in the event, it was neither Holmes nor Roberts that took control. Instead, one of the cardinal rules of diving was broken that day when the reins of the operation were handed over to a non-diver, a man who had no business accepting those reins, leaving men like Roper, Olson and Pelzel to forever wonder why those who knew better abandoned their primary duty.

In the bell van, BNOC Engineer Ron Wilson also sensed that the rescue was being mismanaged. He thought the initial hours of the accident had been marked by a lack of coordination. Supervisor Reid seemed to be taking actions without any apparent direction. "It was not very obvious who was in charge," Wilson would later testify. But it was clear to him that Holmes was not the leader. Holmes only "occasionally" appeared in the dive van. Indeed, it was only when Wilson began hearing Masterson's voice over the intercom, "explaining to Brian Reid what the next sequence of operation would be," that he realized Masterson had taken charge.[62]

Described as a "brilliant" engineer and "the moving force" behind Infabco Diving Services, Masterson relied heavily upon his diving experts to conduct operations as best they saw fit. But nothing they tried had been effective, and now his company was suddenly threatened with a diving emergency that was sliding downhill very fast and under the watchful eye of BNOC. To keep the situation from sinking further he set up a command post on the bridge and began receiving information from the *Stena Welder* then delivering orders to various people on the *Wildrake*.[63]

On diving jobs, non-divers who interfere with operations are a menace to safety because they lack the training and knowledge to understand the consequences of their actions. Throwing the wrong valve can lead to a life-threatening situation, but so can giving the wrong order. And it is for safety reasons that every beginning diver is taken aside and given the first rule on the job: "Don't touch anything unless you know what you're doing." Those words are of such paramount importance that they form a sacred mantra that divers chant and pass on to each other throughout their careers.

When Skip first met Steve Lobban on the *Gulnare*, he told the young tender, "If you don't understand, ask. There are too many things out here that can screw up. You've got to ask."[64] Skip instilled that basic rule of safety in others because he understood its importance. But for reasons that still perplex the diving community to this day, Roberts and Holmes ignored that basic rule and allowed a businessman to insinuate his objectives and priorities into a critical rescue mission.

Both superintendents, no doubt, were in a tough position; Masterson was their boss and opposing him probably would have cost them their jobs. But a higher value was at stake, and they had the law behind them.[a] Had Roberts taken command, he might have made mistakes, but those mistakes—if he made any at all—would never be as wrong as the ones Masterson was about to commit.

At the inquiry, the company owner testified that Holmes was running the rescue,[65] and that he restricted himself to the role of a mere "liaison" in charge of "overall policy."[66] But policy and actions are inextricably twined. Masterson may not have put his hand on a valve or operated a winch, but when he began giving

---

a Although the regulations lay the responsibility of the dive on the supervisor, and not on the superintendent, any actions taken by Holmes and Roberts to thwart Masterson's control almost certainly would have been supported by the courts.

orders to Reid and Holmes he was effectively influencing the outcome of a diving operation, a responsibility he had no legal right to take on. Whether intended or not, his influence and position had the effect of undermining the authority of his own experts, and as a result, a fairly straightforward recovery operation would become muddled by political and economic damage control.

The transfer of power to Masterson was not a sudden grabbing of the reins; it was an abdication. At a time when Richard and Skip most needed a cool head to get them back alive, the people who knew better abandoned their duty and began deferring to a man who had no expertise in diving.

Looking back, Project Engineer Chuck Alexander recalls a man under a great deal of stress who stepped forward to make the hard decisions "because he had to take charge." In his view, Masterson was simply filling a void left behind by two impotent superintendents.

"Brian was a take-charge kind of guy," Alexander recalls. "If he saw that it wasn't being done right, he would have stepped in to do it right in his eyes, and I think that's what he did."[67]

MacNeill: *"And was there one man there who took overall charge?"*

Masterson: *"Yes, I think so."*

MacNeill: *"You think so?"*

Masterson: *"Yes."*

MacNeill: *"And who was that?"*

Masterson: *"I think, as far as the overall charge was concerned, that was myself."*

MacNeill: *"Say that again?"*

Masterson: *"As far as the overall charge was concerned, that was myself."*

—— Brian Masterson, testifying at the inquiry.[68]

Chapter Fifteen

We've Found Them!

Every time we wanted to talk to the diver or he wanted to talk to us, he had to come back to the bell and take off his mask, talk to the surface, the surface would inform Don Cox, and Don Cox would inform us, so there was quite a long delay in these conversations.

—— Peter Holmes, Infabco Superintendent[1]

Since his arrival on board the *Stena Welder* at 0353, Don Cox had been helping the Wharton & William's crew reassemble a diving system that was "in bits" and untested. No one yet knew how the electrical junction box would perform or whether the recently installed main umbilical and bell tethers were free of defects. Then, the clump weight winch had broken down. And if it wasn't apparent that Murphy's Law was in effect, it was about to be made clear to everyone very soon.

At 0540 the end of the guide wire arrived by Zodiac and was attached to the clump weight. As the weight was being lowered to the working depth, Reid unspooled the guide wire from the bell handling davit. Nine minutes later, Eddy Frank, 23, and Phil Kasey-Smith, 32, were in the bell and blowing themselves down as fast as they could on a 4% mix.[2]

Eddy Frank had grown up on the southern bank of the river Tyne in the small working-class town of South Shields near Newcastle, an area famous for salt panning, coal, and ship building. In 1976, he was working in a shipyard as a system's engineer. His father was also working in the shipyard and it just so happened that he was involved with a major refit of a diving ship called the *Oregis*. One day, Eddy was at work when his father approached him and asked if he wanted to get into diving. He said he'd been having a beer with the Comex diving superintendent and there was a need for a tender. Was Eddy interested?

Eddy had visited the *Oregis*; he'd seen the decompression chambers and diving bell and felt the mystical aura of this thing called saturation diving. Despite having never been a sport diver, he said yes. Later, when the Comex superintendent took one look at the five-foot-nine, 200-pound fireplug from South Shields, he said, "We need this guy," and Eddy was on his way to becoming a deep-sea diver.

For Phil Kasey-Smith, it was the chance delivery of some equipment to the Welsh town of Amlwch (*Am-lock*) that brought him into the brotherhood. He had grown up in the tiny village of Micklehurst, near the Saddleworth Moors, a place famous for the Moors Murders when Myra Hindley and Ian Brady tortured, then murdered three young children in 1965.

At age 15, Kasey entered the army and was sent to one of the army apprentice colleges in the city of Carlisle near the border between Scotland and England. After three years as a specialist in vehicle mechanical and electrical fitting he volunteered for the airborne forces and spent the next nine years with the 16th Parachute Brigade jumping out of C-130s. In the late 1960s, he parachuted into Yemen to cover the pull-out of British forces from Aden.

In late 1974, Kasey was accepted to go to Iran as a guided weapons launching system's technician and was expected to leave the following January. At the same time, his re-enlistment date was approaching and he was conflicted about his future. He had already served twelve years and if he served another ten he would get a pension. But Kasey felt he was stagnating in life. "I couldn't envisage me spending the rest of my life doing this," he says looking back.

In September, he left the army. And it was during this period of indecision, while delivering some construction equipment to the town of Amlwch, when he fell in with a group of divers that happened to be staying at his hotel.

In the lounge one night, he began talking to them and found out that some of them were ex-army. They, in turn, discovered he had been in the 16th Parachute Brigade, had tried scuba and "wasn't adverse" to it. Suddenly it was: "Come on—we'll see what you can do," and for the next two years he was laying 25-ton concrete blocks in black water during seven-knot spring tides building a breakwater.

It was hard grueling work; there were no hot water suits and the wages weren't very good. Near the end of 1977, Kasey left to dive in the North Sea because the man running the project onshore, a non-diver, "didn't understand, and wasn't prepared to understand," a lot about the diving side of the project. "We were just a necessity that he had to put up with basically." In Aberdeen, he was hired by Wharton & Williams and sent out to the *Stena Welder* where he met Eddy, the fireplug from South Shields.

Now inside the rescue bell and pressurized to 460 feet, Kasey and Eddy reviewed the dive plan Supervisor Mike Allen had given them. It sounded simple enough: find the bell, connect the guide wire, and when the crane wire came down, connect that.

Just after six o'clock that morning, Allen tracked the bell out to the end of the gantry. At 0609 the tiny capsule plunged through the splash zone and began sliding down the clump weight guide wires.[3] There was no big discussion as to who should be the first to dive. When Kasey said he didn't mind locking out first, Eddy agreed and the matter was settled.[4]

At 0616, the bell came down hard on top of the clump weight at 430 feet.[5] Allen lowered the clump weight and followed it with the bell. At 460 feet he stopped the descent.[6] The two divers equalized and opened the inside door. Kasey finished dressing in and discovered he had no comms to his mask. He tried them on the other umbilical and found them very distorted.[7] Eddy quickly helped him switch over to the standby gear. It was obvious the problem with the electrical junction box had not been resolved. Normally the dive would have been aborted, but no one would have suggested such a thing. There were two divers that needed their help.

At 0644 Kasey entered the water expecting to see a halo of light surrounding the bell, but saw only black space. The bell lights had failed.[8] The bad comms and no bell lights was going to complicate things tremendously. In the dark, the former paratrooper drifted down to the clump weight and spent the next several minutes re-establishing the bell at 480 feet. Shortly before 0700, he notified topside that he was beginning his search.[9]

From the corner of the clump weight he followed the guide wire down to the seabed, where he found yards of it strewn about in the mud.[10] He went to the end of his umbilical and began a quick sweep using a circular search pattern. The water at 500 feet had the clarity of gin, but with no lights, distinguishing objects in the dark was like trying to see through black fog. Kasey was just dealing with shadows. His comms were so bad that he felt it would have been better to have none at all. Bits of words and sentences, mixed in with clicking sounds, interrupted his concentration and forced him to stop what he was doing to try and piece together what Allen was saying to him.[11] Live boating a diver in the dark in the vicinity of a seabed obstruction was bad enough, but with virtually no comms, Kasey couldn't always tell if the fragmented messages he received were normal commands or a warning that *his life* was in danger. None of this mattered to Kasey, because, as the roar of the thrusters rained down upon him, the only thought drumming in his head was—*find that bell.*

Minutes ticked by, and as he plodded across the moonscape, he strained his eyes to detect any dark mass sitting above the mudline. The *Wildrake* bell was not equipped with a strobe flasher, which would have emitted bright explosions of light for dozens of yards in all directions. And because the bell transponder had been hastily removed before the dive, Richard and Skip were now paying the price for that reckless decision.

Looking back along his umbilical, Kasey saw no sign of his own bell. Then, at 0708, an enormous bunker-like object loomed before him. He'd found the

SALM base. The discovery was relayed across to the *Wildrake* with a return message that the last known position of the *Wildrake* bell was about 20 feet away.[12] Seconds later, Supervisor Allen lost all contact with Kasey.[13]

Unaware that someone was searching for them, Richard and Skip decided to release the drop weights before the cold made it impossible. Each weight rested inside the lower bumper ring of the bell and was secured by a single rope wrapped several times between two pad-eyes. The ropes were located in the 8 o'clock and 1 o'clock positions. They thought that if one of them could squeeze an arm out under the trunking, he might be able to cut the ropes. They had to try because they had nothing to lose.

They turned a valve on the inside hatch and bled the pressure back until the seal broke. Stiff and unbearably cold, they pulled the door open and blew the trunking dry. Water and gas gurgled out of the trunking which was 14 inches deep. From the lip of the trunking to the edge of either weight was another 18 inches away. Skip was probably the one to make the attempt because he knew he had long arms. Growing up, his mother would lengthen the sleeves of his shirts by three inches,[14] and his sister Sheila used to joke that "his knuckles dragged on the ground." Roger says no one in Al's Corner would box with him because of this extra long reach.

Using a tool or perhaps his knife, Skip chiseled away at the seabed and pushed the soil aside until an access hole had been mined to the outside. To extend his arm out, he had to force the upper part of his body into the trunking. It was difficult, but he managed to get his knife on the rope securing one of the weights. He hacked and sawed at it and eventually it fell free. The second rope, however, was never cut.[15] Exactly why remains a mystery. One theory put forward by a government investigator was that the bottom door hinge prevented the diver from reaching the second line.

"You couldn't get to that line behind that hinge," Department of Energy inspector Roy Giles recalls.[16]

But photographic evidence and blueprints of the bell do not support that scenario.[17] Perhaps the cold had clouded Skip's mind and he thought he'd cut the line when he had not. Perhaps something other than the bottom door obstructed his access to the second rope. What is known is that, some time between 0710 and 0755, Richard and Skip sealed the bell, closed all the inside valves, and opened the release mechanisms to both weights.[18]

In the bell van, Reid noticed the needle on the internal depth gauge dropping back and wrote the observation down—"Skin valve closed to internal gauge"—and then drew a big question mark after it.[19] The time was 0710. He tried to communicate with Richard and Skip over the radio, to tell them that a diver was searching for them, but there was no response. The umbilical was losing a lot of gas and at 0720 he dropped the supply pressure to 300 psi and then made a note in his log:

0720 - Tried to talk to divers in the bell, but couldn't reach them. No report from "Stena's" diver either.

From inside, Richard and Skip heard the chains drop from the release mechanisms, but the bell went nowhere. Without a bell stage, the weights had no place to drop to. Shivering in an ice-cold steel ball, they must have felt forlorn and abandoned. Suddenly friends, family, West Springfield, Santa Barbara—everything they knew and loved—must have seemed beyond all hope and a lifetime away.

More than three hours had passed since they'd lost hot water, and the temperature of the atmosphere was probably close to the temperature of the water. Their skin would have been extremely cold to touch. Symptoms of confusion, disorientation, lack of coordination, and slurred speech were present. All higher brain functions were becoming impaired, including all forms of memory and decision-making. Richard and Skip were entering the classic hypothermic phase.

We couldn't understand why we hadn't brought in other ships, or another ship to assist in the rescue.

—— Dave Olson
Wildrake Life-Support Technician[20]

Below deck, the mood in the sat van had turned grim. For a period of time, Legere and Olson insulated the other divers in the chamber from the accident.[21] They didn't want to deliver the ominous words: "We lost the bell." But eventually Palmer, Lahren, Mitchell, and Boyd were told that Richard and Skip were on the bottom freezing to death as they spoke. Lahren kept asking what was going on, but Olson says, "there wasn't a whole lot we could tell him."[22]

In the intervening hours, Palmer would feel the stifling and suffocating force of being locked up and helpless to act. "There was nothing we could do," he says looking back. "We were in sat. What do we do? Just twiddle our thumbs? We couldn't do anything. So, we just stayed out of the whole thing. We were there, but we weren't there."[23]

As time progressed, that feeling of disassociation and frustration would spread throughout the ship largely because there were no hourly updates on the rescue and too few radios to provide it. Indeed, even Reid in the bell van felt like he was standing in dark.

"Someone would come in and grab a radio, they would run off with it, so I would be left with nothing," he testified later.[24]

Without a radio, Reid effectively became what Gunter would later characterize as a "broken link" in the communications chain. Cut out of the loop and never in sync with the conversations going on between Masterson on the bridge and Holmes on deck,[25] Reid says he was never "one hundred percent sure of what was going on at the time."[26] All he could do was stand and watch the nightmare unfold about him.

For the deck crew, the lack of information was even worse. "You were getting everything secondhand, or thirdhand down on the deck level because nobody really was telling even the guys in the dive control room what was being done," Dave Lane recalls. "That was all being done between the *Stena* vessel and the bridge....And then we got a garbled version of that down on the deck."[27]

Pelzel remembers Roper going to the bell van on several occasions asking for information, only to see him return shaking his head. "Little bits of information were coming out," Roper recalls, "but, again, there was no central—like 20-minute update or hourly update. It was badly orchestrated—badly controlled."[28]

Meanwhile, Gunter was beginning to question why Masterson, a man with no diving experience, was directing the rescue and repeatedly ignoring advice from his seasoned divers. He thought the company owner and Holmes should consider alternate solutions for getting Richard and Skip out of their death trap. But every time a recommendation was made he says it was "put on the shelf."[29] One of the suggestions he put forward was to call for more outside help.

According to Gunter, when the *Gulnare* crew first arrived, Roberts asked Masterson if he'd put out a general alert or an SOS to bring in other vessels to assist, and Masterson replied that he had not. Gunter thought this was a mistake and expressed his opinion. But the discussion ended with Masterson making the comment that he felt the situation was "not really as serious as it seems to be."[30]

As time passed, Gunter grew more fearful for Richard and Skip's safety. Hot water was not getting to the two divers; there had been a delay in putting the rescue bell into the water; and it was obvious that the captain of the *Stena Welder* was having difficulties maintaining position. It was also clear to him that Masterson had sidelined Roberts and Holmes. Uneasy about the whole operation, Gunter couldn't understand why the rescue was being limited to a single vessel with no DP. He thought there should be a backup for the *Stena Welder*, and when he heard rumors that the *Uncle John* and the *Seaway Swan* were working in the area, he went to Roberts and urged him to persuade Masterson to call one or both of these vessels to come to the rescue.[31]

Unbeknownst to Gunter, Masterson had already received a similar request from Captain Mellor to bring in the *Uncle John*, and rejected it. Likewise, the captain of the diving vessel *Talisman* had contacted the *Wilarake* from the Brent Field and offered his assistance, but Masterson declined that offer, too. His reason—"we felt that we did not need another vessel at that time."[32]

Roberts did speak to Masterson about bringing in another vessel, but the end result of that conversation did not give Gunter much comfort. At the inquiry, the company owner said he discussed the prospect with the BNOC office in Aberdeen, and subsequently called the *Seaway Swan*. But instead of asking them to hurry to the accident scene, he told them to standby, that he would get back to them.[33] He did not speak to the *Uncle John*, but thought that someone in the BNOC office had.[34]

When asked why he rejected the comments and suggestions from the *Gulnare* crew, Masterson replied, "there were a lot of suggestions put forward that would have changed our operation, and we had already started the mode of operation we were in, and it was very very difficult to change."[35]

This would be the mindset he would cling to until the very end.

Inside the rescue bell, Eddy had just received word from Mike Allen that communications were lost with Kasey when his bell partner popped up through the trunking. The time was 0721.[36] While the two divers rushed to put a fresh mask on-line, the *Wildrake* exercised another move to starboard with its crew slacking off an additional 60 feet of bell umbilical.[37]

After changing out the hat, Kasey pulled it on, but the comms were "still very bad."[38] Just before he exited the bell, Eddy recalls Allen issuing a strong warning over the bell speaker: "Before you do anything now, you're going to have to come back to the bell and let the surface know before you do anything."[39] And then Kasey vanished through the trunking at 0737.

On the seabed he quickly located the SALM base and began searching the area near the crude line as advised. Having dived on the base for two years, he knew the area well. The crude line goosenecked up corner number two,[40] but the bell was no where to be found. It wasn't there because, astonishingly, the people on the *Wildrake* forgot they'd moved Richard and Skip 60 to 90 feet away from the base more than three hours earlier. Kasey was searching in the wrong area!

At approximately 0745 he switched to the other side of the SALM base, where the ballast line trails away from corner number six. Five minutes later, after nearly an hour had been squandered, he found the *Wildrake* bell umbilical draped across the 16-inch pipeline. Kasey looked in both directions and saw no sign of the bell. He followed the umbilical until it began rising off the seabed towards the surface. Realizing he'd gone in the wrong direction, he turned and hurried back hoping he had enough slack in his umbilical to reach the bell. He passed the SALM base, and then, a moment later, saw the faint outline of an object standing upright in the dark. He'd found them! Kasey's heart was in his throat. Quickly, he pulled his knife and pounded on the side of the bell. There were frantic movements inside, and instantly a head came down in one of the ports, and then two hands appeared giving the thumbs up. They were okay, but to Kasey, the movements were unmistakable in their meaning: "Get us out of here as quick as you can."[41]

0755 - "Stena" diver says divers in Wildrake bell are OK.

Now five and a half hours after the accident, Richard and Skip were almost certainly hypothermic with their core temperatures at or below 35 C. Their shivering was at its zenith, and their reflexes exaggerated. When they tried to talk, they slurred their words, and they were thinking more slowly. After failing to release the drop weights, their spirits must have been at their lowest point as they huddled together in the freezing cold.

The rapping noise came as a complete surprise, and then seeing a diver's mask in the porthole caused a shock wave that must have given them hope. They had been found.

The diver gave a thumbs up in response before disappearing. And then Richard and Skip heard him clanging around on top of the bell, checking out the pad-eye. They were sounds of music. And then there was silence. The diver must have assessed the situation. His next move, they knew, would be to raise the bell off the seabed. Richard and Skip were freezing cold, and they were shivering uncontrollably, but they felt great. They just needed to hold on a little longer. They were going to make it. They were going home.

Chapter Sixteen

Hookup

I was absolutely convinced at that point, even given the three, four, five hours it took to get the system back together—operable—that provided they didn't screw up anymore on the Wildrake*, we would get them.*

—— Don Cox, Infabco Supervisor[1]

Kneeling on top of the *Wildrake* bell, Kasey was the first person to discover that the bell wire connection had failed. He also noticed something odd about the shackles. Even in the dark, their condition was so remarkable that it would remain one of the clearest images in his mind years later.

"The pin on both shackles, and the bow on both shackles, was really worn through. So they'd been using the system like that for a long time."[2]

What happened next is unclear, but in a sworn deposition by Captain Mellor to the court of inquiry, he reported that Kasey advised the best way to recover Richard and Skip was "to simply re-shackle the main lifting wire" to the bell. But when this suggestion was passed to the *Wildrake*, it was turned down.[3]

Whether Mellor's statement held any credence or not produced some startling and varied responses by several witnesses at the inquiry. When Masterson was asked if that suggestion had ever been communicated to the *Wildrake*, he said, "Yes, I think so."[4] He also acknowledged that the bell wire was chain-stopped to the umbilical, and that it would have been in a direct line, more or less, with the bell pad-eye. Why didn't they follow the diver's advice and re-shackle the main lift wire back onto the bell? he was asked.

"Because the rescue attempt had been based on the fact that we thought that the bell wire had been severed in some way or had broken."[5]

When he was reminded that he had already received a message from the *Stena Welder* that the wire was not broken and that the best method for recovery was to re-shackle the main lifting wire, he hedged his earlier testimony and said, "I can't remember." And when the examiner continued to press the importance of the matter, he said, "It didn't come to me and I didn't hear it, no."[6]

When Supervisor Brian Reid was asked if it was even possible to re-shackle the main wire, he said, "Looking back on it now, it could have been, yes."[7]

And when asked if that would have been "the quickest and most direct method of recovering the bell at that stage," Reid answered, "Definitely."

"Why wasn't it done, do you know?"

"I have no idea."

When Superintendent Holmes was questioned about this, he said he was under the impression that the suggestion had been made, but insisted that it was never made to him.[8]

As with Brian Reid, Holmes was asked if such a thing was possible?

"If it [the lift wire] was within reach, he could have made a temporary connection to it," Holmes answered. "There were shackles on top of the bell that he could have used, yes."[9]

Holmes also speculated that the rescue diver might not have seen the end of the bell wire,[10] which is certainly plausible, given the earlier attempt by the crew to raise the bell on the chain stops which could have caused the lift wire to ride back up the umbilical, if not completely separate from it.

Craig Roberts had a diametrically opposed view. While he agreed that it would have been easy for the rescue diver to find the end of the lift wire, there would have been a "slim chance" for him to reconnect it,[11] a surprising response given the fact that Infabco was contracted to reinstall a 550 metric ton loading buoy to its base. When the solicitor questioning him proposed a simple method for reattaching the lift wire using one of the two shackles already on top of the bell, Roberts answered, "I was looking at the drawing, thinking about that—yes, I guess he could."[12]

Why Kasey's suggestion was not quickly agreed to, no one can be certain. But there is no doubt that re-shackling the original lift wire was an obvious and straightforward strategy. And yet, for all the experience and talent on board the *Wildrake*, the obvious was lost in the belief that the lift wire had "broken," when in fact it was first reported by Richard, and later by Kasey, that it had not.

Five minutes after Kasey's discovery, Brian Reid closed the topside gas supply to Richard and Skip.[13] They had used a lot of gas and bubbles were coming up on the portside of the ship leaving Reid to theorize that the supply hose had broken.[14] For Richard and Skip, this meant the only remaining gas they had to vent the bell was what they carried on board the bell—six bottles of mixed gas.

Meanwhile, Kasey had returned to the coils of guide wire lying about the seabed and parts of the SALM base.[15] He asked topside to recover the slack to clear the tangled mess from the base. At 0803, the crew began recovering the wire. As Kasey watched the wire snake towards the surface he had to give precise commands with critical timing to make sure the wire did not hang up on the pipework. Under ordinary circumstances this recovery could have been accomplished quickly. But with distorted comms, no bell lights, and a ship constantly pulling him off the job, he struggled for the next hour starting and stopping the recovery process no less than seven times.[16] And every time he gave a command it had to be relayed over a "tortuous link" to Mike Allen to Holmes and eventually to Reid in the bell van.[17] If Allen couldn't understand what Kasey was saying, then Kasey had to race back to where the bell was hovering, climb his umbilical, squeeze his body inside the trunking, pull his mask off and shout orders into the bell speaker. Once he delivered his message, he pulled his mask back on, dropped into the water, swam down to the seabed, and raced back to where he thought the job was, which was forever shifting about because the ship could not be held in a fixed position.

Constantly repositioning the *Welder* would have been a waste of time for this reason. Much of the time Kasey was fighting to stay on the job at the very end of his umbilical, which on occasion trailed over the top of the SALM base. He could only hope the *Welder* would remain within a favorable radius of where he wanted it to be. But there were times when he was suddenly and violently yanked off the job, dragged along the seabed, or over the base itself—causing him to dodge bits of debris and pipework—to be deposited as much as 30 feet away. When the ship eventually re-centered itself, he would crawl back to the job only to start the work all over again. And if he couldn't reach the job, then he was forced to dash back and order a move.

Courtesy of Marcus Taylor Ltd.

The *Stena Welder* during the rescue.

Had the *Welder* been on anchors, its bell would not have been flying around beneath the ship. It would have been stable, with its clump weight hanging below the bell and running in the same direction as the ship. Kasey would have been able to use the orientation of the weight to determine direction and distances to objects underwater. But on this dive, he was never certain where the job was. All he knew, when he left the bell, was that out there, in a sea of darkness, he had to get two desperate divers out of the water as quickly as possible.

Added to his frustration was another vexing problem. Kasey was losing hot water. With the *Welder* rolling on a sea 500 feet above, the clump weight had oscillated back and forth like a weather vane until it eventually gained enough momentum to spin the bell completely around causing the guide wires to cross like puppet strings, squeezing the main umbilical in a viselike grip, the effect of which was to restrict the hot water flow to Kasey. But the burly Englishman was built like an ox and his mind so focused on his mission that he blotted out all thoughts of the freezing water that was now slowly and insidiously pulling him down into a state of hypothermia.

In the *Stena Welder* bell van, Supervisors Mike Allen and Don Cox huddled around the gas panel feeding Kasey. For nearly an hour the fragmented sounds of a diver struggling against overwhelming odds had penetrated Allen's brain through his headset. Kasey's labored breathing, and the commands he gave, told the story—"Up on the guide wire," and then a few minutes later, "All stop on the wire." And then the sequence was repeated again and again and again. Patiently the two supervisors sat quietly, knowing that their diver had his hands full, but Cox and Allen were increasingly alarmed by the loss of another precious hour of Richard and Skip's time. This was the guide wire phase and there was still the heavier crane cable to connect. It was obvious to them the rescue was not going fast enough. It was time to change course.

Cox contacted Holmes and suggested that the *Welder* lift the stranded bell and hand it off to the *Wildrake*.[18] He explained that they had a 30-ton davit a short distance from their dive station which would make the hookup easier for Kasey. They could raise the bell off the seabed, cut the umbilical away, and make a vertical lift to shallow water where the bell could be transferred to the *Wildrake*.[19] There would be no need for a guide wire to lead down a crane cable.

When Holmes was asked at the inquiry whether the *Stena Welder* was capable of performing this task, he said, "I don't know, I have never been on board the *Stena Welder*...I wasn't conversant with its equipment."[20] Instead of finding out, he went to the bridge to discuss the idea with Masterson.[21]

After listening to Holmes, Masterson consulted with a variety of people—Reid, Burgess, and "various experts" in Aberdeen who commented on the plan. "Everybody's opinion was being attempted to be taken into account," he said at the inquiry.[22]

Ultimately, Cox's suggestion was rejected.[23] He was never told why,[24] but at the inquiry Masterson testified that if the weather had blown up during the transfer, and the *Wildrake* bell was recovered onto the deck of the *Welder*, this would have marooned Richard and Skip on the wrong ship because the two diving systems were incompatible: the *Wildrake* bell could only be mated to the *Wildrake* chamber system.[25]

Holmes, like Masterson, also expressed concern about the weather. He said it was "not very good" during the rescue attempt,[26] and that transferring the bell from one ship to another would have been "a difficult operation."[27]

Although Masterson did not mention Roberts as one of the people he consulted with, Craig was made aware of Cox's suggestion.[28] At the inquiry, he was asked about the viability of the plan.

"From your experience, would that not have been a straightforward and speedy method of recovering the bell?"

"Do you want my personal opinion on this?"

"I want your opinion based upon your experience as a superintendent?"

"As a superintendent, yes."[29]

As for the weather that day, far from threatening to blow up, it was in fact calming. Supervisor Mike Allen testified that it was "not very rough,"[30] and in the *Wildrake* Daily Progress Report, Kurt Baukau noted that the sea state was 2-4-foot waves, down from the 8-10-foot waves reported at the start of the dive. Despite his earlier testimony, Holmes was eventually forced to admit under cross-examination that the weather was fine with no indication that it was going to deteriorate.[31] Masterson also had to concede that, throughout the rescue attempt, they had been transferring people between the *Wildrake* and the *Stena Welder* by Zodiac and had "no problem at all."[32]

When it was suggested to Craig Roberts that weather conditions were quite calm, he agreed. "Yes, maybe four or five feet swell," he answered.

"Nothing which would prevent a transfer of the bell thereafter over to the *Wildrake*?"

"At that time, no."[33]

They knew they were in deep trouble.

—— Rescue diver Phil Kasey-Smith

On the seabed, Richard and Skip were huddled together, trying to hold death back. With signs of rescue outside, they were coping, but only barely. It had been an hour since they'd seen the diver in the porthole, and morale was sinking again because of the overpowering cold. They needed to get out of there fast because they were now past the stage of feeling pain. Their brains were being slowly anesthetized: amnesia and clouding of consciousness would soon set in.

Dramatic behavioral changes such as disrobing and failure tc protect themselves, could occur. Hallucinations were possible. It was difficult to focus, speaking nearly impossible. As they became more lethargic, their conscious levels declined. And then they stopped shivering. It was the intermediate step to death. The urge to fall asleep was overwhelming, but they knew they had to stay awake to stay alive.

When Richard looked into Skip's eyes he saw his own. Both men must have been thinking about their families and whether they would ever see them again. Was this how it was going to end for them?—in a diving bell on the bottom of a dark sea, with no way to say they loved their families; no way to say good-bye?

Skip was curled up mesmerized by the refection on the porthole as though it were a campfire, thinking about home and what he'd do when he got out of this mess. *Get up to Maine for awhile. Sit down and have a beer with Roger. Start on the cabin? No, Florida. Lay out on a hot, hot sandy beach for several months. How could he ever purge the cold from the marrow of his bones?*

The buzz of the scrubber motor stole his thoughts away and he began to reflect on his childhood. *Purring crickets on a summer's starry night, dark glittering sequins always bright...star light, star bright... How does that rhyme go? Stargazing...sleeping bags on the garage...the telescope...deer hunting...Al's corner...chestnuts...family...friends... family...Sheila, Roger, Ma...*

They were fighting hard against sleep, but, oh, it was painful to stay awake. It wasn't time to die...it wasn't time to die... it wasn't time to...

They were swimming their hearts out.

—— Mike Allen
Wharton & Williams Supervisor[34]

At 0902 Kasey managed to drag the guide wire to the top of the *Wildrake* bell and connect it.[35] When this information was passed tc the *Wildrake,* Chris Roper says the confidence level of everyone suddenly rose, the collective thinking being: *Okay we got it. We're tied in now, there's still time to save them.* But privately his fears had been mounting because Richard and Skip were probably getting close to the edge.[36]

At 0907 Kasey ordered topside to take the slack out of the guide wire. When this was accomplished, he ordered them to send the crane wire down. Then he checked on Richard and Skip and noticed that their hand signals had deteriorated into "very very frantic" movements, which indicated one thing: despair.[37] In dive control Don Cox recorded the latest events in the dive log.

> 0908 - Lowering on crane from Wildrake which is attached to guide wire...Diver reports divers in bell seem OK.[38]

As Kasey waited for the crane hook he could tell from the angle of the guide wire that the *Wildrake* was not directly overhead. After six minutes the *Wildrake* crew stopped lowering the hook which had a light attached to it. Although Kasey never saw the light, he noticed that the guide wire had gone tight with a sharp dip to it. He followed the wire back and found the headache ball on the seabed against the side of the SALM base approximately 60 feet from the *Wildrake* bell.[39] On the end of the hook were several heavy 10-foot wire slings.[40]

He gave orders to clear the ball off the seabed. His plan was to raise the crane hook off the bottom, re-tighten the guide wire, then bring the hook back down in an attempt to work the end of the slings closer to the bell. But after several unsuccessful attempts, combined with trying to drag the boulder-sized headache ball along the seabed, it was clear the crane hook wasn't going anywhere.

Five hundred feet above, Cox and Allen listened with growing apprehension to what sounded like a repeat performance of Kasey's earlier struggle with the guide wire. Another hour was slipping away. The accident was now almost seven hours old and Richard and Skip had been without hot water for more than five hours. Inwardly, every tick of the clock sounded an alarm in their brains. They knew that if they didn't do something fast, they were going to have two dead divers on their hands. Once again, Cox contacted the *Wildrake*, but this time he suggested they be allowed to perform a "wet transfer,"[41] a method of rescue that had been on the minds of divers for years, but never attempted.

The idea was to get Richard and Skip out of a bell with bad integrity and into one with good integrity. Cox wanted to suspend the *Wildrake* bell on the *Stena Welder* davit, and with the two bells only 30 feet apart, Kasey would be able to drag Eddy's umbilical over to the *Wildrake* bell, enter it, stabilize the divers with hot water, and bring them across one by one into the rescue bell, thus eliminating any further need to handle the damaged bell at all.

Although Peter Holmes was never asked at the inquiry whether this suggestion ever came to him, Masterson acknowledged that a wet transfer had been considered, but that he took no part in the discussion.[42]

As with the previous request, this one was also denied leaving Cox with a bitter memory. "We had the means to do it, and the whole idea was just, 'No, we're not going to do that, we're going to lift it with the crane on the *Wildrake*—get our own bell back with our own equipment.'"[43]

On the bridge-wing of the *Stena Welder*, Captain Mellor stood in an orange raincoat listening to Cox over the radio trying to alter the course of the rescue. For nearly six hours Mellor refused to leave the Poscon. His original plan, to come in and make a quick dive, had been shattered, and the longer he continued to live-boat Kasey and Eddy, the more concerned he became for their safety. Mellor had been captain of two other diving support vessels. He understood diving procedures, and he was "totally convinced" that the strategy for this rescue was "unnecessarily complicated." But when he expressed these sentiments to his BNOC representative, he was told "the decision as to how to operate was to remain with Mr. Masterson on the *Wildrake*."[44]

Frustrated, Mellor could only wait for what he believed would be a bad outcome. And as time wore on, he began to suspect that Masterson was attempting to limit the role of Wharton & Williams for business reasons. "I mean if we ended up having to recover their bell, in any shape or form," he says looking back, "...it wouldn't look very good for them in any future contract negotiations to get the contract. So, we believed that that was first and foremost—paramount—in the minds of the Infabco people on board the *Wildrake*."[45]

At 0927, after spending eleven minutes hassling the headache ball, Kasey concluded he was never going to get the wire sling attached; the distance to the bell was too great. Instead, he estimated a shortfall of 40-50 feet, then raced back to the bell to instruct topside he needed more slings to make up the difference.[46] According to Captain Mellor, the 2W crew had suitable slings on board and were prepared to pass them down to Kasey, but when this was suggested to the *Wildrake*, the offer was rejected.[47]

Over on the *Wildrake*, word that the rescue diver needed more slings sent the crew into an immediate flurry of activity. Already lying on deck were four slings and it was proposed to send these down to the diver. These slings, however, were rated at 5 tons, which Tim Newson considered to be inadequate for lifting the 7-ton bell directly out of the water. Newson didn't want to waste time stopping the bell in shallow water to re-rig it with heavier slings before bringing it on deck.[48] Once they had a purchase on the bell, he wanted it brought directly on board without further delay, so he turned to John Brook and said, "Hang on, I'll get the required strops,"[49] and then he ran inside the shelter-deck hatchway, bounded down a staircase to the chamber room, raced past the sat van, turned down a short corridor, only to find the rigging store locked.[50]

Eventually he found a bo'sun who found the seaman who had the key to the locker. Newson selected two, 12-ton, 10-foot slings and two, 20-ton shackles, then struggled back to the dive station and tossed the heavy gear on deck. "There are two," he said to Brook, "I'll be right back with two more. Hang on."[51] And then he repeated his race to the rigging locker, but found the door locked again.

Exasperated and furious that someone would do such a thing during a major, ship-wide diving emergency, he began to search for the man with the key. At the same time, some of the divers contemplated using a pair of bolt cutters on the lock,[52] but before that took place, Newson came across Kurt Baukau, Wilhelmsen's Offshore Manager, who had a key to the locker.[53] He quickly grabbed another set of slings and shackles and hurried up to the dive station, where he discovered that the crane hook had already been sent down to the diver, and that, instead of waiting for the stronger slings, the crew had mixed two, 12-ton slings with two 5-ton slings.[54]

A mile away, on the *Gulnare*, Bob Gemmell was sick with worry for Skip. The recovery was taking too long, and there was nothing he could do. Unable to escape his feelings, he went to his room to be alone. It was quiet, and he lay down on his bunk, but he couldn't keep his mind still, and then he recalled a feeling he'd had almost a month earlier. At the time, he remembered pushing it out of his mind because he couldn't understand what the feeling meant, but now it made horrifying sense to him.

He was at home in Eugene, Oregon packing his bags to come back offshore, when all of a sudden he felt this "big cloud" come over him...

> And I felt like, my God, something's going to happen. I mean it just kind of went through me...like fear, and...it was just profound. It was like something was going to happen, and I really thought it was going to be me. I really did. It kind of made me nervous a little bit. I just had this overwhelming feeling that this was not good. It was like somebody saying, 'Don't go, this is not a good day to do this.' But you just go ahead and do it anyway. I don't remember it lasting and I never recalled it again until after...while we were into the accident.[55]

Like a bolt of lightning, Bob understood that his premonition hadn't been about him at all that day; it was about Richard and Skip. And then he realized that they weren't going to make it.

> *All you could think about is the people you know in your dive bell, who were your dive friends, and they're dying a slow death down there; and you don't care, or you are not really aware of, what the other guys are going through.*
>
> —— Rod Pelzel, Infabco diver[56]

On the back deck of the *Wildrake* every passing minute was torture for the Infabco crew. And while the wind still blew, and the ship still rocked, they had nothing to do but stew in a mixture of withering hope and inescapable doom that cast them into a kind of perverse limbo. Roughly one hundred feet apart, the 2W and Infabco divers looked across at each other, just barely able to make out each other's expressions. None of the *Wildrake* divers understood what Kasey's problems were on the seabed. No one said to them that he had no comms and very little hot water. No one knew he was working without bell lights. They all imagined the recovery to be a simple job—down and back, over and done. As Rod Pelzel would say years later, "Shit, man, it was easy as...drop the line, hook it up, come up on the bell."[57]

But the passing hours betrayed that vision. At one point the frustration became so intense that Dave Lane says some of the men began making bitter comments, tainted with obvious resentment and anguish, in the direction of the rescue divers: "What kind of fucking divers are they? They can't even find a bloody wire. *Stena*'s a bunch of wankers."[58]

But some of the *Wildrake* divers knew that Kasey was fighting more than just the elements and a bad dive plan, and they began to speak their minds. After witnessing a chain of command that was "upside down and backwards," that ignored suggestions that made good sense while executing those that made none at all—Chris Roper and Randy Gunter gathered below decks with a handful of the other divers, and began discussing how they were going to solve the weak-leadership problem to get Richard and Skip out of the water alive. Roper says the general consensus of the group was: "We should be fucking running this show. We should take it over."[59]

Intent on changing the game plan, Gunter went to the bridge and advised Craig Roberts that control of the rescue operation should be taken out of the hands of Masterson and Holmes. Roberts, in turn, talked to Masterson, which resulted in Gunter being ordered off the bridge.[60]

"We left," Gunter said later at the inquiry, "and then Mr. Roberts came down and told us that we could find ourselves in very deep water in the event if we were to go and tell Brian to look at it another way."[61]

Gunter was warned "to leave everything in Mr. Masterson's hands," and that if he did not do as he was told, then "they were prepared to take whatever action was necessary," including the application of a mutiny charge against him and the others.[62]

Recalling that incident, Roper says, "Craig did tell us to basically, 'Chill out guys. Brian's running the show and if you do anything to take that power away then you are in some serious trouble.'"[63] Roper realized Roberts had decided to let Masterson and Holmes control the operation. And without Craig on their side, Roper knew he and the others had little chance of wrestling the operation away from Masterson.

Shortly after 0947, Kasey received the crane hook with the extension slings. He spent the 20-minute wait on the seabed in the dark, his body racked by cold, unwilling to think about his declining condition. He found the end of the sling chain and dragged it from the silt towards the shadow in the dark, and after ordering the *Stena Welder* 20 feet to port, managed to climb the bell, relieve the strain on the wire with a piece of rope and slam the shackle pin home through the emergency pad-eye.[64]

1010 - "Stena" diver has slings hooked up to Wildrake bell.

On orders from topside, Kasey released the guide wire which was recovered to the surface.[65] At 1015 he checked Richard and Skip again and reported they were still alive. He was now ready to plumb the bell for the lift, but the crane wire and headache ball were lying against the SALM base. To clear the rigging and effect a vertical lift, he had to move both ships simultaneously. The *Stena Welder* had to move to starboard while the *Wildrake* had to move to port. It was a critical move and with good comms, it probably would have gone according to plan, but for a miscommunication. When Kasey ordered the *Stena Welder* to starboard, topside interpreted it to mean the *Wildrake*.

At 1022 both ships began moving in tandem. At the same time the *Wildrake* crew began to slowly raise the crane wire.[66] Gradually the headache ball started bouncing on the seabed traveling away from the SALM base. But the cable wasn't slacking as Kasey expected; it was getting tighter! Everything was moving in the wrong direction. Instantly, he shouted into his mask, "ALL STOP! ALL STOP!" But the command came too late. As he was being dragged away by the *Welder's* engines, he watched the image of the *Wildrake* bell disintegrate into the surrounding gloom.

At 1028, the ships were stopped, but by this time the *Wildrake* bell was lost again.[67] Holding back his frustration, Kasey began another search for Richard and Skip.[68] Twelve minutes later, at 1040, he collapsed on the seabed, exhausted, and with no more adrenaline to give. The lack of hot water had finally taken its toll. His arms and legs were so badly paralyzed that he could neither walk, nor crawl, nor climb into the bell. Eddy had to pull him back inside.[69]

> *What struck me...was the overriding lackadaisical approach that seemed to be played into this—there seemed to be so much time wasted. They didn't seem to appreciate what hypothermia could do to a diver. A diver understands it. Your hot water has only to go off for a minute and you've got problems haven't you? You immediately start to feel uncomfortable and you start to shiver within minutes. And they had these guys down there all this time, wet, sitting in a cold wet bell, with no hot water, and they thought they had all the time in the world. They did not seem to appreciate the enormity of the problem.*
>
> —— Dave Lane, Infabco diver[70]

Richard and Skip were now too cold to be afraid. Breathing from their manual CO_2 scrubbers, their core temperatures were probably below 32 C, a level when cardiac arrhythmias are possible. Venting the bell was impossible, not because they had used up their on board gas,[71] but because their muscles had stopped functioning. They were extremely rigid and unable to speak, stand, or help themselves. They were becoming cyanotic, with pupils dilated. Brain activity was severely depressed. Whatever conscious state they were in was growing more and more clouded as each minute passed. It's very unlikely they were aware of anything. They were approaching the stage when it's believed that thought itself is lost. "Among the few survivors," one expert says, "no one ever remembers anything."

Soon they would lose consciousness. Richard and Skip were dying.

Diver E. Frank locking out—no comms with diver, only bell speaker.

—— *Stena Welder* dive log.

Shortly before 1100, Eddy plunged through the trunking. His mission was to plumb the bell and give "a visual" as it left bottom. But first he had to find it. His eyes adjusting to the darkness, Eddy circled the entire radius of his umbilical, but found neither the *Wildrake* bell nor the SALM base. At 1102 he returned to the bell to give his report. The ships were moved 50 feet, but, again, in the wrong direction, away from Richard and Skip. At 1112, he was back in the water continuing the search.

Meanwhile, Masterson had reached a crucial decision. Because Richard and Skip's condition was thought to be critical, he was giving Mike Allen five minutes to find the bell and then he was going to commence the lift.[72]

Five minutes went by, then another, until the ultimatum stretched to twenty minutes. Eddy searched frantically for a silhouette, a shadow, anything that would indicate a bell sitting upright against the horizon; but he saw nothing. The ships had migrated too far off location, and by 1134 he was back inside reporting that there was no sign of the lost bell.[73]

Under pressure from Masterson, Supervisor Allen asked Kasey if the *Wildrake* bell was clear to lift? And Kasey assured him that it was, that "it wasn't near the base."[74] Allen ordered his divers to remain in the bell.[75] Cox then informed Masterson that the bell was clear to lift.[76] Moments later Captain Mellor began moving the *Stena Welder* to starboard to allow the *Wildrake* over its bell to carry out the recovery. But the *Wildrake* did not move to port to make a plumb lift as Mellor expected; it remained in its present position.

At 1140, BNOC diving consultant Jim Burgess saw the lift begin and made a note of it in his log:

Picking up on crane wire w/bell on it.

Masterson, Holmes, and Roberts were aware that the *Wildrake* was not above its bell. The purpose for attaching the guide wire was to lead down the crane wire at an angle, because the *Wildrake* was not over its bell. For the past six hours, the ship had made a series of recorded moves to starboard, but never to port. There was a working transponder on the SALM base that indicated where the ship was in relation to the base and the bell. And yet no effort was made to shift the vessel back in the direction necessary to obtain a vertical lift, as is standard practice. Instead, Holmes and Roberts allowed Masterson to order a dangerous "side pull" on the *Wildrake* bell.

While Brian Reid operated the hydraulics to reel in both the umbilical and the original bell wire, the crane driver—some 60 feet away—was hoisting in his wire to raise Richard and Skip to safety. Because he had no radio, Reid was watching the crane wire to synchronize his actions. When Reid saw the crane stop, he stopped.

When it hoisted, he hoisted.[77] Although the *Wildrake* had a brand-new intercom system which linked the two posts, it didn't work. It was Holmes standing on deck that was directing the crane driver with a portable radio.[78]

Anticipation and hope were running high as the lift progressed. From the bridge, Chuck Alexander saw the wire rising out of the water and thought, *"Oh good it's coming up,"* and then, very soon, it went taut and everything stopped.[79] The time was 1150.

Also on the bridge, Randy Gunter had been listening to the radio transmissions with Craig Roberts and Tom Almon when he heard the crane driver report he had a "heavy strain" on the crane wire, and that he had stopped to await further instructions.[80] Gunter saw the crane wire leaving the top of the boom and running at a sharp angle towards the *Stena Welder*. At this stage, "a great deal of discussion" followed between the two ships as to why the lift was halted. Moments later Gunter was ordered off the bridge.[81]

At the dive station, Rod Pelzel saw the crane boom "flex" under the load, and wondered what the crane's rating was, and why it was having such a hard time lifting the bell. It didn't make any sense to him; *"The bell ought to be a feather,"* he thought.[82] And then word began circulating that it must be "stuck in the mud."[83]

Arriving on deck to check on his gas quads, Gary Legere says a "big controversy" broke out over how much lift was required to overcome the suction force holding the bell to the bottom. Legere was astounded by this talk. *"It's a little bell,"* he remembers thinking to himself.[84] They had a 30-ton crane pulling on an object that weighed just over a ton in the water.

Roper, who had been standing at the port-side railing, was less concerned about the bell being stuck in the mud than about the angle on the crane wire.[85] It was not running plumb. Nearly everyone saw it. Chuck Alexander, Steve Lobban, Ron Wilson, Dave Lane, Brian Reid, and even from the *Welder,* Don Cox and Captain Mellor saw it. Masterson, himself, would later testify that the pitch was somewhere in the region of 45 degrees.[86]

Suddenly, a bolt of fear swept across the dive station. They had two men in a bell on the end of a 5-ton sling under heavy strain, and somewhere on the seabed was an obstruction the size of a small house. Roper says there was "all sorts of concern about dragging the bell across the base—snagging the base."[87]

Normally blind lifts are never performed because they are fraught with danger. Yet this lift had commenced without verifying the bell was off the bottom, and now the crane driver was reporting a heavy strain on the wire. Something was preventing the load from coming up. As diver John Brook would say years later: "Common sense dictates now—even after all this time—that the next move was to go and find out what was going on on the bottom."[88]

But Masterson decided to continue the lift without that crucial inspection. At the inquiry, Roberts and Gunter testified that the crane driver took a series of "picks" on the bell—taking a strain and slacking off, taking a strain and slacking off—in an attempt to jar the bell loose from whatever was restricting its freedom.[89] To Dave Lane it was obvious there was "too much strain" on the wire. And when the bell wouldn't dislodge itself, he heard a lot of people "shouting and waving" and commenting, "Got too much weight," and "It won't come up." And then a new theory appeared out of nowhere: not that the bell was stuck in the mud, but that the umbilical must be snagged on the SALM base and was holding it back.

Indeed, Masterson believed the strain on the crane wire was caused by the bell umbilical snagging on the SALM base, and that the solution to free it was to continue pulling with the crane until the umbilical broke away.[90] When asked at the inquiry why he didn't stop the lift when he saw the angle on the crane wire, Masterson, a chartered engineer, explained: "To be able to get an angle like that on that type of load line, you don't need anything more than 2 tons. It was going to take more than that to be able to sever the umbilical."[91] When it was pointed out to him that the strength member in the umbilical would require more than 10 tons to break it, he disagreed with that view and suggested that another part of the umbilical was possibly caught on the base.[92] When asked why he didn't take the time to simply reposition the ship for a vertical lift, he said, "Because we would have needed a diver on the seabed again to be able to tell us where we were."[93]

1212 - Holding a strain on crane wire.

After trying unsuccessfully for twenty minutes to jar the bell free,[94] the crane driver was ordered to take a strain and maintain it. According to Ron Wilson the wire began "vibrating quite considerably."[95] Rod Pelzel says the load on the crane was so great it brought the whole side of the ship over. "It was just preposterous," he recalls. "Obviously [the crane driver] didn't have a clean pick on the thing."[96]

By this time Masterson had come down from the bridge to the dive station, took Holmes's portable radio from him, and ordered the crane driver to continue to come up on the load.[97] But every time the load approached 10 tons, the crane stalled; and with each failure, it had to be restarted.[98] When the crane man reported he could no longer raise the wire, but could achieve a lift by raising the boom, he was ordered to do so.[99]

Standing nearby and watching the lift was tender Steve Lobban. With barely three months of offshore experience under his belt, he didn't fully understand everything that was going on at the time. But, when he saw the crane driver holding the control handles back and "maxing out" the load, it became obvious to him that the bell was hung up on something. "I remember them booming forward, booming back, pulling, and obviously there was a problem, but no one was going to stop and sort it out. They were just pulling for all it was worth."[100]

From the *Stena Welder,* Captain Mellor was stunned by what he was observing. It never occurred to him that Masterson would attempt the lift at an angle. All at once he turned to the BNOC representative and said, "There is *no way* they can do a safe lift at 45 degrees!" And he went on the VHF radio shouting, "FOR GOODNESS SAKE STOP HEAVING. STOP! LOOK AT

THE ANGLE!"[101] But he says Masterson replied, 'Heave on everything—ignore that—heave on everything!'"[a]

At 1218, after nearly half an hour of "pulling for all it was worth," in an effort to get that "extra pound of pull," something unforeseen happened. The crane driver suddenly began whipping the boom back and forth in huge arcs like a fisherman trying to free a snagged line in a trout stream.[102] Holmes would later testify that the driver used a "slight slewing motion,"[103] but other witnesses described a more savage action. Dave Lane says they had the wire up tight on the winch, and were beating the boom backwards and forwards "like a bloody terrier." They just kept "yanking, and yanking, and yanking," he says.[104]

At that point, Pelzel saw Randy Gunter "jumping around," and going "berserk," shouting, "These assholes. What do they think they're doing?"[105]

Meanwhile, Roper noticed the crane wire tracking across the water, and by the time the boom had slewed for the third time, he was climbing the crane pedestal to tell the man to stop. It was a difficult 30-foot climb. He had to wend his way through a series of caged ladders and narrow gangways.[106] Just before he reached the crane cab, the air was broken by a deep resonant—boom. The crane wire danced like a rubber band, and then it seemed to relax and come to a vertical position beside the ship.[107] From the bell van, Ron Wilson saw this as a sign that the bell had finally cleared the bottom.[108]

Approaching the crane driver, Roper asked, "What's going on? What's happening?"

"I was given the order to slew," he was told.

Indeed, Masterson had instructed the man to slew the boom back and forth "as quickly as he could."[109] Distraught and chalk-faced, the crane man continued to recover the wire.

Suddenly the air became deathly still, and strangely hushed. It was as though the weight of the world was pressing down on this tiny patch of water. Most of the crew had gathered at the ship's railing. Dave Lane, still in his hot water suit, was leaning over the gunwale next to Tim Newson. Gunter, Lobban, Arscott, Sowards, Pelzel, Brook, Almon, McKenzie, and O'Connell were all grouped together next to the umbilical basket. Further aft, Masterson stood alone wearing a white hard-hat, staring down at the water with a grave look on his face.

For two long minutes, Reid and the crane driver continued the recovery, both wires extruding out of the water like strands of liquid steel, rising up to deliver a solemn verdict. It was the culmination of hours of exhaustion and frustration; either Richard and Skip were on the end of that wire or they were lost forever.

---

a At the inquiry, Masterson denied hearing Mellor over the radio (Walker/Guiel FAI, p. 596). But Sub Sea International Diving Superintendent Ronnie Morrison, stationed a few miles away on the *Atlantic II*, recalls overhearing the conversations being broadcasted from the *Wildrake* and at one point heard someone say, "Heave up on the bell." *Source*: Ron Morrison, Personal Interview, May 14, 1996.

Courtesy of Marcus Taylor Ltd.

Brian Masterson (left) at the side of the *Wildrake*.

Courtesy of Marcus Taylor Ltd.

Inspecting the open socket where the clevis pin fell out. From left to right: Rod Pelzel, Dave Roberts, Randy Gunter, Mac McKenzie.

The first wire to break the surface was the original bell wire operated by Reid.[110] Swaying high above, the empty socket gave painful and glaring evidence that the clevis pin had simply fallen out. The wire, it was plain to see, had not broken after all, and just above the socket, wrapped around a "dead head" collar, was a tangled collection of battered and distorted chain stops—every one of them ripped from their umbilical Dee-rings.

No one said anything and eye to eye contact was avoided as the crane driver continued the lift.

Standing behind Tom Almon, Rod Pelzel was still feeling hopeful, but also afraid to admit what he'd just witnessed. "*This is it,*" he said to himself. "*Either this is happening and it's going to work or they're dead.*" And in Pelzel's heart, whatever happened with that wire when it suddenly danced, simply caused the bell to release, and now Richard and Skip were on their way up.[111]

But those thoughts started to erode as the headache ball cleared the surface followed by a chain of shackled slings. And then Pelzel saw the last sling behaving strangely, as though it was a fishing line floating freely in the water. At 1220 his feelings of elation and anticipation suddenly vanished into gut wrenching horror when the feathered end of a 5-ton broken sling emerged from the North Sea without the *Wildrake* bell.[112] Shock and anguish spread like a raging fire. Pelzel's world suddenly collapsed into self-confession: "*Fuck! That's it. They're gone.*" At first no one said anything. Then fear, galvanized by the dangling wire, sent screaming curses across the deck unveiling the sickening reality that Richard and Skip were still on the bottom, still living their nightmare, still praying to God for salvation.

They had not been saved.

Roper just stared off into space. He felt tired and hollow, as though someone had shoved a fist into his stomach. He realized now that Richard and Skip were not going to make it. They had used up all their time. Almost instantly, he turned his feelings of anger towards Masterson for what he had done.[113]

Dave Lane says the divers now realized "it wasn't going to happen," and they began to leave the dive station. "They all knew it had gone bad," Lane recalls, "and it was going to be quite awhile before they actually managed to put a wire on it again."[114]

In the bell van, Burgess recorded the event in his log:

> 1220 - Bell cable on surface. Note: pin missing from bell cable connection.
> Crane wire on surface without the bell. Cable broke!!

Four minutes later, a message was sent to the *Seaway Swan* in the Brent Field "to proceed out here and help."[115] As Chuck Alexander recalls, when the wire broke, it was time to do something different: "We're in deep trouble now. We need something better than what we're doing. Those decisions [to call in the *Seaway Swan*] were made ashore, the platform, or someplace."[116]

In fact, they were made in Aberdeen.[117] Ten hours after the accident, BNOC officials finally came to the conclusion that the *Stena Welder* was not going to succeed. They also knew that, now that the *Swan* was urgently needed, it was hours away.

Chapter Seventeen

Lost Again

I've got to live with the memory of seeing them two guys in the bell for the rest of my life.

—— Rescue diver Eddy Frank

It had taken six hours, marked by confusion, argument, and threats to put a new line on the bell to save Richard and Skip; and in a heart beat, the progress of those six hours was blown to smithereens because a man with no diving qualifications and no vision of what happens underwater had taken command of the operation. Once again, the *Wildrake* bell was lost. If anyone had a reason to try something different—to let the *Stena Welder* recover the bell—it was Masterson and Holmes. But neither man considered changing course. Instead, they decided to repeat the same strategy all over again,[1] but with one important difference that had been previously overlooked. They decided to get Richard and Skip some hot water.[2]

Taking the end of a spare hot water hose, the *Wildrake* crew rigged it to a 35-ton shackle, then connected the shackle into the umbilical. In theory, the heavy shackle (with the hose and crane attached) would slide down the umbilical to the bottom. To lead Eddy to the hot water hose, a "messenger line" would be run between the two vessels. One end would be connected to the shackle on the umbilical, while the other end would be run down one of the guide wires of the rescue bell.[3] In this way, Eddy would lock out, take hold of the messenger line, travel the length of it, find the hot water hose, and take it to Richard and Skip.

While the *Wildrake* crew worked to effect this plan, Captain Mellor began maneuvering his ship back into position.[4] Angry and upset, he felt sick that Kasey had not been allowed to recover the bell. Mellor thought the entire strategy behind the rescue operation had been a complete farce culminating in a fit of panic. And now, instead of pulling Kasey and Eddy from a successful recovery, he was being forced to repeat the fruitless plan all over again as though he was trapped in some pathological loop of time.

Mellor wondered how much more the two men could endure before they became victims themselves. He tried to avoid these thoughts, but their safety weighed heavily on his mind. For sixteen years the sea had been his life, as it had been for his father and grandfather before him. He loved the sea, but now the thirty-four-year-old captain felt tired and wasted, and he didn't want to be part of it anymore. His mind was paralyzed by the senseless act he'd just witnessed. He stood staring at the *Wildrake* as a fresh wind whipped across his face. The sound of waves were lapping against his ship, and in the background a radio message broke the airwaves—the *Seaway Swan* was steaming north—but Mellor took no notice. He just continued to stare at the orange and white ship off his portside, his eyes fixed bitterly upon her bridge, knowing that somehow his life at sea was over and that he would never be the same again.

Five hundred feet below, Kasey and Eddy watched the water swirl and ripple in the trunking as the bell shuddered to its new position. They were cold and wet, and their faces drawn with exhaustion. More than an hour had passed; the *Wildrake* seemed to be taking a long time in raising the bell, but the two men knew better than to bother Allen for information. So they waited, pensive and deaf to the sound of the scrubber motor, hopeful that they would soon be able to stand down on news of a successful recovery.

More time passed and then feelings of hope and expectation dissolved into shock and disbelief as the bell speaker blurted out the disembodied words: the *Wildrake* bell was not on the surface. Before Eddy could react internally, he was ordered into the water to begin another search. Six hours of work had to be retraced with even greater urgency. From the day his father got him into the business, Eddy had always enjoyed diving, but now he seemed to be stuck in a bad dream. As he pulled his mask back on and splashed through the trunking, he hoped there was still time to save the two divers.

He made several lock outs to the seabed with no results, and then on his third exit, as the ship came to a new location, he landed almost directly on top of the SALM base. At one of its corners he saw the silhouette of the lost bell and said to himself, "*I've found it!*" His frustration broken, Eddy now felt he could do something to get Richard and Skip out of there, and he immediately began to assess the situation.

The bell was wedged hard up against the SALM base between corners five and six where the ballast line and an eight-inch guard rail climbed the base and formed a natural V-trap preventing the bell from glancing off and swinging free of the steel structure. Deep drag marks in the seabed led to the corner where white paint marks were visible on the piping, left there presumably by the bumper frame of the bell.[5]

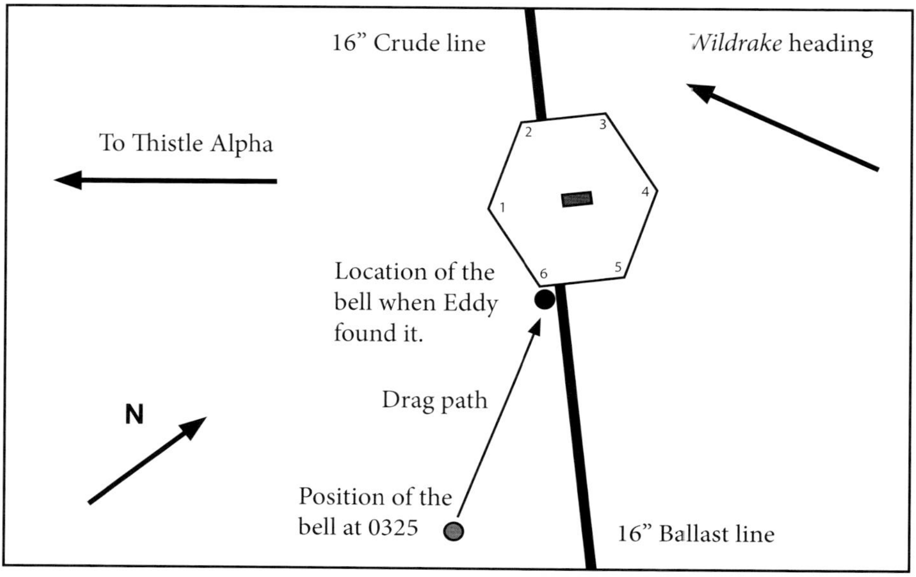

The unsuccessful "picks" that Gunter and Roberts witnessed were probably due to the bell jamming under a short horizontal cross-brace near the top of the base. With no chance for the lift to succeed, the wire sling broke, causing the bell to fall about ten feet to the seabed. It was now sitting upright and tilting outboard at a slight angle because part of the lower bumper ring was resting on a pile of sandbags. The broken sling was draped over the top of the bell. And then Eddy saw the drop weights. One was loose and lying out of position, while the other was still in its cradle. He also saw the ropes. Painfully he understood that the guys inside had tried in vain to save themselves.[6]

Dragging his umbilical behind him, he pressed his mask against a porthole and tapped hard on the side of the bell with his knife. A faint glow of yellow light was radiating from within and it took a moment for Eddy's mind to register the vision he saw. Skip was lying curled up on top of the hatch in his survival suit. He wasn't moving. Richard was slumped back in his seat with his arms at his side. His survival suit was pulled down about his waist exposing his white top. He was drawing hard for breath, and at the tapping sound, his eyes moved, and very very slowly he glanced up with a glimmer of recognition.[7] Richard was right on the edge of death.

And then Eddy was dragged away as the *Welder* moved off location. After all the worry, the stress, the lack of hot water, the swimming back and forth to communicate, the disorientation of being live-boated, and being led to the dying men with the promise of being their savior—cruel fate withdrew that promise and left Eddy with a vision of hell that would never wash from his mind, not even by the tranquilizing effects of time.

Back in the bell, he reported what he'd seen, but when he began suggesting how they should extricate the bell from the SALM, he was told to get back in the water, the *Wildrake* was sending down a hot water hose. On the clump weight, Eddy spent the next eleven minutes waiting for the messenger line. When the shackle jammed sliding down the *Wildrake* umbilical,[8] he was ordered to continue searching for the lost divers. For the next five minutes he slogged across the bottom, but couldn't find the bell. At 1304, he was back in the trunking suffering from leg cramps. Supervisor Mike Allen made a note of Eddy's condition.

> 1304 - E. Frank was tired to the point of distress and had to be helped into the bell on his return.

At 1315, doctors Morven White and George Shirriffs arrived on board the *Wildrake* by helicopter.[9] According to Gary Legere, he managed to convince Masterson and Holmes earlier that morning that they should seek medical help to treat Richard and Skip for hypothermia, "because at that point we were still really hopeful we were going to get them."[10]

On the *Welder*, Mike Allen's concern for Kasey and Eddy was now a dam ready to burst. Both divers were physically exhausted, suffering from severe cramps in their arms and legs and could only tolerate very short periods in the water. The two men were constantly flushing the bell to keep the atmosphere clean. Allen knew it was only a matter of time before he had one diver unconscious in the water and the other physically unable to rescue his partner. He gave his two divers a short rest period and asked them to analyze the CO_2 content.

At 1337, Kasey took over for Eddy and was back in the water determined to save Richard and Skip.[11] Ten minutes later, after a fruitless search, he popped his head through the trunking. His leg cramps had returned and once again Eddy had to help him inside. It was then that Allen decided, "Enough is enough." He had to get them out of there and put a fresh team in. He took the radio from Cox and spoke to Masterson explaining that his divers were being endangered "beyond acceptable limits."[12] He said if he didn't put a new team in, then the rescue operation could well end up with four people in trouble instead of two. Masterson agreed with Allen's assessment,[13] but seven minutes later, after Kasey and Eddy had already sealed the bell, he was on the radio again making a plea for one more lock out.[14]

Reluctantly, Allen told Masterson he wasn't going to order it, but would leave the decision up to the divers. If they said no, then he was bringing the bell up no matter what. But Kasey agreed to go back out and for the next 13 minutes searched the seabed and eventually found the *Wildrake* umbilical heaped over the SALM base. At the very end of his lifeline, he discovered the bell and the broken sling, and then, just as it happened with Eddy, he was dragged away, never to see the *Wildrake* bell again.[15]

It took another 12 minutes to recover Kasey and Eddy to the surface. Both men were severely depressed. "*Why? Why did the lift go the way it did? Why weren't we able to recover the bell? Why did we fail?*" they wondered. For years, Kasey would question his capabilities as a diver, blaming himself for the broken wire, never knowing what actually happened on the surface. These were feelings he locked away deep inside, and in his darkest moment he would ask himself, "*Is it my fault?—this thing about lifting the bell.*"[a]

At 1501, the *Stena Welder* bell reached the surface. It had been in the water for nine hours, and it was then that the deck crew discovered the twisted guide wires and the reason for the lack of hot water.[16] It took another ten minutes to sort the wires out to bring the bell on deck. Inside the chamber system, relief divers Michael Mangan and Tony Slayman were waiting.

> *You know, I thought about it a long time afterwards...'What should I have done.' I mean that stuck with me for ages, years...because it made me step back and take a look at myself.*
>
> —— Craig Roberts, Infabco Field Superintendent

On the *Wildrake*, Craig Roberts's composure was disintegrating. The plan to reconnect a new hot water hose had failed, and now, the *Welder's* diving system had to undergo repairs. There would be no quick bell turnaround. It was the last thread of hope he had clung to, and now it was gone. All the years of training, all the fast solutions, all the talent and experience had been used up, and no power on earth could turn the clock back to save his friends. Unable to escape from this awful nightmare, and with the reality of losing Richard and Skip crashing down upon him, Roberts found a quiet place on the ship to be alone, and broke down,[17] letting the tears and the grief spill out uncontrollably, forever burdened with the question: "What should I have done?"[18]

An hour passed. Allen's crew worked feverishly to repair the *Welder's* clump weight system. Bell checks were performed, on board gas checked, and because there was no time to fix the fault within the oil-filled junction box, diver communications were not corrected.[19] Inside the main chamber, a feeling of "lost hope" permeated the atmosphere as Kasey and Eddy briefed Mangan and Slayman. Meanwhile, Masterson advised Cox that he was moving the *Wildrake* 50 meters forward to realign the ship with the SALM base.[20]

At 1600, the launching sequence began.

The bell was rolled out and lowered into the water. Again, the divers would have no comms and no bell lights. The hot water problem would not return, providing the bell didn't spin. Thirty minutes later, Mangan and Slayman came to a full stop at 482 feet, and then Mangan locked out of the bell.

---

a Phil Kasey-Smith was never called to testify before any court on the matter of the *Wildrake* accident.

Raised in Boulder, Colorado, Michael Mangan had never been a water person. He had a machining and welding background, but the pay was terrible. After a buddy introduced him to scuba, he thought diving might be a way to earn more money. In 1973, he graduated from a training school in Oakland, California, then spent the next five years working for nearly all the diving companies in the Gulf of Mexico. When he arrived in Aberdeen in 1978, Mangan was an experienced gas diver.

Now twenty-six years old, he dropped through the trunking at 1639 to begin his search. Thirty minutes later he found the bell. Richard and Skip had been without hot water for more than thirteen hours. Fearful of what he might see, Mangan looked inside and saw the two men toppled over each other and face down.

1708 - Bell found on side 5 - 6 of SALM base.

West Springfield, Massachusetts August 8, 1979

August 8, 1979, had started out like any other day for Sheila Borsari. Her husband, Rick, had gone to work that morning while she stayed at home to tend their three-month-old daughter, Rebecca. It was just an average weekday—until the telephone rang.

Putting the receiver to her ear, she heard a pause, a faint beep, and then static hissing on the line. Instantly, Sheila recognized the sounds of an overseas connection and thought it might be her brother calling. But it was the starched British accent of Bob Lloyd, Infabco's general manager, introducing himself.[21] An uncomfortable pause followed. Questions were asked to confirm her relationship with Victor Guiel, and then the paralyzing words poured out of the receiver. They seemed incomprehensible at first. There had been an accident, she was told, and Skip's bell was lost. Two diving companies and a submarine[a] were supposed to be on the scene.[22] Lloyd said he would call back with more information later. In the meantime, he gave her a few telephone numbers for reference, and then he hung up.

In an instant, dark frightful images began to fill Sheila's mind. The call was real, and she knew it was serious. Skip had talked obliquely about the danger, and now all the feelings of love she had for her older brother began to overwhelm her heart.

She picked up the phone and dialed her husband's work number, but Rick was out of the office. She called her brother Roger, but he wasn't home either. She was less than a mile from her mother and father, but they were at work, too. Everyone was at work. In a big house on Fairview Avenue, Sheila was alone with the walls and the silence and her fears. She walked over and picked up

---

a It's likely that Bob Lloyd mentioned to Sheila that the 'semi-submersible' *Seaway Swan* was steaming to the accident site. The word submersible was probably misinterpreted to mean submarine.

Rebecca and pressed her tightly against her chest and looked out the window. It was a beautiful hot autumn day, but Sheila could only see the edge of an impending nightmare.

Santa Barbara, California

Twenty-five hundred miles away, Jeanne Walker was helping her daughter, Marisa, down the front step of their small Spanish style rental on Alisos Street. It was almost 10 o'clock in the morning and the sun was burning off a coastal haze in the Santa Barbara sky.[23] Marisa was 15 months old, and mother and daughter were going shopping when a car suddenly pulled up to the curb. Jeanne recognized Ted Roche as soon as he got out of his car. He was one of Infabco's supervisors and part of the *Wildrake* crew.

On leave for the birth of his second child, Ted had just returned from visiting his wife and their one-day-old daughter in the hospital when he received a telephone call from Bob Lloyd.[24] He wanted Ted to break the news of the accident to Richard's wife.

Inside the house, Jeanne was told the lift wire had "snapped," that the bell was on the seabed with Richard and his partner inside. They were okay and a rescue was in progress. Ted seemed optimistic about the recovery and expressed as much to Jeanne.[25] But he wasn't aware of the condition of the bell umbilical, the loss of hot water, nor the duration of the attempted recovery thus far.

Thistle Field, North Sea

Michael Mangan now began the task of reconnecting the *Wildrake* crane wire to the bell. The *Wildrake* was repositioned and the crane wire was lowered. When Mangan found it, it was "piled in a heap" on top of the SALM base.[26] He spent the next hour racing back and forth to the bell to order several ship movements and a series of short, 20- to 30-foot lifts to recover more than 100 feet of slack to the surface.

At 1816, he shackled the wire to the *Wildrake* bell,[27] and then he was ordered to close all the external valves and cut the umbilical away to prevent it from fouling on the base during the recovery.

While Mangan set about to complete these instructions, preparations were being undertaken to receive Richard and Skip back into the system. Life-support Technicians Greg Hansen and Tom Ulrich, were flown from the *Gulnare* to the *Wildrake* to assist Olson and Legere.[28] Inside the sat system. Dave Mitchell had been thinking about the problem of opening the bell. It was likely Richard and Skip would be lying on the access hatch, and Mitchell had come up with a plan to force the door and was now instructing topside on how to fabricate the equipment he needed.

Santa Barbara, California

After several phone calls, Ted Roche managed to gather more information about the accident by talking directly to Tom Belcher who had been monitoring the accident from the *Gulnare*.[29] In Ted's customary, hard-driving way, he fired one hundred obvious questions at Belcher concerning the status of Richard and Skip's hot water, their communications and gas supply, and the weather conditions. As he talked, he could not disguise the look on his face from Jeanne.

When he put the phone down, she asked him why he was so worried; the bell had hot water and gas, didn't it? But Belcher had given Ted a more complete picture of Richard and Skip's situation and he had to tell Jeanne that the umbilical had been damaged. He knew Richard and Skip's equipment situation in the bell; he knew the depth of the water; and he knew that the drop weights had been lashed.[a] But more than anything else, it was the cold and the length of time that worried him the most. All he could say to Jeanne was that hope could only save them now.

At that point, perhaps sensing the worst was about to happen, Jeanne called Doug and Patti Campbell and asked them to come over.

Thistle Field, North Sea

In the sat van, Dave Olson was informed the bell would be on the surface very soon. To facilitate a quick entry into the bell, he wanted one of the doctors blown down immediately so he would be in place and ready to act when the bell was equalized with the saturation complex. Nor did Olson want to get caught in a scramble performing duties that should be dealt with beforehand when Richard and Skip needed urgent medical attention. But when he contacted Masterson on the bridge to ask when the doctor should be saturated, he was told he would be notified "when necessary."[30] It was an incident that Olson still remembers with bitterness today:

> When your sat van calls up and says we would like to sat the doctor before we recover the bell so that we are not running 27 different operations at once, I'd have thought somebody would have listened. We got hired because we were supposed to be the 'Pros from Dover.' We had beaucoup experience, both in

---

a "I remember thinking about that during the accident," Ted says thinking back. "And I remember in my mind's eye I see white nylon, multiple passes of white nylon." Not surprisingly, Ted's counterpart, Supervisor Frank Montafia, would have the same thoughts. "I'm not making excuses of course," Frank said during interview, "but, again, this was a major mistake [lashing the drop weights], and it could have been a pivotal reason why things worked out the way they did. And I look back on that very clearly. I wasn't responsible, nor did I put those ropes on, but I didn't raise objections then that I should have....As soon as I heard about the accident, that was the first thought that came to my mind."

> running chambers and plumbing chambers and building them—and that got torpedoed....I mean, these guys could be dying on you as they were laying in the bell. Why not have the [doctor] in place and everything ready to go, so that—boom—soon as we're locked on we're ready to do a bell entry?[31]

Later, Olson would view his failure to prevail as one of the many "incredibly frustrating comedy of errors that just continued to the end." And today he still castigates himself for not being more steadfast with Masterson: "I should have been more forceful and said, 'Look asshole, this is what we have to do.'...Why would I be calling Brian Masterson to find out when to lock in a doctor? Where in the goddamn hell is the superintendent? Where's Craig?"[32]

On the bridge, Olson's plan for a quick bell entry was being advocated by another source. When Dr. White learned that the bell was close to being retrieved, he advised Holmes and Masterson that "it might be wise" for him to be in saturation when it arrived on board.[33] But he was told "that possibly there would be too much going on, on deck, with the bell being connected up," to have him blown down at the same time.[34]

Meanwhile, Mangan had carried out his instructions for preparing the recovery of the *Wildrake* bell. By 1914, he had closed the external valves on the bell, cut the umbilical free, and secured the drop weights to the bell frame to prevent the bell from accidentally surfacing. Taking another look through the porthole, he thought he saw a slight movement from one of the divers and reported it to topside.[35]

During the next six minutes he directed the movements of the *Wildrake* to put a slight angle on the crane wire. In this way, when the lift began, the bell would swing away from the SALM base. At 1920 he was in the rescue bell and ordered the lift to proceed.

Over the bell speaker he could hear a steady relay of information—that topside was taking the load, that the crane driver was reporting no problems, that the bell was coming up.[36]

At 1929, after 17 hours, the strategy to recover the *Wildrake* bell was finally achieved. Richard and Skip were on the surface.[37]

Chapter Eighteen

Primeval Screams

I came out of the water and looked at people in positions of responsibility quite, quite differently from that point forward.

—— Lacey Lahren, Infabco diver[1]

From the side of the *Wildrake*, Dave Lane quickly jumped onto the crown of the bell and connected the davit wire.[2] At 1937, Richard and Skip were hoisted out of the water.[3] Like stalks of wheat, more than a dozen men leaned over the ship's railing to get their first glimpse of the bell. It looked as though it had been rolled down the side of a mountain.

Once covered with a fresh coat of paint, the framework and on board gas bottles now exhibited long collision scars. The bottom door hinge assembly, made from heavy plate steel, was badly distorted.[4] A bell light cable had been severed; plumbing lines for the emergency gas bottles were mashed and crimped; the through-water comms transducer had been fractured;[5] a metal spool for winding manila rope was twisted. There was "considerable indentation" to the bottom bumper ring.[6] The video camera that Richard held in his hands at the time of the accident was nowhere to be found; its headless cable hung down, torn and shredded. Below the bell, one of the drop weights dangled vertically at the end of a wire strop, the other was still in its cradle.[7] Randy Gunter noticed the release mechanisms had been activated.[8] Inches away, strands of frayed white nylon rope hung from makeshift pad-eyes. Only later would he understand the significance of those ropes.

Courtesy of Marcus Taylor Ltd.

Bringing the bell on board.

Courtesy of Marcus Taylor Ltd.

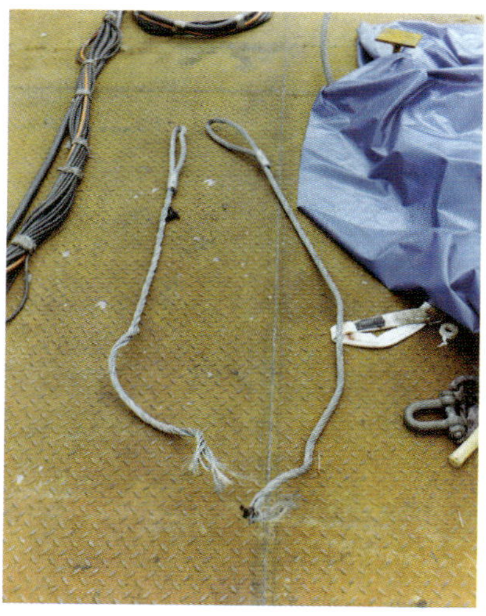

The broken 5-ton sling.

The damaged bottom door.

As the bell cleared the ship's railing, some of the crew were still hoping that Richard and Skip had managed to stay alive, but the more experienced believed their fate was a foregone conclusion.[9]

Standing nearby, Rod Pelzel watched as the steel casket was slewed inboard. It was the most solemn moment of his life. In his heart, he still could not believe this was happening. All of the hours of anxiety and frustration, the desire to find a way to save his friends, and the realization that there had been absolutely nothing he could do to change anything, had culminated to this moment of truth and madness. "*Why is this going on here?*" he kept asking himself. "*Why is this going on here?*" Exhaustion and lack of sleep had undermined his sense of reality. He felt like a stranger in his own body flashing between two time dimensions—one minute sharing laughter with Richard and Skip, the next, lining up to attend a funeral service at sea.

Courtesy of Marcus Taylor Ltd.

Landing the bell on the docking station.

But was this a funeral?

In this wide-eyed twisted dream of his, the final outcome was still unknown and subconsciously denied. Down past imaginary rows of pews, the casket was open and waiting. It was the moment of reckoning and Pelzel didn't want to know the outcome, but there was no time for him to turn and walk away. As the bell passed high above his head, his eyes drifted up to see a vision of unimaginable horror that shattered his consciousness and swept away all remaining sensibilities. There, pressed firmly against the lower porthole, was Skip's contorted face staring down at him in judgment.[10] For an imperceptible moment Pelzel's mind stumbled, transfixed by the death mask emanating from the halo-shaped port. In his last dying moments, Skip seemed to be delivering a clear and unmistakable message: "Hey, thanks—thanks for fucking me up."[11]

At this point, John Brook says some of the divers began to lose it. "He was obviously dead. And his face—I mean, all these guys just freaked out. I mean, we all did."[12]

With a profound sense of sorrow, the deck crew walked trance-like through the rest of the ongoing operation. The drop weights were removed, the bottom door was cut off, and at 1946, the bell was mated to the sat system.

From below deck, Gary Legere came to the dive station and looked through one of the portholes. Inside, equipment was strewn everywhere. Skip was lying on his side across the hatch and looked as though he had been beaten up. A large gash marked the side of his face. Brook noticed it, too. Richard, half-sitting, was slumped over Skip's shoulder.[13]

Standing nearby, Dr. Shirriffs overheard someone in the crowd utter that one of the divers had moved.[14] He thought Richard and Skip were dead,[15] nevertheless, Dr. White decided to go into sat to make a clinical determination.

West Springfield, Massachusetts

At 2:30 in the afternoon, Sheila called Scotland for an update on the accident and was told the bell had been recovered, that they had "spotted movement inside," and that a doctor was being put into sat.[16] In an hour, she would receive a call with the condition of her brother.

With a sense of dread and helplessness, Sheila waited. All her life, Skip had been the kind of brother a little sister could ever want. He'd protected her and demonstrated his love countless times, and now, in his time of need, she couldn't gather him up and bring him to safety; all she could do was bear the misery and wait for the next call.

Santa Barbara, California

Almost from the moment Richard began living in Santa Barbara, Doug Campbell had been his friend. Doug's father, Fos Campbell, owned Kilovac Corporation where Richard had once worked. It was Fos who had taken his son on a tour through the factory one day and introduced him to Richard. After that, Doug and Richard became fast friends and roommates during college.

Both men shared common goals in life, and after Doug married Patti, the two families became land partners with Doug's parents at Hollister Ranch. Now, at this time of crisis, Doug and Patti arrived to find Richard's wife in "a mess."

"It was horrible. Just horrible," Doug recalls. "She was obviously very, very upset. Marisa was just a small child at this point, so we tried to take care of Marisa, which we did."

And then Scotland called. The bell was coming on board. The voice on the phone said he would call back with the condition of Richard.

Thistle Field, North Sea

On the other side of the world, the "scramble" that Dave Olson wanted to avoid was now coming to past. The bell was married to the system, but the doctor, and the equipment Dave Mitchell requested, were outside the system.

At 1951, Olson and the other technicians began pressing Dr. White down on air in chamber two. At 29 feet they switched to helium and continued the pressurization for a while, then suddenly stopped the descent. Worried that the speed of the blow down might produce helium pockets and cause the doctor to pass out, Olson switched from helium to a mix of HeO_2 (3.75 %).[17]

While the doctor was being committed to sat, Olson's team began lowering the system depth from 460 feet to 540 feet to match the depth of the bell. Five minutes later, at 1959, the entrance lock and bell were equalized. Inside, the entry team were ready, but as precious minutes ticked by, Dr. White and the equipment were still sealed behind the door of chamber two.

At 2009, after eighteen "extremely hot and uncomfortable" minutes, the door popped open.[18] Dr. White quickly passed the tools to the divers.

Working in cramped quarters, Palmer, Boyd, Mitchell, and Lahren took turns inside the tunnel using blocks of wood, a crudely fashioned telescopic pipe, and hydraulic power to force the bell door.[a] It was difficult work. With Skip lying on top of the hatch, the door could only be pumped open fractionally in stages. Outside the system, Dr. Shirriffs made an entry in his notebook:

> Desperate measures to open the hatch.[19]

Meanwhile, inside the entrance lock, the seal on the bell door had been broken sending an icy concoction of blood and sea water down upon the group. A gas hose connected to the pressurization line was quickly inserted into the bell, and 6% HeO_2 was blown in to increase the oxygen level and flush out the carbon dioxide.

At approximately 2030, Lahren managed to squeeze his upper body into the bell.[20] It was like a freezer inside and the atmosphere was still foul with carbon dioxide. Both Richard and Skip were out of their hot water suits. Skip was wrapped in his space blanket and lying on his left side with his hands cupped in front of his face.[21] A small oral-nasal mask connected to a manual carbon dioxide scrubber was lying inches away.[22] Richard was leaning against Skip and was not in his space blanket.[23] His manual scrubber was at his side.

Both divers appeared to be dead.

The ex-Marine, who'd been wounded during the Siege of Khe Sanh, began checking Skip's vital signs: pupillary responses, heartbeat, and rate of respiration.[24] He found no signs of life. With the door only partially open, he was unable to reach Richard. After seven or eight minutes, the contaminated atmosphere forced him down the tunnel gasping for breath.[25] When Dr. White asked about the condition of the divers, Lahren said Skip was dead.[26]

For Lacey, entering the bell and finding the collapsed bodies of Richard and Skip would have a profound affect upon his career. In a week, the former marine would come out of sat never to return.

---

a The top door of the entrance lock operated by a hydraulic ram. Using this, in combination with the pipe and wood blocks, they were able to raise the inside door of the bell.

"I decided that I wasn't going to let anybody have that kind of control over my fate, ever again," he says, looking back on the incident.[27]

After Lacey, David Boyd climbed the tunnel to examine Richard, came back, and told the doctor that he could find no signs of life.[28] Shortly before 2100, Dr. White entered the bell with his medical bag and conducted an examination which consisted of checking Richard and Skip's cardio-vascular, respiratory, and nervous systems.[29] It is not known whether he measured Richard and Skip's core temperatures.[a] That question was never posed at the official inquiry.

After 15 to 20 minutes, Dr. White convinced himself that Richard and Skip had died, then rejoined the divers in the entrance lock and communicated his findings to Dr. Shirriffs over the radio. Dr. Shirriffs went up on deck, and with a light, inspected the interior of the bell through a porthole. Peter Holmes and Brian Masterson were standing nearby. At 2110, he turned to the two men and officially pronounced Richard Arthur Walker and Victor Francis Guiel Jr. dead.[30] They were the 44th and 45th divers to die in the North Sea.

Almost two years later, Dr. White was questioned at the inquiry whether he'd considered "trying any resuscitative procedure of any kind" on Richard and Skip.

"No, I did not consider it," he answered.

"Why not?"

"I felt that the situation did not warrant it," he replied.[31]

Nor did he consider rewarming the bodies before determining death.

That disclosure would cause a well-respected member of the British medical community, a physician with a strong background in hyperbaric medicine, scholar of North Sea diving accidents, and Senior Lecturer at the University of Dundee, to find fault with the treatment Richard and Skip received.

"In the case of the *Wildrake*, I was very critical of the medical management," Dr. Philip James says. "Death should never be attributed to hypothermia without an attempt to rewarm."[32b]

Indeed, there is even a saying among doctors that "You're not dead until you're warm and dead." And in his book, *Hypothermia and Cold Stress*, Dr. Evan Lloyd cites no less than seven studies conducted prior to 1979 which find that "the only certain diagnosis of death in hypothermia is failure to recover on

---

a Later, Dr. White would be the focus of a false rumor that, because access into the bell was restricted, his examination consisted of only reaching in and feeling Richard and Skip's pulse. Lahren and Palmer dispute this charge: "He got completely into the bell," Lahren says. "Initially we forced the hatch open with a device engineered by Dave Mitchell. Subsequent to my entry, the device was adjusted to allow better access."

b Additional research supporting Dr. James: Golden, Surgeon Lt. Commander F. St. C. *Recognition and Treatment of Immersion Hypothermia.* Proceedings of the Royal Society of Medicine, 66: p. 1058-1061; Lloyd cites studies by Kugler-Podelleck *et al.*, 1965 and Siebke et al., 1975.

rewarming."[33] This point had become "so well recognized" in Britain, Lloyd says, that in 1974 the National Health Service issued an official circular which advised that "before brain death can be diagnosed the doctor must check that the core temperature is normal."[34]

Meanwhile, rescue divers Michael Mangan and Tony Slayman had just been recovered to the *Stena Welder* sat system. A few minutes later, word followed that the *Wildrake* divers were dead. All four 2W divers were deeply affected by the outcome, especially Kasey. At first he couldn't feel anything as he rehashed the accident in his mind, still mystified that the *Wildrake* bell wasn't where topside said it should be, that the sling had broken, and that the *Stena Welder* was not allowed to make the recovery. "*This should have been a fairly straightforward operation. Why did it fail?*"

For Eddy, the accident would haunt him for years and forever bury the glamour of deep-sea diving. Eventually he would trade in his fins for a job as a topside systems technician. And for the next two decades he would find himself getting hot under the collar whenever he saw safety being compromised. On the outside, his crew mates saw him rant and rave like a lunatic, but on the inside Eddy had a reason for feeling the way he did. "You cannot just forget them," he says about the accidents he's seen. "I mean, they stick in your mind; they're with you all the time."[35]

Back on the *Wildrake,* Gunter and Roper had now been awake for more than thirty-four hours. Exhausted beyond feeling, they couldn't escape the reality that just a few feet away two of their friends were lying dead inside a diving bell. Roper refused to go near it. Instead, he retreated to be alone, still angry at Masterson.[36] Gunter, on the other hand, vented his anger more predictably. After witnessing a plethora of wholesale blunders culminating with Skip's battered face in the port, he began lashing out at the *Wildrake* crew with charges of: How could you let this happen? And for tender Ron Francisco, the accusations were particularly harsh.

Under the impression that Francisco had been the last person to check the bell wire connection before the fatal dive, Randy cornered the young man and accused him of killing Richard and Skip.[37] And while Francisco had completed the external bell checks on the previous shift under the supervision of deck foreman Dave Lane,[38] John Brook would later admit in court that he had performed the final checks.[39]

"[Francisco] got the big blame of not checking the pin," Kurt Baukau recalls. "And he was running around at night, and he took more and more of that shit, and he was going around and crying...and he was desperately trying to find somebody who would take sympathy or listen to him. I felt so sorry for the boy."[40]

Don Cox, who by now had returned to the *Wildrake*, said Gunter's efforts to "browbeat" Francisco was the nearest he'd ever come to blows with anyone offshore.

It wasn't morally right, he said, "and I made it known in no uncertain terms that if he did it again, then it would be fisticuffs at five paces on the back deck."[41]

Just before midnight, some of the *Gulnare* divers were flown back to Thistle Alpha where Superintendent Tom Belcher was waiting to hear what happened. In appalling detail, Rod Pelzel took him aside and spoke nonstop, recounting the horror of the botched rescue and how one shocking display of reckless disregard spawned another: ruination of Richard and Skip's lifeline, failure to plumb the initial lift, snagging the bell on the SALM base, failure to lock out the rescue diver to inspect the problem; then mindlessly trying to break the bell free with the ship's crane as though the payload were a lump of steel, not flesh and blood.

Belcher just sat there, frozen in disbelief, as Pelzel streamed on, struggling with his emotions, his mind poisoned by having witnessed the senseless wasting of two human lives. Grief-stricken, he recounted the crane boom whipping from side to side and the broken cable emerging from the sea without the bell. Finally, at the point of describing Skip's face in the porthole, Belcher and Pelzel broke down and cried.[42]

Later that night, Rod went to his room, opened his small red diary, and recorded the saddest event of his life:

> August 8 - Dick and Skip killed. Called over to Wildrake to assist in bell recovery. Divers were given the most incompetent support from incompetents in position of authority. There was no one on board who had the expertise to raise [the] bell successfully, that would or possibly could step out and take control of [the] situation.[43]

In the weeks ahead, Belcher and Pelzel would accompany Richard's body back to California, but Pelzel would never return to the North Sea again. He had just witnessed the cream of the crop "basically massacred, senselessly," he would later say:

> It was like, man, I don't need this anymore. I have no compunction, no desire whatsoever to be involved in this anymore because they had no control. When you're in that pot somebody else is calling the shots, man, and with all their training and all their knowledge, they were trapped, which made it an even more dramatic story when the bell was finally lifted and seeing the attempts that they'd made to free themselves to get the drop weights off.[44]

Pelzel, Gunter, Roper, Legere—in one way or another they would all be emotionally scarred by the events of that night—one of the men fighting back tears seventeen years later during interview as he described seeing Skip's face in the porthole. For Gary Legere the passage of time would never erase the sense of shame and guilt he felt for what happened to Richard and Skip:

> In general most of the people involved with the accident were pretty messed up for a long time afterwards. They might not talk about it, but they were. It's really weird for this to come back into my life. You know I thought I had it all

> kind of figured out, but you never do; it's just what happened. It does well up a lot of emotion to me because it was a highly emotional—that one guy, that rep—he was kind of going around [saying]...'We're all sticking together; we all really worked hard; we all tried to do our best.' I was kind of looking at him thinking, *'Well, we didn't.'* You know, that's the truth of the whole matter. No one did their very best on that job.

West Springfield, Massachusetts

By now, Rick had been located and was home with Sheila. Shortly before 6 p.m., she called Scotland and heard a distant voice utter the shattering words that her brother was dead. After that, nothing else registered in Sheila's mind.

The burning sensation began in her throat, and migrated down her windpipe to her chest. Suddenly, the air became too hard to breath and the pain too much to bear. The god that Sheila worshipped—who held her tiny hand on the way to countless Saturday matinees, who dug clams with her in Penobscot Bay, who taught her how to shoot a rifle and throw newspapers and read constellations in the night sky, who protected her from "bad boyfriends" and brought back the world from his travels—was no more. Racked by inconsolable grief, she ran into the backyard with tears streaming down her face, tormented by visions of Skip dying a slow death on the bottom of the sea. These would be the images that would endure, and for years, sleep would be her only defense against the demons that haunted her every waking moment.

Rick dialed Roger's house again.

Across town, Roger picked up the phone to hear his brother-in-law crying on the other end. Get over here now, he was told, and Roger, not knowing what was going on, didn't make it into Rick and Sheila's house before they met him on the back-porch landing to deliver the devastating news. More than the words, it was the look of utter despair in their swollen, tear-filled eyes that spoke the truth. In a daze, he suddenly found himself breathless on the ground with disembodied animal sounds emanating from his mouth. To the human mind, the concept of death is transcendental and painful, not because the mind feels the suffering of those that have died, but because death marks the beginning of eternity. And in that brief instant, when Roger realized he would never see his brother again, he was stripped of his identity and became a lost soul.

Minutes later, Joey Graveline walked into the house. He carried with him a replica of a 45 caliber 1775 Pennsylvania long gun he'd built for Skip. "And I could tell there was something really important going on just by the look on their faces." When Joey asked what was going on, Roger turned around with the look of the damned on his face. "Vic's dead," he said.

Joey dropped the gun, walked outside, and sat on the steps and cried for the next several hours. He couldn't believe Skip was gone. It was as though someone had told him his best friend never existed and it drove him to the brink of insanity.

Later, when he got his mind back he would wonder about the date—August 8—it was the same day his grandfather died, his daughter was born, and now, it marked the death of his best friend.

In a trance, he drove home, took a shower, changed his clothes, and returned to Rick and Sheila's to stay the weekend. Broken and shattered, the gang from Al's Corner walked around like ghosts, overwhelmed by grief, their eyes red raw from the tears that never stopped. The devastating news seemed unreal. No one really knew what was going on. There were no details, no understanding why Skip was dead, and so they consoled each other as best they could.

When the weekend was over, Sheila told Joey to keep the gun.

Santa Barbara, California

On Alisos Street, it was almost three in the afternoon when the telephone rang. Ted Roche took the call, listened quietly to the voice on the other end, and then asked about Richard and Skip's condition. In an adjoining room, Jeanne heard the question, and from the expression on Ted's face, she understood what the answer was. But even then, her mind refused to believe what her senses already knew. Even then she was certain that there had been a dreadful mistake, that, somehow, they had the wrong man. Even when she asked if Richard was okay and heard Ted telling her, no, that her husband was dead—even then, she couldn't believe that Richard had been fated to die.

Jeanne Walker, tall and beautiful, had married the man she loved. She had borne him a darling child, shared his dreams, and expected to be at his side the rest of her life. And now, at the age of twenty-four, she'd been made a widow in the blink of an eye. Suddenly, she began screaming and beating on Ted's chest; and in the last remaining seconds of sanity, before the walls fell in on her, before Doug and Patti whisked Marisa away from the spectacle of seeing her mother lose her mind, the sound of Jeanne Walker's screams suddenly changed in quality. For these were primeval screams welling up from a mind lost in madness, and for a very long time, they were screams that could not be quieted.

Chapter Nineteen

How Could This Have Happened?

DIVING BELL: TWO ARE DEAD

—— *Press and Journal*, August 9, 1979

Since the discovery of oil in 1969, Aberdeen's Press Corps had reported on everything from OPEC's stranglehold on the nation's economy, to the pace at which the oil bonanza should proceed, to government predictions on self-sufficiency. The public, accustomed to the usual diet of sports scores, marriage announcements, and the odd sighting of Nessie, now looked on with a sense of curiosity and awe as the oil rush began to seep into the social fabric and eclipse the slow pace of everyday life. Oil was jobs and wealth; oil was Britain's economic salvation.

But as the years passed and the death toll climbed, the public's sense of curiosity and awe began to sour into a mixture of trepidation and déjà vu. Mind-numbing headlines became all too familiar: more dead at sea. News accounts were dramatic but typically superficial because accident scenes were inaccessible, oil operators tempered the facts, and reporters, unfamiliar with the industry, didn't know the right questions to ask. Still, the dead did not go unreported. From downed helicopters to capsized rigs to catastrophic explosions, the human cost was tallied in one tragic article after another.

On Thursday, August 9, the *Wildrake* accident became front-page history and Richard Walker and Skip Guiel took their place among the fallen—faceless men with no discernible past and no way to speak for themselves. Instead, Infabco Diving Services did all the talking.

At a press conference the night before, journalists gathered to listen to a series of misleading statements designed to paint the false impression that

Infabco had pulled out all the stops to save the two helpless divers. According to a company spokesman, when the alarm was raised, two ships—the *Stena Welder* and the *Seaway Swan*—"raced to the scene" and were involved in the recovery operation, and that the cause of the accident was likely due to "mechanical failure rather than any human error."[1]

There was no mention that BNOC had sat on the sidelines for eight hours before calling the *Seaway Swan*. No mention that the *Stena Welder* crew spent critical hours putting their diving system back together, that offers of assistance from other vessels had been rejected. No mention that the man in charge, Brian Masterson, had no diving experience or that more desperate hours were lost after a reckless attempt to break the bell free from the SALM base.

BNOC understood that the potential for scandal in this case was enormous. And nothing focuses the British mind more keenly than scandal. Scandal creates bad publicity and generates a lot of finger pointing. Two men had just suffered an horrific death under circumstances which would shock even the most unsympathetic ear. Public outrage could prompt regulators to call for stiffer legislation, forcing oil companies to disrupt their normal and accepted practices. Fatalities were simply bad for business. So the true dimensions of an accident were rarely handed to the press on a sliver platter—the rationale being: if everyone kept their heads low, then, in time, the whole matter would blow over; these two unknown divers would be forgotten, and it would be back to business as usual.

And that strategy probably would have succeeded in the *Wildrake* case except that Infabco had pushed the practice of cutting corners too far. In an unprecedented move, the Department of Energy and the Crown prosecutor in Aberdeen would attempt to use the *Wildrake* accident to set an example for future violators, and in turn inspire Infabco Diving Services to raise the meaning of infamy to new heights.

Far out at sea, saturation technician Gary Legere woke early that morning to find Richard and Skip still lying in the bell under pressure on the back deck of the *Wildrake*. Angry that the two divers had not been taken care of,[2] he separated the bell from the chamber complex, decompressed them, and had the bell rolled onto its side. Then he and Greg Hansen removed the bodies and placed them in a pair of makeshift coffins which were subsequently moved to a landing below the helideck.[3]

At 0800, BNOC took the *Wildrake* off-hire.[4] For Brian Masterson, this was bitter news. The SALM contract should have been a springboard to bigger and better things. But nothing had gone right, and now government investigators were on their way, sure to be asking questions about an operation that unexpectedly metastasized into a dual fatality. An ordinary man might have been grief-stricken, even defeated by the tragedy, but Masterson was no ordinary man. He had more pressing matters weighing on his mind.

With the diving system in total disarray, the company owner entered the sat van to consult with Dave Mitchell, who's opinion on diving matters he respected. In the small control room, Don Cox was standing behind his boss when he overheard him ask Mitchell how quickly the system could be repaired so they could get back to work.[5]

Inside the dimly lit chamber, Mitchell must have thought he'd been committed to a lunatic asylum when he heard Masterson's question. The subject seemed to be completely out of place. Less than twenty-four hours after the tragedy, Masterson was discussing work as though nothing had happened. As Legere put it bluntly years later, "The guys aren't even cold yet, and he's planning the next dive!"[6] When Masterson suggested Mitchell and the others remain in storage while the system was made ready for diving operations, Lacey Lahren says everyone in sat emphatically answered, "Absolutely not!"[7] They wanted out to fix the system before anyone got back inside.

This attempt to breathe new life into the crew was registered on deck as well. Dave Lane saw Masterson appear at the dive station attempting to give the divers a pep talk, using words to the effect of, "Come on lads, let's get on with it. We've got work to do. Let's try and get it all sorted out." Suddenly, it dawned on Lane that Masterson might actually be thinking he would sail into Lerwick, drop off the bodies, pick up a new umbilical, then sail straight back out again.[8]

But the crew had had enough of Brian Masterson. Disgusted with his offensiveness, they ignored him and wandered back to their cabins. Their only inclination now was to get off the death-ship as quickly as possible.

At two o'clock that afternoon, Senior Diving Inspector Roy Giles of the Department of Energy boarded the *Wildrake*.[9] With him were DOE Inspector Lawrence Piper, several representatives from BNOC, and police officers William Bailey and William Adams who joined the investigation to determine whether there had been any malicious intent to cause loss of life.[10]

Courtesy of Marcus Taylor Ltd.

DOE Diving Inspector Roy Giles (in coveralls) and Gary Legere survey the bell.

The first person Giles spoke to was Chuck Alexander, the SALM project engineer, who gave him a brief account of the accident and introduced him to Masterson and Holmes.[11] Holmes then led the ginger-haired Welshman around the diving system with officers Bailey and Adams trailing behind. In the bell van, they discovered that the audio tape of the fatal dive had been erased. Holmes would later explain at the inquiry that the tape was actually a continuous loop and "had fed itself back, through, and wiped itself clean."[12]

Courtesy of Marcus Taylor Ltd.

Parts of the bell umbilical lying around on deck like "spaghetti."

On the back deck Giles noticed the badly tattered bell umbilical lying about like "spaghetti."[13] Near the stern of the vessel the bell was on its side and sea fastened to the deck. Its painted undersurface, from 12 o'clock to 9 o'clock, had been scraped to bare metal, indicating that it had been dragged along the seabed for a considerable distance.[14] He noted the modifications to the drop weight system and surmised that, in trying to save themselves, Richard and Skip were able to cut one of the nylon ropes, but not the other.[15] He also calculated that, had they been able to release the weights, the bell had enough positive buoyancy to return to the surface.[16] Despite Holmes's assertions to the contrary, Giles concluded that the bell had been operating without a secondary means of recovery,[17] and that the lift wire connection assembly had been compromised. When he asked to see the swivel, no one could produce it.[18] And he discovered that Skip's medical certificate was out of date—another breach of the regulations. Everywhere Giles went, he found one safety violation after another. In fact, even before he stepped on board the *Wildrake*, he knew that Infabco Diving Services had broken the law.[19]

Among the current diving regulations were two prohibitions: diving contractors were not allowed to 1) conduct operations below 410 feet (125 meters), nor 2) use saturation techniques, unless details of the operation and the type of procedures and techniques to be used were first submitted to the Secretary of State for approval.[a]

This formality was to give DOE inspectors time to organize an offshore inspection (something they hoped to conduct prior to commencement of

a The Submarine Pipe-lines (Diving Operations) Regulations 1976, Section 7(d,e).

operations, but rarely achieved), and to put the diving contractors on notice that the inspectors "were watching."[20] But to the DOE's embarrassment, a thorough search of its records revealed that Infabco had never applied to the Secretary of State for all of the 14 months it had been operating in the Thistle Field.[21]

In a conference room, as officers Bailey and Adams took statements from the witnesses, Giles listened quietly, took notes, and asked questions regarding the rescue operation.[22] As each sorry detail poured out of the mouths of the crew members, Giles became convinced that this was not a "pure accident," but a case of gross criminal negligence ending with a bungled rescue attempt due to "the headless chicken syndrome," the likes of which he'd never seen before.[23]

At 1600 hours, Don Cox, Chuck Alexander, Tim Newson, and the rest of the BNOC installation team transferred to the French Diving Support Vessel *Talisman* to complete the job of reattaching the SALM to its base.[24] The *Wildrake* would proceed to the port city of Lerwick in the Shetland Isles. Where the *Talisman's* offer of assistance during the rescue attempt had been rejected, now the ship's commercial value loomed large enough for BNOC to obtain a spot-charter without difficulty.

In the coming days, *Talisman* divers would search the seabed for the missing clevis pin and confirm that the *Wildrake* bell had indeed collided with the SALM base.[25]

Courtesy of Marcus Taylor Ltd.

The following morning Randy Gunter lay in his bunk still possessed by the tragedy. His body needed sleep, but his mind wouldn't leave him alone. He kept imagining himself with Richard and Skip,[26] trapped on the bottom, tormented by crippling cold, fighting to stay alive—*My God, what were they saying to each other?*

Over and over, Gunter tried to piece the events of the accident together, but he couldn't make sense of it. *Why was the rescue limited to a vessel that had its diving system under repair? Why hadn't BNOC called in more ships?* It didn't make sense.

Unable to sleep, he dressed and walked from the accommodation area to the stern of the ship where waves of cold air made his eyes water. Gunter felt tired and lonely. His hometown of San Diego was a long way off. In the early morning light, he watched the dark gray horizon slowly rise and fall off the fantail of the *Wildrake*. The ship was steadily cutting a path towards Lerwick. Nearby, the death bell lay on its side and the two shackles that had been substituted for the swivel were still attached to the pad-eye. Gunter walked over and began inspecting them closely. He saw the grooves that Roper told him were as deep and as wide as his thumb.[27] The drop-weight ropes, torn and shredded, hung from make-shift pad-eyes. Gunter just shook his head in disgust.

How could this have happened?

Courtesy of Marcus Taylor Ltd.

Indicating the grooves.

And then Gunter's eyes fell upon a small gap between the hull of the bell and the flotation collar. Wedged in a crevice was a piece of wire.[28] Without thinking, he reached in and pulled it out. It was the remains of a broken welding rod, and it was bent, as though it had been wrapped around a circular object. Like a bolt of lightening, Gunter instantly grasped its significance.

Clevis pins and some types of shackle pins are designed to be locked in place with split pins (also called cotter pins). But on the job, split pins frequently become lost or broken or simply thrown away as a "nuisance factor" because they require tools to insert and remove—tools that divers rarely carry on their person.

Welding rods, on the other hand, are readily accessible and easy to bend into any shape by hand. So for many temporary, non-life-threatening applications, where the diver judges there is little or no risk of injury, welding rods are, and have been for decades, the accepted standard as "mousing" to secure a shackle pin.

Courtesy of Marcus Taylor Ltd.

Damage to the shackles by wear.

But for permanent applications where human life is at stake, where every effort should be made to preserve—not compromise—the integrity of a vital connection, using a welding rod as a "keeper" in a shackle assembly, known to be undergoing constant shock loads and vibration, is a manifestly bankrupt rigging practice because welding rods are prone to break.

As an experienced diver, Gunter knew this. And the instant he pulled that piece of wire from behind the flotation collar, he believed he had solved the mystery that had been nagging his conscience. He quickly left to wake Roper, and together they retrieved the blueprint of the clevis mechanism and discovered that the fragment of wire closely fitted the curvature of the clevis pin.[29]

The two divers believed they were holding a strong piece of evidence in what might evolve into a criminal matter, given the serious blunders they'd observed during the rescue.[30] But in the aftermath of the accident, they didn't trust anyone on board the *Wildrake*, in particular Brian Masterson, Peter Holmes, and Brian Reid in whose hands they thought the evidence might suddenly "disappear."[31]

Eventually, Gunter's discovery found its way to the official inquiry, where the Crown attorney tried to prevail upon Supervisors Don Cox and Brian Reid to admit that they had condoned the use of a welding rod as mousing for the clevis pin. Both men vehemently denied the assertion, Cox even going so far as to say that he had actually seen a split pin in place prior to the fatal dive.[32]

Notwithstanding those denials, this line of questioning was irrelevant given the fact that the Crown introduced photographic evidence clearly showing that the surviving shackle pins had been secured with wire, not split pins.[33] This wasn't conclusive proof that a welding rod had been used for the clevis pin, but it was compelling evidence indicating that the crew had no compunction about compromising the security of one of the most vital elements of a diving bell.

How then did the clevis pin come out of the open socket? Did it break? Or did it simply fall out? The missing clevis pin would have provided the answer, but it was never found by the *Talisman* divers, and the Crown never called an expert to the witness stand to offer an opinion on the subject. So the true mechanism of failure has always remained a mystery. It does seem unlikely that the clevis pin broke in half. While it had undergone repeated shock loading, even reasonable estimates fall far short of the amount of force required to shear the 41 millimeter pin.

Engineers also point out that the requisite shear force would have deformed the holes in the lugs of the open socket.[34] But Gunter had inspected the lugs after the accident, measured them with a micrometer, and found the holes to be undamaged and perfectly symmetrical.[35] In addition, there were three pins connecting the bell to the main lift wire (two shackle pins plus the missing clevis pin): if simple overloading had caused the bell connection to fail, then logically the shackles pins would have sheared first because they were smaller in diameter.

That leaves the more plausible theory that the clevis pin, just a smooth bolt with no threaded nut on the end, simply fell from the open socket after the keeper broke free. Several years after the accident, an article written by David Hughes appeared in *Underwater Systems Design* explaining how a split pin could fail in a bell shackle assembly, an arrangement that Hughes characterized as "dangerously prone to failure."[36]

As the bell is snatching on the end of its lift wire, the shackles separate and come together with such terrific force which, over time, causes grooving to occur. And with repeated separations, Hughes says the pin does not always "exactly hit the center of the groove of wear, and a very considerable sideways jar can result."[37] And it is this lateral force which eventually shears the split pin. It is also worth noting that, on the *Wildrake*, the lateral force would have been exacerbated by launching the bell over the side of the ship, rather than through the moonpool.

Beyond Hughes's analysis, this mechanism for failure received help from another source. The arrangement of the shackle assembly itself had an inherent flaw, in that it was the size, shape, and orientation of the shackles to each other, that allowed the lugs of the safety shackle to impact directly against the split pin (or welding rod) of the clevis pin. And just as these repeated, violent collisions produced grooving in the shackles, they also would have worn away the head of a split pin (or welding rod). With the keeper gone, it was just a matter of time for the clevis pin to follow from lateral movement.

One remaining factor triggered the accident. As bad a rigging practice as this jury-rigged assembly was, nothing undermined Richard and Skip's safety more than a crew that abandoned its duty in performing external bell checks. The first item on the checklist was to ensure that the lift wire shackles were moused and secured. The fatal dive was three and a half hours long when the clevis pin fell out. According to an engineer, that amount of time would not account for the degree of grooving found in the shackles.[38] This means that, for a very long time no one was actually climbing the *Wildrake* bell and taking a hard look at those shackles, because if they had, then they would have seen what rescue diver Phil Kasey-Smith saw when he found the bell on the seabed.

Meanwhile, in Lerwick, Richard's personal effects had been passed out of the chamber to be shipped home to California. Blue jeans, harmonica, pocket knife, underwear, shaving kit—there was nothing unusual about the items, except a blue spiral-bound notebook. It was Richard's diary. In a matter of days this seemingly nondescript booklet would echo the last words of a man on a collision course with destiny.

That day, August 10, 1979, just two days after he died on the bottom of the North Sea, Skip's lawyers closed on the purchase of his land in Maine.[39]

Chapter Twenty

The Diary

I think they would have been happy to just hide the bodies and say, 'Well they must have fallen overboard.' I really think they would have done that if they could have.

—— Jeanne Walker[1]

In Santa Barbara, Jeanne Walker was still struggling to put her emotions back together when Richard's personal effects arrived in the mail. Since his death, she'd spent each tearful day lost in abject misery dissociated from the unbelievable reality that the man she loved would never walk through the door again. With no mental picture of what had happened, or why, Jeanne was now left in a vacuum, emotionally wasted from expelling her grief and cursed by unrelenting nightmares that came in the night.

Doug and Patti Campbell did their best to ease Jeanne's pain, but ultimately they had to leave her alone with her fears and the horrible task of looking into her daughter's eyes, full of innocence and perfect love, to explain she would never see her father again. Too young to understand, Marisa couldn't grasp the concept that she'd been condemned to a fatherless childhood. The memories of the Daddy she knew—who played with her for hours and swung her around and took her to the zoo—those memories would eventually fade in time, much like the fairy tales he'd once read to her from long ago.

While Jeanne was explaining this to her daughter, Richard's body was still in Scotland. He was under an autopsy blade and she decided there would be no memorial service until he was home.

And then the package from Scotland arrived.

When the contents of Richard's yellow canvas duffel bag were dumped out onto the floor, the musty smell of his clothing instantly filled the room. Among the pile of dirty work clothes were a few books, a drafting compass, a ruler, drawings of their future home, Richard's passport, an empty billfold—and a blue notebook. Jeanne recognized the diary immediately.

She opened it and flipped to the beginning of Richard's last tour, July 19, when he'd made an entry at the airport in Los Angeles. His handwriting brought pain to her heart, but it was the words he wrote that caused her eyes to burn and fill with tears. She'd always known Richard loved her, but to read the affirmation of that love, to know in raw and absolute terms the anguish he suffered when he was away from her, was both precious and agonizing to her.

The next page showed a series of logged entries dealing with work: what Richard saw and did, his arrival on board the *Wildrake*, the surface dives he made, the day he went into sat (listing Skip G. as his bell partner), the duration of his bell runs, and complaints about no hot water, no lockers, and no tools to work with. Some notations Jeanne couldn't fully comprehend. Richard used unfamiliar words and phrases like SALM, stabbing a tow shackle, and pulling a haul-down wire through. And on August 5, he mentioned something about "Big Wednesday." This would later turn out to be the last movie he viewed through the chamber port, a Jan-Michael Vincent film about California surfing.

On August 7, the day of the fatal dive, Richard had scribbled "WOW"—shorthand for Waiting On Weather. When Jeanne turned the page the short entries disappeared. The next two pages were filled with her husband's longhand as though he were writing a story. But Richard wasn't writing a story; he was expressing his core thoughts and feelings about his safety, his family, and about an unforgiving industry he felt shackled to. They were words of horrifying prophesy, and for the first time in her life, Jeanne realized that the man she knew to be experienced, competent, and quick thinking was afraid for his life.

> 7 August—On location at Thistle. The boat heaves a lot. Roger says there are conger eels all over. We will find out today. Poor topside management. Two guys in here are both nuts (ungood nuts) and dear god, I want out. I've really got to scare up another job after this one. I'm no longer impressed. They have made no effort to get out [my] gear from the Gulnare—anchor up to dive first. Dick around in the swell for 24 hours. I sure would like to have my white turtleneck sweater and, of course, some ear drops. Not impressed. Too many Brits/political hassles. It just leaves my stomach twitching. I miss my little one. It really does carve a hole right out of me. I'm still lost after all. I wish all were more solid and happy. Oh god, please help me to exercise my talent and will to pull out of it. I don't even know if I'm gonna get out of here alive. I never know. It's dangerous and unhealthy. All for $$$. Marisa would never understand. I don't either.

I always say, just that little bit more, and I'll be ahead enough, but at home I forget, get lazy and next thing you know I'm begging for it again. I must grow up and stop this madness and be with my three girls.[a] How I love them all.

And then, because Richard was feeling the need to switch to a safer career, he ended the passage with a list of new career paths he was hoping to pursue when he got home.

Welder/fabricator

Record a hit

design an invention

Paint poetry

go back to sensitivity

Maybe this time.

With tears streaming from her eyes, Jeanne turned the page. Richard had written "8 August" for the next day's entry, but the page was blank. After that, Jeanne's nightmares grew worse.

At 3 o'clock in the afternoon on August 18, an open-air memorial service was held for Richard at San Marcos Trout Club, the small private community where he was first introduced to Jeanne. The radiant sun and the panoramic view of the Pacific Ocean stood in stark contrast to the reason for the solemn gathering. A loudspeaker system was set up on the promontory where Richard and Jeanne once lived in a forest service lease house. One by one, Richard's friends came to the microphone to say good-bye to the Renaissance man who never lost sight of the person he was and who he wanted to be. Some, like Mary Ellen Fletcher, still paralyzed with grief, found it impossible to hold herself together or comfort Jeanne through the grieving process.

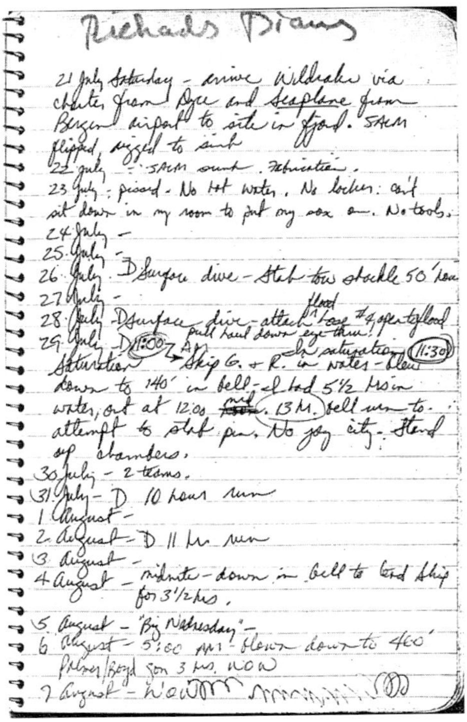

Richard's Diary

Courtesy of Jeanne Walker

a Jeanne, Marisa, and the family dog, Jozie.

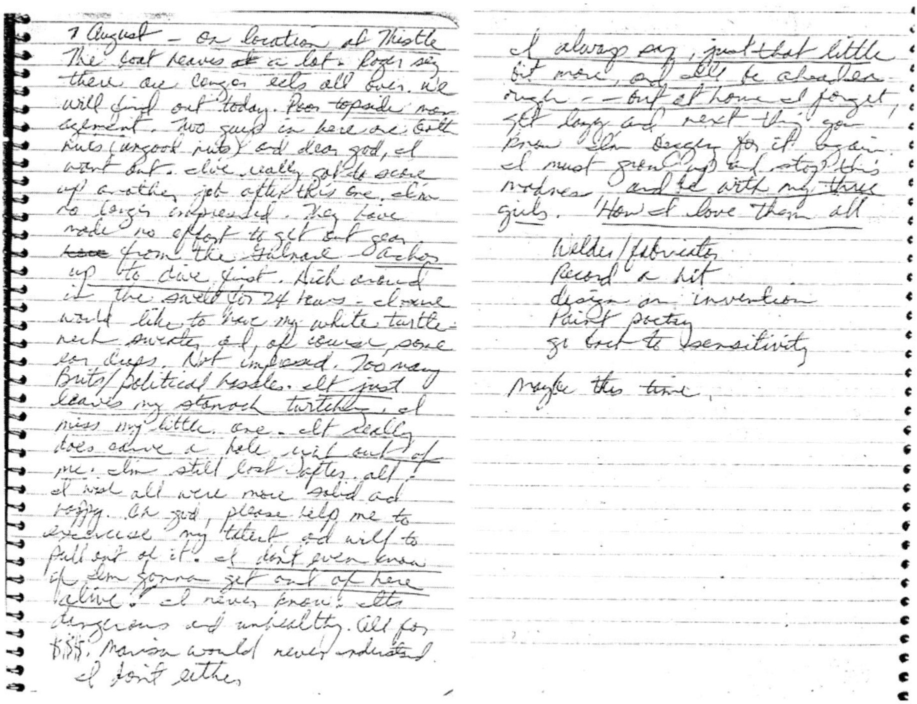
7 August – On location at Thistle. The boat leaves a lot. Roger says there are conger eels all over. We will find out today. Poor topside management. Two guys in here are both nuts ([illegible] nuts) and dear god, I want out. I've really got to secure up another job after this one. I'm no longer impressed. They have made no effort to get out gear from the Galveston Anchor up to dive first. [illegible] around in the sweats for 24 hours – I sure would like to have my white turtle-neck sweater, and, of course, some ear drops. Not impressed. Too many Brits/political hassles. It just leaves my stomach twitching. I miss my little one. It really does carve a hole right out of me. I'm still lost after all! I wish all were more solid and happy. Oh god, please help me to exercise my talent and will to pull out of it. I don't even know if I'm gonna get out of here alive! I never know! It's dangerous and unhealthy. All for $$$. Marisa would never understand. I don't either.

I always say, just that little bit more, and I'll be clear enough — but at home I forget, get lazy and next thing you know I'm begging for it again. I must grow up and stop this madness and be with my three girls. How I love them all

Welder/fabricator
Record a hit
design an invention
Paint poetry
go back to sensitivity

Maybe this time.

Among the seventy grim-faced mourners were Jeanne's parents, Doug and Patti Campbell, Doug's parents (Foster and Dorothy Campbell), Ron Walker (Richard's stepfather), Tom Belcher, Rod Pelzel, Ted Roche, Craig Roberts, and Brian Masterson.[2]

When Masterson appeared in the crowd of mourners that day, stories and rumors concerning his role in the accident had preceded him through the grapevine. Doug Campbell had received Richard's August 6 letter by that time, and, like a voice from the dead, it echoed his concerns about the *Wildrake* and his desire to quit the job. At the same time some of Richard's friends in the diving community had been asking hard questions about the rescue and puzzling over why it took so long to recover the bell.

Meanwhile, Jeanne had been making her own inquires and hearing a lot of excuses that didn't add up. Separating fact from fiction was difficult for her, but one haunting theme refused to go away: the rescue effort had been bungled. Mingled with the excuses were Richard's last words. For him to write gut-wrenching statements about his safety, and for some of the people close to the event to brush her off with responses like, "Nobody killed him; he just died,"[3] was simply unacceptable to her. When she pressed Craig Roberts for details she found him evasive and unable to speak. He became upset by her questions and his voice shook as he generally defended Masterson's actions.[4] Craig had been Richard's classmate and a family friend, and yet he was not coming forward with plain talk to calm Jeanne's anxieties and lessen her fears. If anything, it fed her suspicions that Infabco had done something wrong and was now trying to cover up what it did.

So Jeanne found herself on the other side of a drawbridge raised by a closed society that possessed all the facts to an event that took Richard from her. Like so many young widows before her, it would take her years to learn the truth; years that would drag her through four court cases and a kaleidoscope of lies and disinformation to knock her down, to make her go away, and to keep her from learning why Richard died that day. But this twenty-four-year-old mother was not about to play the shrinking violet and let the lies and the cover-up reconcile the last words of her dead husband. If anything, the tortuous journey that would be demanded of her would only galvanize and strengthen her resolve to go to war over Richard's memory.

The conclusion of that fight was nearly a decade away. On this particular day, high up in the Santa Ynez mountains under the glare of an afternoon sun, Brian Masterson was paying his respects, expressing his condolences, and doing the official thing. Doug Campbell, who would become an instrument of Jeanne's quest to find justice, sensed the tension in the air.

"Jeanne didn't want to have anything to do with him. She didn't even really want him there, but nobody wanted to make a scene either. And there was no confrontation with Masterson, but there was certainly an undercurrent that this guy was to blame. And there were a lot of divers there who felt that way....But I think everybody was still in the grieving stage at that point. And it was just like, 'Well, we'll deal with that later—deal with Masterson and what happened and all that later.'"[5]

On the day Richard was being eulogized in California, Skip was being laid to rest at St. Thomas Cemetery in the heart of West Springfield. The funeral procession from St. Ann's Church was huge. More than 200 people gathered to honor the man who "didn't have an enemy in the world," and who, for so many years, added meaning to their lives.

For the Guiel family, waiting for the return of Skip's body had been horrible, especially for Sheila and Roger who expected to say good-bye to their brother one last time before they buried him. This would have given them the "true feeling" that he was really gone. But the funeral director had told them not to look at the body; there had been too much damage from the rapid decompression and the autopsy. So the coffin went into the ground that day without Sheila or Roger ever fully accepting that Skip was dead. It was a decision that cast them into a kind of purgatory for the rest of their lives.

Somehow, Skip always knew the sea would take him. Like most divers, he rarely discussed the close calls or the dangers of his job with his family. But on one occasion, he explained to his brother-in-law, Rick, what every sat diver recognizes as his greatest worry: topside makes mistakes. Skip knew that the majority of divers killed in the North Sea did not die by their own hand. And in making that point, he turned to Rick and said, "If I die, somebody else is going to kill me."[6]

Skip's death was hard on his friends, especially Joey Graveline. For Joey, twenty-five years of history were shoveled over that day. From building cardboard forts in the tall grass to working on cars, his memories of Skip would never be nourished by new ones. There would be no more surprise visits from his friend standing bigger than life in his Navy pea coat at his doorstep; no more chances to make him laugh; no way now to reassure the little boy who wandered into his backyard so long ago, that they would always be together. Joey knew that the best friends in life are the ones you grow up with, but now he had to find some way to cope with this terrible loss and go on without him. Not long after the funeral, someone made a toast to Skip, and Joey had to leave the room.

"To this day I find it very difficult to go to his grave," Joey says, with tears in his eyes. "I think I've probably been there, in the 17 years, seven or eight times. Just find it very difficult to go there—not being able to see the body—it's very difficult to admit that he's gone."[7]

Draped across Skip's casket was a large red and white diver's flag. Conrad Rivest, Skip's dive instructor, had brought it to the service. To him, the flag represented his protégé because this was the same flag they'd used on their dives together. After the funeral, Sheila and her mother took it down past the locks and threw it into the Connecticut River so it would flow into Long Island Sound.

Infabco divers Chris Roper and Bob Gemmell attended the funeral, too. They'd brought Skip's body back from Scotland. Throughout the rescue attempt and during the funeral, Bob had kept his emotions under tight control. He shed no tears. This silent mourning continued for several more months; and then one day, the ticking time bomb exploded. He and Maggie were talking to relatives at the kitchen table, when the conversation turned to the accident. All of a sudden, memories of Skip came flooding back in a single, overpowering instant. "Man, I just lost it," Bob says. "I mean, I wept like I lost my brother."[8]

After that, he made one more trip back to Thistle and discovered he just didn't have it anymore. "I didn't have the enthusiasm. I didn't have the desire. All the glamour, all the excitement, everything was gone." Bob had followed Skip to the North Sea and then to Infabco. He'd gone into the industry blind and made it this far. But Skip's death represented a bad omen, and the *Gulnare* team wasn't what it used to be. He says the supervisors wouldn't exercise discipline and run the wankers off; they'd give them a second chance. And so he lost faith in the men who were supposed to protect him. With a wife and two young boys, he couldn't take the chance anymore. It was time to get out of diving and go back to machining, and so he did.

Today, when Bob thinks of those days, he thinks of the good times—taking classes at dive school, diving on the weekends in California, and living in Stonehaven. He remembers the small, but deeply personal things Skip did for his boys, like taking them for walks, giving them motorcycle rides and making presents for them. They were the good times and now they've left Bob with a wish that can never be fulfilled—he wishes Skip was still alive.

For Chris Roper, Skip's death destroyed all the mental barriers he'd constructed to deal with the rigors of the North Sea. Having witnessed the abrogation of authority during the rescue caused him to internalize Skip's death and mobilize him into action. Determined not to let the true circumstances of the accident be swept under the rug, he set about to help the families come to terms with the facts.

On the evening of the funeral, Roper met with Sheila, Rick, and Roger and explained in detail how the rescue operation went wrong and who was to blame. He handed over a copy of the *Wildrake* dive log, photographs from the accident scene, notes from the bell van in Brian Reid's handwriting, and copies of the external and internal bell check lists. He didn't know it then, but in time, this information would become invaluable to Sheila's legal representatives.

The next day, at 10:30 in the evening, Brian Masterson and John Nadeau showed up on Sheila's doorstep to offer their condolences.[9] Nadeau, who lived in Rhode Island, had called earlier that day to say he was bringing Masterson over after his arrival from California. After listening to Chris Roper's account, Sheila sensed her conversation with Masterson was going to be one of those moments in life that begged to be memorialized on tape, but Roger and Rick wouldn't allow it. They felt he would answer more freely if the conversation wasn't recorded.[10]

For the next five hours, they gave Masterson the opportunity to explain what happened and why Skip died. Masterson did most of the talking while Nadeau played a supporting role. None of the family members had any idea that the superintendent sitting before them had modified the bell and authorized the disabling of a system that could have saved Skip's life. Nadeau wasn't volunteering any information. He just sat there and kept quiet. "As far as we knew," Sheila says, "he just came up from Rhode Island because he knew Skip from diving."[11]

As the hours wore on, the gap between Masterson's version of the events and Chris Roper's got wider and less convincing. Sheila, Roger, and Rick were given the impression that the accident was "nobody's fault," that "everything was done according to the right way and couldn't have happened any other way," and that Skip's death was simply the result of "a freak accident."[12] They countered with Roper's version and eventually emotions began to heat up as Roger and Rick became irate over Masterson's denials. Finally, at 3:30 in the morning, the two visitors left.

It would be another two years before the Guiels learned the depth of the violations and the level of incompetence and mismanagement that caused the unbearable pain in their hearts. Only then would Sheila recall the meeting with Masterson and Nadeau and realize that everything she'd listened to that night "was just bullshit," that the two men "were covering their butts."[13] Only then would Roger say, "Had I known what I know now, they wouldn't have made it out of there."[14]

Two days later, on what would have been Skip's twenty-ninth birthday, the skies blackened and hurricane-force winds swept through West Springfield tearing the roof off a local K-Mart store.

Roger said it was Skip "kicking his way into heaven."[15]

Photo by Author

Skip's grave site.

Chapter Twenty-one

I Can't Buy Richard's Life Back

[Jeanne] was put in a situation where she really had no choice....She had to get on her white horse—her white charger almost—and do battle with the forces of evil for her husband's memory, if for no other reason than for her own sense of well-being. She just could not let this go by. She had to fight the fight....She just had no choice.

—— Doug Campbell[1]

Almost immediately after Richard's death became public knowledge, personal injury lawyers began calling Jeanne, pressing her to begin legal action right away. But Jeanne didn't want to speak to a bunch of ambulance-chasing attorneys. She was still in shock and still uncertain whether Richard's death warranted a law suit or not. All she had at the moment was a strong feeling that her husband had died because of something that was being kept hidden from her.

So Doug spoke to the lawyers, and like members of a Sunday chorus, the song they consistently sang was: you'd better get an investigation moving in Scotland before the facts are lost, before witnesses disappear, and before people have a chance to concoct a story.[2] Doug advised Jeanne not to make any hasty decisions regarding attorneys, but he could see that, more than legal advice, Jeanne needed to know the truth about what happened to Richard. By coincidence, Doug's mother and father were scheduled to meet with their company lawyer, Bob Parker, on a business trip to Switzerland just days after Richard's memorial service. So Doug approached his father, Fos, with the idea of soliciting Parker's help in conducting a fact-finding mission on behalf of Jeanne. When Doug and his father explained the situation to Parker, he readily accepted the assignment, and at his own expense. Several days later they left for Europe.

While Parker and the Campbells were gathering information overseas, Jeanne began to join forces with the family of the man who died with her husband.

She learned that "Skip G.," listed as Richard's bell partner in his diary, was Skip Guiel from West Springfield, Massachusetts. She obtained his sister's phone number, and one night in early September, she called Sheila. The conversation began awkwardly, but it quickly developed an electric quality to it when the two women compared notes and realized their suspicions about the accident were shared. Sheila related her discussion with Chris Roper and the meeting she had with Masterson and Nadeau, while Jeanne revealed the contents of Richard's last diary entry and the rumors about the bungled rescue mission. Together they had enough suspicion to accuse Infabco of taking the lives of their loved ones.

This was Jeanne and Sheila's first contact. And over the next twenty months they would unite to become what Sheila's husband called the "Dynamic Duo," speaking on the telephone, sharing information, never meeting, never putting a face to each other, but always profoundly aware of what the other woman was feeling as the pain they endured became immeasurable and mutual.

In late September, Jeanne wrote the first of many long letters to the Guiels describing how her trust in people was "rapidly diminishing." She was unsure if Roger and Sheila even wanted to discuss the accident, but she had to talk to someone and the Guiels were the only people in the world who could possibly understand her pain and suffering:

> I hope I am not upsetting you further by discussing what must be so painful to you too. Please let me know if you would prefer I not continue to discuss it. My feelings are not likely to improve for many years, and I don't even foresee it then.
>
> In the meantime, I am becoming obsessed with finding the truth, and, I suppose, the guilty parties if such exist. And I must see this never happens again. And I am beginning to search for a justice to be done, although nothing will ever be right again.[3]

On October 2, Jeanne sat down with Bob Parker and Fos Campbell to hear the results of their investigation. Parker turned over his 24-page report outlining his meetings with officials of British National Oil Corporation, insurance brokers, legal experts, diving managers, Brian Masterson and the man who would ultimately prosecute his company. Of particular interest to Jeanne was their meeting with Malcolm Williams, co-owner of the diving company that tried to save Richard. Williams told Parker that, of all the diving fatalities he was familiar with, this was the first that "shouldn't have happened."[4] There were too many anomalous circumstances surrounding the tragedy. He questioned why the bell separated from its lift wire in the first place? And why was there a double shackle arrangement on top of the bell? Williams told Parker this wasn't just unusual, it was worse than strange because everyone in the industry knows that shackles wear against each other under dynamic loading. He said there were photographs to document this.[5] And then Williams gave his visitors a brief description of the rescue culminating with the odd behavior of Brian Masterson.

Sometime between 4:30 and 8 o'clock that morning, the *Welder* specifically asked for back-up from the *Uncle John*,[6] but for reasons which were not clear to Williams, Masterson made the following decisions:

1) he blocked the request for back-up,
2) he insisted on using the *Wildrake* crane to recover the lost bell, instead of allowing the rescue vessel to make the recovery, and,
3) although Masterson knew the crane driver didn't have a vertical purchase on the *Wildrake* bell, he insisted on starting the lift without first allowing the rescue diver to verify it was off the seabed.

As a result, the bell snagged on the SALM base, and when the lifting sling broke, more critical time was lost in reattaching a new line. Williams emphasized that his divers worked to the limits of their physical endurance, but with the main umbilical damaged and no hot water getting to the bell, time simply ran out on the trapped men.

Even without a technical background, Jeanne knew this was devastating news and it confirmed her worst fears that Richard had been the victim of negligence and incompetence.

Parker also told Jeanne that the police and Department of Energy were conducting parallel investigations into the accident, and that the prosecutor in Aberdeen, a man by the name of Malcolm MacNeill, was waiting on the findings of these investigations to determine whether he would press charges against Infabco. MacNeill had said he saw nothing to indicate that Richard and Skip were at fault for what happened to them, which, to Parker, implied that Infabco must have furnished Richard with unsafe equipment or procedures to work with.[7]

Shortly thereafter, Jeanne shared Parker's report with Sheila. It was more bad news and every time the two women managed to gather another piece of the *Wildrake* puzzle, the physical picture of Richard and Skip's last hours grew darker in their minds.

For Sheila, Skip's death had given rise to a kind of emotional paralysis that would keep her numb and disconnected from life for years. Never viewing the body, there was nothing for her to come to terms with. In one sense, Skip had become a missing person, an apparition that haunted her consciousness. There was no escape, no drug, nor remedy strong enough to kill the pain. Every waking day she was forced to cope with the elemental conflict between that part of her mind which managed reality and the irrational part that still expected him to emerge from the diving bell. She knew Skip was never coming home, but the utter conviction of that certitude didn't make the suffering go away or allow her to move on. Instead, it pushed her further away from herself. Peace of mind—if it would ever come—was years away.

In one respect, Sheila was more fortunate than Jeanne: she had the love of her husband Rick, the rich vitality of her Italian heritage, and dozens of friends and relatives to help her through the trauma. Jeanne, on the other hand, was virtually alone. She'd never been particularly close to her parents, and now half her life had been reduced to a handful of ashes scattered across her property on Hollister Ranch.

Jeanne didn't know it then, but she was slowly dying inside. Her death wouldn't happen in six months or a year, but in time the woman who married Richard Walker would disappear into a colorless world where anger and bitterness embraced years of depression and distrust. The smile and the laughter and the joy that once filled her heart would be replaced by an overwhelming urgency to wage war against the company that killed her husband. Roger and Sheila would fight this battle, too, but Jeanne would do it with such intense ferocity that it would become obvious to them she was on a private crusade. Infabco had ended her marriage; now she vowed to bring an end to them.

Across the Atlantic the *Wildrake* accident had become a public relations nightmare for Brian Masterson. Accounts of the tragedy began popping up in industry magazines, with one startling editorial appearing in the October issue of *Roustabout* that added an ugly twist to the story. The headline read:

Story that took five attempts to write

CONTINUING SAGA OF BNOC BRIBES PROBE[8]

The article described how a leading Sunday newspaper, a financial weekly journal, BNOC's security division, and the local police were looking into allegations that Brian Masterson had bribed BNOC officials in order to win his contracts in the Thistle Field. This rumor, which apparently had been floating around Aberdeen for some time, had suddenly grown legs due to the investigation into Richard and Skip's deaths.

According to the article, when Masterson returned from the United States, BNOC security men were waiting at the airport to question him, and still later, Grampian Police invited him to headquarters to ask how he had obtained the contract for the SALM re-installation.

The police were unable to prove any wrongdoing and Masterson was released, but the article revealed that "several other oilmen" were implicated in the scandal and that, at the end of September, at least one of them was "hurriedly removed" from BNOC on direct orders from Chairman Lord Kearton himself. When BNOC was questioned about the allegations of corruption and the sudden departure of its employee, the oil company issued a statement claiming that the man's contract "had expired" and that he was due to leave the company anyway.

As a pro-oil publication, *Roustabout* did its best to defend Masterson, emphasizing that he deserved the benefit of the Scottish verdict "Not Proven" (vis-à-vis Not Guilty). "The real culprits," the article said, "are the faceless men who level these 'charges' and immediately scurry back into the woodwork."

And then the magazine issued a public challenge to Masterson's accusers:

> We don't say Brian Masterson is innocent: we don't say he's guilty. Firstly because we cannot pursue the depth of investigation which the 'charges' merit. Secondly, because we don't know the accusations. And, finally, because we do not know the accusers.
>
> We DO invite the faceless men to tell us their tales...backed up with facts! If there is a fire let's find it: if there's not, pack it in!

But the suspicions about Masterson and his true relationship with BNOC would never go away. Two years later, they would resurface with even greater clarity in a British documentary entitled *The Last Dive*.

In the United States, Jeanne and Sheila had decided to initiate separate wrongful death lawsuits against Infabco. Sheila retained a lawyer in Springfield, who engaged a New York law firm with an office in London.[9] Meanwhile, Jeanne selected the Law Offices of Gerald Sterns in San Francisco.

Sterns, whose main practice was limited to personal injury and wrongful death litigation, had made a name for himself handling high-profile international catastrophic accidents, particularly those in the aviation industry. In 1974, a fully loaded Turkish Airlines DC-10 took off from Orly field in Paris. Nine minutes later, Flight 981 slammed into the Ermenonville forest 23 miles away killing all 346 passengers and crew. No other air crash in history had produced so much carnage. When it was later discovered that a defective cargo door had blown off during the flight, Sterns represented about 80 English families against the American aircraft manufacturer, McDonnell Douglas, and General Dynamics, the subcontractor who made the door. In a landmark decision he was able to bring what was essentially an overseas case into the US court system and prevail.

In early November, Jeanne wrote to Sheila about her decision to hire Sterns. At this stage she was beginning to see that Richard's life was being reduced to the cold-hearted business of collecting dollars.

"I am learning something about the way [lawyers] operate," she wrote, "and really dislike some of them." One lawyer Jeanne interviewed told her, "Quite frankly Mrs. Walker I usually handle cases where there are 3 or 4 children involved and a husband who was earning much more."

But Sterns seemed "genuinely concerned" and didn't come across "like a money-hungry arrogant funeral director," she wrote. Jeanne suggested Sheila consider having Sterns take her case as well, and then she remarked on what she hoped the law suit would achieve:

> So, [Marisa and I] will pursue this to the end now, and it is a burden lifted to have put it in other hands. I have found I am not capable of doing as much,

> even legally, having access to the information an attorney could get. It does mean though (I think) that most of what is done is at his discretion. The suits filed are organized and written by him; the decision on settlements and damages are now his, and not mine. At least partly maybe that is for the best.
>
> I think the hardest part for me is learning that mostly these big companies (oil etc.) think hardest about their financial statements, not their moral obligations, not human lives, and so you must fight them for money which seems terribly wrong to me. I can't buy Richard's life back. They cannot make up for what they took in any way, and they'll never hurt me either. And really I'd rather be able to attack someone, but how do you scratch out the eyes of a corporation? So, there are lawyers to do that (I guess), and they'll try to get them where it hurts, and if it hurts enough they'll see accidents are more expensive than research, training, safety precautions and it will be less likely some other men will suffer so much as they....
>
> It seems that very few lawyers would want to try a law suit in Britain, but if jurisdiction could be established in the States they might be far more interested. I have a feeling that Sterns feels fairly confident that can be done—he has not said so, of course, but I suspect that is a very big consideration.
>
> Anyway, perhaps you are no longer thinking of legal action, I realize [your] situation is very different. Partly because for us the financial problems are very direct—all the commitments we made and plans etc.—were depending on Richard's salary. We are really forced to either sell what we have or sue them. They (Infabco) have changed their corporate mind about helping us—which they originally said they'd do—and they never got any life insurance for their employees—so we're pretty much forced to do this on that level, not to mention my need to feel I can somehow punish them, and perhaps protect other men.[10]

That December Sheila wrote to Sterns to ask if he would be interested in merging the two cases to which Sterns replied that he was.

"The combining of the cases with one attorney," he wrote, "will probably result in a saving of expenses to all concerned, and this can be significant since the expenses in matters like this can be very substantial. We normally handle these cases on the basis of a contingency fee and advance the expenses necessary in the investigation and prosecution of the case."[11]

But it would be another seven months before Sheila and Roger would decide who should carry Skip's case to its final conclusion.

When Bob Parker and Fos Campbell met with Procurator Fiscal Malcolm MacNeill in Aberdeen several months earlier, neither man understood what the title meant nor that Scotland and England had separate and distinct legal systems.

The origin of MacNeill's office dated back to the Middle Ages, when the king asserted his power in the outlying burghs by vesting a portion of his authority in the "shire-reeve," or sheriff, who maintained law and order in the region, collected royal taxes, attended to military affairs, and served as both prosecutor and judge in judicial matters.[12] As the number of sheriffdoms spread, so, too, did the king's influence.

By the sixteenth century, the office of procurator fiscal had been established. Appointed by the sheriff, the fiscal's job was to assist in prosecuting criminal offenses and collecting fines which were paid into a special fund called "The Fisk."[13] As time passed, he took over the duties of prosecutor entirely while the sheriff retained his role as judge. During the mid-1800s, control over the fiscal shifted from the sheriff to the Lord Advocate, the chief legal officer in Scotland. In 1907 Parliament removed the remaining political link between the two offices when it placed the sheriff's right to appoint the procurator fiscal in the hands of the Lord Advocate.[14]

At Inverlair House in Aberdeen, Malcolm Torquil MacNeill had been in charge of his office for the past ten years. Similar to a district attorney in America, he was responsible for investigating and prosecuting all crimes committed within his jurisdiction, an area which included the entire northern UK sector of the North Sea down to the border between Scotland and England.[15] In addition, he was required to investigate all sudden, suspicious, and unexplained deaths and to present evidence relating to the fatality before a sheriff at a Fatal Accident Inquiry. It was due to these responsibilities that MacNeill was knee-deep into the task of deciding whether or not to prosecute Masterson's company.

His file on the *Wildrake* accident was thick with reports from the Department of Energy, Grampian Police Department, and eyewitness accounts from the *Stena Welder* and *Wildrake* divers. Commander Warner and Roy Giles of the DOE had personally traveled to Aberdeen to discuss the severity of the violations. Violations which MacNeill would later say were "being painted in black and white by the police and in color by Jackie Warner and Giles."[16]

Warner's views notwithstanding, MacNeill believed the evidence before him supported a *prima facie* case of criminal liability against Infabco Diving Services.[17] And so in the spring of 1980, this mild-mannered public servant quite literally found himself at a well-worn legal juncture. Under Scots law he had to decide whether to prosecute the diving contractor by either Indictment or by Summary Complaint—two distinct legal procedures that would enhance or reduce the punishment inflicted upon Infabco if it was subsequently convicted.

Of the two options, summary procedure would be the more expedient path and much less injurious to the defendant. There would be no jury trial. Masterson's attorney would enter a plea to the charges—either guilty or not guilty. In more than eighty percent of the cases tried in summary courts the defendant pleads guilty.[18] And if Infabco followed that trend, MacNeill knew that Masterson would walk away after paying only a token fine of no more than what the statute permitted, in this case £400.[19]

But if MacNeill chose the other route, prosecution by Indictment, and Infabco contested the charges, then the case would go before a jury. If the jury found the company guilty, then the sheriff could impose a discretionary fine of any amount.[20] This was the extent of the law. There would be no potential for anyone to go to jail unless MacNeill named the person directly in the indictment and obtained a conviction.[21] In that event, the sheriff could assign jail time for up to two years, impose a discretionary fine, or both.

In the ten years as procurator fiscal of Aberdeen, MacNeill had never prosecuted a diving company by Indictment before. All his previous cases had been by summary procedure. But this time the evidence was overwhelmingly in favor of spending the added time and public expense of bringing Infabco before a jury. This is what MacNeill wanted, this is what the DOE wanted, and this is what the violations demanded. But the decision to conduct a solemn proceeding (trial by jury) was not his alone to make.

In Scotland, all cases for Indictment must be referred to the Crown Office in Edinburgh for review by members of an elite legal staff known as the Crown Counsel.[a] This process of "seeking Crown Counsel's instructions" allows the Lord Advocate and the Crown Counsel to "in effect shape what constitutes the criminal law in Scotland without recourse to parliament or the judiciary."[22]

So MacNeill bundled his evidence off to Edinburgh to await their decision. Crown Counsel would either concur with MacNeill's recommendation, bump the case down to summary procedure, or even direct him to *no pro* the case (not to proceed).[23] Whatever the decision, it would be binding on the fiscal.[24]

But Edinburgh agreed with MacNeill, and by mid-June 1980, he'd written to Jeanne and Sheila informing them that the Crown was going to prosecute Infabco.[25] MacNeill explained there would be two important court dates. At the first appearance, called the "pleading diet," Infabco would enter a plea to the charges. If the company pled not guilty, then the Court would be adjourned to the "second diet" or trial. If there was a plea of guilty, then the fiscal would enter a statement of facts for the prosecution, and the defense would enter a plea of mitigation of penalty, with the sheriff subsequently deciding the extent of punishment. In any event, MacNeill promised to let Jeanne and Sheila know when the case would be called into court, and then he wrote that, if the circumstances of Richard and Skip's deaths were not made clear at the criminal trial, he would petition the Court to conduct an inquiry.[26]

In early July 1980, Sheila paid Gerald Sterns a retainer and signed the authorization papers allowing him to represent her and Roger and their mother against Infabco. Skip's father was now out of the family picture. Shirley had filed for divorce just nine days before Skip was killed.[27]

Barring any settlement with Infabco, Sterns was prepared to take his case into court. Infabco's liability was unmistakably clear, but not altogether shocking to the San Francisco lawyer. He'd seen plenty of these screw-ups before. "It's the

a Crown Office approval is not required for summary prosecutions.

way the system works," Sterns says. "People are trying to make money, they cut corners on safety and poor bastards like Richard Walker pay the price."[28]

Unable to bring the victims back, Sterns goal was to try to collect the greatest amount of compensation from the people responsible to make the relatives' lives as comfortable as possible. But Sterns knew from experience that human life was worth very little in terms of dollar recovery in Britain. The original Fatal Accidents Act of 1846 allowed the courts to compensate dependents when the breadwinner of the family had been killed, but it did not provide damages for mental anguish of the survivors or for non-economic losses such as those associated with the loss of care, comfort, guidance, nurturance and protection—all the intangible things that make up a relationship. These so-called "loss of society" damages were not recognized in the UK. Awards were based solely on the ability of the dependents to prove actual financial loss.

Worse yet, Sterns knew that non-dependency cases, such as the Guiels, had virtually zero value in Britain. "Under British law, you must have a dependency situation to sue for wrongful death. You have to have someone supporting somebody, which of course we had in Walker, but we did not have in Guiel."

The underlying principle of this view is that the law should not allow people to profit from someone else's death. After the 1974 Turkish Airlines crash, author Moira Johnston wrote a riveting account of the disaster in *The Last Nine Minutes* in which she cites a BBC exposé on the value of non-dependency cases resulting from air crashes. The niggardly amounts awarded by the UK courts could not have been less comforting to the relatives of the victims:

> A father had received only five hundred pounds—roughly $1,200 dollars—for the death of his single son, a typical payment for a non-dependency case...A man who had lost his wife had received three thousands pounds. In a crash near Manchester, where seventy-two people had been killed, more had been paid by insurance companies for the lost plane than for the people who had died.[29]

Even in cases where dependency was established, compensation was shockingly low. "A widow with two children had received five thousand pounds," Johnston points out. "How long could a widow live on five thousand pounds?" she asks. Moreover, in order to prevent overlapping of damages, the courts factored inheritance into the calculation of an award. Thus, children who lost both parents saw their settlements reduced by the total value of the assets they could expect to inherit.

Britain was not where Sterns wanted to take his case. And this was the challenge he faced: although Scottish jurisdiction was evident, he had to find some way to clamp the long arm of US law onto Brian Masterson and drag him back to America to answer for the deaths of Richard and Skip.

On July 30, 1980, Sterns filed wrongful death complaints on behalf of the Walker/Guiel families in the Superior Court of Santa Barbara and in the United States District Court in Los Angeles (the federal court system). Throwing every basis for jurisdiction into the 16-page complaint, he cited General Maritime Law, Article III, Section 2 of the United States Constitution, Federal Common Law, the Jones Act, and the Death on the High Seas Act, arguing that by recruiting Richard in Santa Barbara (via its agent, Craig Roberts), the relationship between Richard and Infabco had been founded in the US, and that Infabco had in effect been doing business for profit in the United States, and therefore should be held accountable on US soil for the harm it had caused.

The defendants[a] were liable, the complaint read, because they were negligent, careless, and in violation of their duty to provide Richard and Skip with a safe and seaworthy vessel, a competent crew, and necessary and appropriate rescue services. Furthermore, the defendants knowingly placed into service a diving system which was used without prior inspection for defects. As a consequence, Richard and Skip were required to perform their duty under "unsafe conditions and circumstances." The absence of the prior inspection, and the fact that the diving system was "in a dangerous and defective condition," constituted the basis for placing the full weight of responsibility firmly on the shoulders of the defendants.[30]

The defendants "expressly and impliedly" represented the *Wildrake* and its equipment to be of good quality and reasonably fit for ordinary use, Sterns wrote. But "in fact, those express and implied warranties were untrue" and caused the deaths of Richard and Skip.[31]

In conclusion, he charged the defendants:

> were guilty of the most outrageous disregard for the health and safety of Richard Walker and Victor F. Guiel and of fraud, oppression, and/or malice, so as to warrant the imposition of punitive or exemplary damages against them for their conduct, by way of punishment and for the sake of an example.[32]

Above and beyond general damages, special damages, court costs, and for the conscious pain and suffering of the deaths of Richard and Skip, Sterns was asking for punitive damages of not less than $1 million for each of the deceased.

Each of the defendants was served with a copy of the complaint, and in time Sterns would learn who would respond to it.

The first anniversary of Richard and Skip's deaths came and went with no word from MacNeill about the trial date. Then, in late August, he wrote to Jeanne and Sheila informing them that Infabco would be tried in Aberdeen Sheriff Court on September 23, 1980. Two months later, he followed with another note stating that the trial had been postponed until December 15.

In a letter to Sheila in October, Jeanne commented on the new date, expressing her worries that, because the trial would be held so close to Christmas, the Crown might proceed more swiftly than usual or adjourn the proceeding for a long holiday:

---

a Initially ten defendants were cited in the complaint, but in the end, Infabco Diving Services and Møllerodden would receive the judgment of the US Court.

> At least the December 15 date has been given by both US Consul and London Times. It would be a good Christmas present to see IDS [Infabco Diving Services] etc. prosecuted.
>
> I do have an interesting and encouraging opinion about this summer's (past) IDS job. One of the divers told me it went absolutely horrible—that the company was deteriorating (lack of supplies, tools, organization, morale, etc.—many left or contemplated leaving early). So maybe they can be put out of business with some bad publicity and a verdict of guilty.[33]

Roger, Sheila and Jeanne had planned to attend the trial, but in a telephone conversation with MacNeill, Jeanne says the fiscal advised her not to attend. "We were told that our presence wasn't necessary, and we didn't need to be there."[34] On that advice, they canceled their plans to fly to Scotland. It was a decision Sheila would later regret. "Had we been there I know the trial would have gone completely different."

On November 28, 1980, at Aberdeen Sheriff Court, charges against Infabco Diving Services were read aloud before Sheriff Alastair Stewart:

> INFABCO DIVING SERVICES LIMITED...You are Indicted at the instance of THE RIGHT HONORABLE THE LORD MACKAY OF CLASHFERN, HER MAJESTY'S ADVOCATE, and the charges against you are that on 7th and 8th August 1979 in controlled waters at a position in the North Sea...

Thus began the four charges against Infabco Diving Services in the three-page indictment. On the first charge, which was the most serious, MacNeill alleged that Infabco had failed in its duty to provide a diving plant in good working order because (a) the clevis pin had come out causing the bell to become detached from its lift wire; (b) the purported secondary means of bell recovery (the umbilical and its winch) failed to function as intended; (c) the divers could not exit the bell while the bell was resting on the seabed; and (d) the drop weights could not be released from within the bell.

On the matter of the second and third charge, Infabco failed to notify and receive approval from the Secretary of State for conducting diving operations below 125 meters and using saturation diving techniques.

Charge four alleged that Infabco allowed Skip to dive without a valid diving medical certificate. These were all violations of the Submarine Pipelines (Diving Operations) Regulations 1976 and the Petroleum and Submarine Pipelines Act 1975.

Working on his birthday, forty-two-year-old Sheriff Stewart peered over his bench at Masterson's solicitor. James Tierney was a very competent and clever attorney who knew the law and vigorously defended his clients in court. Masterson had chosen well and was now expected to enter a plea on behalf of Infabco. But before he did so, Tierney launched an objection against the relevancy of the charge that dealt with the drop weights. MacNeill had inadvertently cited the wrong regulation in the indictment and defense counsel Tierney had picked up on the error.[35]

Sheriff Stewart, however, was not sufficiently swayed by the challenge and repelled the objection. At this point, Tierney entered pleas of not guilty for all four charges, and then, in what would later be dubbed the "Infabco defense," he dropped a bombshell on MacNeill by proclaiming that Infabco Diving Services was not in fact the employer of Richard Walker and Skip Guiel. In essence, Masterson was saying that MacNeill had charged the wrong company.

In putting forth what seemed to be an outlandish defense, Tierney drew the Court's attention to Section 67 of the Criminal Procedure (Scotland) Act 1975, which stated that if the accused is alleged to have committed an offense "in any special capacity," and enters a plea without challenging the Crown's assertion that it possesses that capacity, then the qualification shall be held "as admitted."[36]

Central to Tierney's defense strategy was the way in which MacNeill had worded the charges. Each charge had been prefaced with the words, "being the employer of." He did so because the regulations that had been breached pertained *only to the employer of divers*. Thus, "being the employer of" became the "special capacity" of the defendant which triggered Tierney's use of Section 67. If Tierney had entered pleas of not guilty without disputing the Crown's claim that Infabco was the employer of Richard and Skip, then Masterson would not have been able to turn around later and say to the Court that MacNeill had charged the wrong company, as he was doing so at the moment.

It seemed to be an absurd defense, but, as MacNeill would soon discover, proving the obvious would not be as easy or as straightforward as he'd expected.

Upon recording the not guilty pleas, Sheriff Stewart continued the case to the trial diet.

The next day, the Aberdeen *Press and Journal* headlined the event in an article:

Diving Firm For Trial After North Sea Deaths[37]

Chapter Twenty-two

The Trial

Nemo me impune lacessit

—— Motto of The Royal Arms of Scotland

Even from a quarter mile away, at the corner of Bridge and Union where a statue of King Edward VII stands overlooking passersby, Aberdeen's prison tower is clearly the most prominent structure on the horizon. Built in 1629, it pierces the sky 190 feet and marks the site of an old tollbooth where Quakers were once imprisoned in the mid-1600s,[1] and where a century later, in a scheme masterminded by local merchants and magistrates, it became the temporary residence of hundreds of children who were kidnapped off the streets and sold into slavery for the plantations in the New World.[2]

In 1868, a firm of Edinburgh architects added the present-day Sheriff Court and Townhouse to the tower section. Designed in Scottish Baronial style and built of Kemnay granite, the courthouse spanned an entire city block, and it was here, on the morning of December 15, 1980, that a legal battle was about to be fought in the austere surroundings of Courtroom No. 1.

It was a large windowless room, as tall as it was wide, and of the six courtrooms in the building, it was the least pretentious. With long rows of plain wooden benches on the ground floor and upper gallery, the decor was old fashioned, as though nothing had changed since the days of Queen Victoria. Along the back wall and wainscoted around the courtroom, dark, almost ebony-colored wood paneling stretched past the jury box on the right and press box on the left. A canopied witness stand stood elevated in the well of the courtroom where the clerk of court was busy making preparations to record the Minutes of Proceedings and where Malcolm T. MacNeill would press an Indictment before a jury against a diving contractor for the first time in his career. Regardless of the seriousness of the crime, all criminal cases appearing before Aberdeen's Sheriff Court were heard in the first instance in this room.[3]

Front and center of the judge's bench, the Accused Box stood empty. Infabco Diving Services was a legal fiction, not a person, and Brian Masterson, due to be called as a witness, was not seated in the courtroom. Across from the jury box, the press benches were packed with reporters from Scotland and one or two national newspapers. Graeme Smith, who reported on the accident 16 months earlier, was there, as well as Norman Harper from the Aberdeen *Press and Journal.*

Harper, with boyish good looks was a junior grade reporter on the verge of promotion to senior journalist. All his previous court reporting had been chaperoned, but on this particular occasion a senior journalist was not available and he was alone. As he waited quietly for the judge to make his entrance, he noticed that half the public benches were buzzing with spectators. Among the curious were representatives from the various diving companies who'd gathered to hear this unprecedented case, because they knew that what would happen to Brian Masterson's company would affect them all.

Had Harper taken notice, he would have seen one of the spectators holding pad and pen like himself. But the man wasn't a reporter, and he wasn't a member of the diving community, either. He was there on assignment for Gerald Sterns.

Colin Sanderson was a freelance public loss assessor and Sterns had employed his services in the past to provide background information on various claimants in personal injury cases, including the 1974 Turkish Airlines disaster. Rather than send one of his attorneys to the trial, Sterns hired Sanderson to be his eyes and ears and to notify him regarding the outcome.

As the Englishman fiddled with his notepad, lawyers began assembling in the well of the courtroom dressed in black gowns. Gathering papers at his table, MacNeill noticed that Masterson had engaged a new defense counsel, a man much older than James Tierney and wearing a black silk gown with falls and white horsehair wig atop his head. His name was James Law and he was a senior advocate, or what is also known as Queen's Counsel (QC).

In Scotland, the legal profession is divided into two principal branches—solicitors and advocates. Solicitors, which Tierney was a member, are allowed to form partnerships and conduct a wide range of general legal services such as buying and selling real-estate, drafting wills, creating companies, and processing deceased person's estates. They can appear before the lower courts (tribunals and district and sheriff courts), but they are not permitted to argue cases before the superior courts (the Court of Sessions and the High Court of Justiciary).[a]

Advocates, on the other hand, generally deal with complicated cases and complex issues of law. They are the virtuosos of the Scottish legal corps and are divided into two groups, junior and senior counsel. Not all junior counsel do so, but after about 15 years of high quality work, a junior counsel may petition to be appointed by the Queen to the rank of Queen's Counsel on the recommendation of the Lord Justice General.[4] "Taking silk," as this is called, then allows the QC to charge higher fees.

Unlike solicitors, advocates (whether junior or senior) are the only members of the legal profession that are permitted to argue civil cases before the Court of Sessions and serious criminal cases (murder, rape, and treason, for example) before the High Court of Justiciary. They receive their instructions through the client's solicitor and are forbidden to consult directly with the client unless the solicitor is present. They work alone, may not form partnerships, nor are they allowed to be compensated on the basis of a contingency fee.[5] And if the advocate commits malpractice either through "wrong advice in law, negligence, mistake, indiscretion, error of judgment, or the mismanagement of the case," he is not liable for damages.[6]

All advocates, like Olympian gods, are based in Edinburgh in a building known as Parliament House just off High Street behind St. Giles Cathedral. And it was there, during the summer of 1980, that James Tierney found the legal wizard he believed would extricate Masterson from his predicament.

As a boy, James Law had been schooled at Kilmarnock Academy just south of Glasgow. After graduation he matriculated at the University of Glasgow where he received both his Master's and law degree (LLB) and then served seven years as a member of the prestigious Crown Counsel in Edinburgh. In 1971, at the age of 45, he took silk, and, for a period of time in the mid-1970s, served as sheriff in the very courthouse where MacNeill was now prosecuting Masterson's company. MacNeill knew Law. He knew his reputation and his habits in court. In fact he'd conducted several cases before him in 1977,[7] and he knew that Law was familiar with the strategies diving companies employed to avoid prosecution.

---

a The distinction between solicitor and advocate is somewhat blurred today. In 1992 the ruling was changed to allow solicitors with more than five years of court experience to apply for 'extended rights' to appear in the superior courts. *Source*: *The Scottish Legal System* p. 237.

The clock mounted on the upper gallery read just after 10 a.m. when a paneled door behind the judge's bench suddenly opened and Sheriff Alastair Stewart, bewigged and dressed in black robe, stepped into the courtroom. An officer of the court immediately shouted, "Court, all rise," and everyone stood up. MacNeill and Law bowed to the judge, he bowed back, and then everyone took their seats.

MacNeill started the ball rolling by making a pretrial motion to amend the indictment on a minor point, and then, in order to give the accused one last chance, Mr. Law was asked to confirm his client's plea to the charges. Once again, a plea of not guilty was entered into the record. At that point, the process of balloting the jury began.[8]

The clerk of court drew 15 names from a large glass jar and they became the requisite quorum to hear the case. The first juror called was a sales representative, followed by a consultant supervisor and then a housewife, until seven women and eight men had taken their seats in the jury box.[9] A summary of the indictment was read aloud and then the jurors were asked to stand. As each of them raised their right hand, the clerk of court administered the oath:

> You fifteen swear by Almighty God, and as you shall answer to God at the great Day of Judgment, that you will truth say and no truth conceal, so far as you are to pass on this assize.[10]

All fifteen answered, yes.

At this stage, none of the jurors knew what the case was about and the sheriff did not instruct them on their duties before they heard testimony. MacNeill and Law made no opening speeches as arguments concerning the relevancy of specific evidence would come later.

In Scotland, a jury can put forth one of three possible verdicts—guilty, not guilty, and not proven. To convict, there must be a majority in favor of guilty. The difference between not guilty and not proven is basically technical. A not guilty verdict implies innocence and is more definitive, whereas not proven signifies that the jurors concluded the prosecutor did not prove his case.[11] In the event that MacNeill prevailed over Law, it would be in the hands of Sheriff Stewart to decide the severity of the punishment, not the jury, and MacNeill would not be allowed to make any sentencing recommendations.

The trial lasted four days. Testimony ranged from the technically mundane, when Sheriff Stewart had to intervene and caution everyone to speak in terms that a layman could understand,[12] to the dramatic eyewitness testimony of rescue diver Eddy Frank describing Richard at death's door, to the embarrassing news that Infabco had been operating in the Thistle Field for 14 months outside the knowledge of the Department of Energy.

To win his case, MacNeill had to overcome two challenges. First, he had to prove beyond a reasonable doubt that Infabco was guilty of the charges laid down in the indictment, and second, because Masterson's solicitor had invoked Section 67, he had the added burden of proving that Infabco was the employer of Richard and Skip at the time of the accident.

On the first challenge, MacNeill had little difficulty proving the charges. More than half his witnesses—including six *Wildrake* divers, several Department of Energy officials, Grampian Police officers, and the Norwegian DNV official who originally certified the *Wildrake*—gave corroborating testimony in relation to the condition of the diving system and its modification in violation of the regulations. DOE Inspector Roy Giles testified that the bell had been operating without a secondary means of recovery.[13] Don Cox and Dave Lane stated the drop weights had been lashed on the outside of the bell,[14] which the Norwegian official said invalidated the certification of the diving system.[15]

The evidence in support of the charges was so persuasive that, during cross-examination, defense counsel Law remained fairly silent with respect to the equipment, and the few times he did raise questions, the issues concerning safety only grew darker for Masterson. When Law ventured onto the subject of the bell umbilical, for example, Infabco diver Tim Newson revealed that, prior to the accident, some of the crew complained about its integrity, feeling it was not strong enough to take the weight of the bell.[16] And John Brook testified that the bell locator (transponder) had been removed prior to the fatal dive, and that the shackles on top of the bell had too much movement in them.[17]

Where MacNeill ran into difficulty was on the issue of employment. Immediately following the accident, all of the divers who gave statements to the police and the Department of Energy claimed they were employed by Infabco.[18] But, by the time of the trial, MacNeill found himself confronted with half his *Wildrake* witnesses flip-flopping on him. Supervisor Brian Reid was the first diver to take the witness stand and it was then that the jury, and surprisingly MacNeill himself,[19] heard for the first time about a company called Offshore Co-ordinators Limited. According to Reid, this was his true employer.

As the examination progressed, the jury learned that Offshore Co-ordinators was a company registered in the Isle of Man but had its administrative office in a bank on the Channel Island of Jersey, just off the French coast. Residing outside British jurisdiction, the Channel Islands were often used by individuals to hide income from Britain's Department of Inland Revenue. And like the financial groups that operate from other offshore tax havens, the bankers on Jersey were highly secretive about their business affairs. When MacNeill discovered that Infabco had been paying its divers from Jersey, he tried to subpoena one of the bank officials, but the response he got was an emphatic: "We can come if we want to, but we don't have to, and [we] don't want to."[20]

So it was established that Reid had been paid by a third party company. When MacNeill asked the thirty-seven-year-old supervisor from whom he received his work assignments, Reid answered from Superintendent Peter Holmes.

On cross-examination, Reid admitted he had no written contract of employment with either Infabco or Offshore Co-ordinators.[21] On redirect, MacNeill submitted a document into evidence showing that it was Infabco, not Offshore Co-ordinators, that issued Reid's letter of appointment as diving supervisor. This letter was dated seven weeks prior to the accident and was signed by Bob Lloyd, the general manager of Infabco.[22] The point MacNeill wanted the jury to appreciate was this: that under the diving regulations, *only the employer of divers* was permitted to appoint the diving supervisor.[23] If Infabco was not Reid's employer, then what was it doing issuing a supervisor appointment to him? Or put another way, if Offshore Co-ordinators was in fact Reid's employer, then why hadn't it issued the letter?

MacNeill asked Reid how he got his job with Infabco.

From Infabco Superintendent John Nadeau, Reid answered.

How did he join the *Wildrake*?

Someone from Infabco took him to the ship.

Later, Reid hedged his earlier testimony by stating he really thought he was working for Infabco because they had the SALM contract and they were giving the orders.[24]

Dave Lane followed Reid with a repeat performance, initially claiming to work for Offshore Co-ordinators, then, like Reid, steering back to reality after further examination by the fiscal. Infabco was really the operating company, he said; Offshore Co-ordinators was just a financial convenience.[25]

Lane was asked if he had ever received any instructions from Offshore Co-ordinators while on board the *Wildrake*?

No.

Did he ever sign an employment contract with Infabco?

No, but he might have signed one with Offshore Co-ordinators, he wasn't sure.

MacNeill put Supervisor Don Cox on the stand and, just as he had with Brian Reid, produced a copy of a letter on Infabco stationery authorizing him as supervisor. It was signed by Bill Hemmings, Infabco's *Wildrake* project manager.[26]

MacNeill asked Cox how he had obtained his job.

From Bill Hemmings.

Did he ever sign a contract with Infabco?

No, replied Cox.

Where did his diving instructions come from?

From Infabco.

On cross-examination, Law asked the witness who he thought his employer was.

Offshore Co-ordinators, Cox answered; they were the ones who paid him. And then Law produced a time sheet signed by Superintendent Peter Holmes. It was headed with the words Offshore Co-ordinators Ltd.

Were Walker and Guiel employed by Infabco?

Cox replied that divers were typically secretive about their employment arrangements but felt that they probably were.

Another hurdle MacNeill had difficulty getting over was introducing documentary evidence which pointed to Infabco as Richard and Skip's employer. Law constantly frustrated these efforts by raising objections with regard to the rules of evidence. When MacNeill put DOE Inspector Roy Giles on the stand, for example, and asked him if he'd received any paperwork relating to Richard and Skip's employment with Infabco, Giles testified that the *Wildrake* project manager had turned over copies of their employment applications to him headed with the words Infabco Diving Services Ltd.

Aware of their content and not wanting the jury to see information prejudicial to his client, Law immediately objected on the grounds that Giles's evidence would amount to hearsay because he was not competent to prove that Richard and Skip had completed the forms. Ultimately Sheriff Stewart ruled in Law's favor, leaving one of MacNeill's star witnesses able to say he had received the applications, but unable to say what was in them.

And when MacNeill began questioning Giles about the contents of an official accident form (OIR/9) that BNOC Safety Coordinator John Kensey had drafted (listing Infabco as the employer of Richard and Skip), Law blocked the inspector's ability to refer to its content because he had not produced the report himself.[27]

When MacNeill put Kensey on the stand to speak to his own report, Law prevailed again, declaring Kensey unqualified to claim that Infabco was the employer of the deceased.[28]

Short of resurrecting Richard and Skip to testify on their own behalf, the idea must have flashed across MacNeill's mind that reason and sanity had been thrown to the wind and that the odds on winning his case were stacked against him.

Changing tactics, he asked Kensey where he got the idea that Infabco was the employer of Richard and Skip?

From Brian Masterson and other Infabco personnel, Kensey answered.

All that week, MacNeill and Law parried back and forth, playing their version of pin the tail on the donkey: MacNeill trying to fix blame, convinced that Infabco was the donkey, while Law tried to show that the beast was anyone but his client.

In the crowd of spectators, Sterns investigator Colin Sanderson listened intently while he furiously sketched and scribbled details of courtroom diagrams and witness testimony into his notebook. After court, he would sometimes interview witnesses and gather additional bits of information to send to San Francisco.

As the trial progressed over the next several days, press reporters captivated the public's imagination with startling headlines.

Meanwhile, just off Courtroom No. 1, Chris Roper sat patiently in the witness room. He had already spent an entire day waiting to testify when MacNeill called Brian Masterson to the stand.

Masterson's defense centered around the assertion that Richard and Skip were self-employed consultants and not under the control and dominion of Infabco Diving Services at the time of the accident. In the witness box for most of the day, the thirty-five-year-old businessman denied that Infabco had operational control over the divers, denied that Offshore Co-ordinators was simply a paying agent for his company, denied knowledge of how Richard and Skip received their monthly paychecks, and strenuously and repeatedly denied that they were employees of Infabco Diving Services. Infabco may have been the diving contractor, Masterson said, but it never had any divers working directly for them.[29] It was simply a management company engaged by British National Oil Corporation to organize the reinstallation of the SALM. To fulfill that obligation, Masterson said his company executed an option to subcontract divers from Offshore Co-ordinators, including Richard and Skip.[30]

But MacNeill produced a copy of the SALM contract and showed Masterson where it stated that Infabco Diving Services, not Offshore Co-ordinators, was responsible for providing a double-shift diving crew. There was even a stipulation in Appendix 2 that laid out the specification of the crew.[31] Masterson, who testified he had been directly involved with the preparation of the agreement, tried to wiggle free by denying the contract even related to diving services, insisting that it solely concerned the reinstallation of the SALM. At this point, Sheriff Stewart interrupted the witness and said he thought the contract dealt with more than simply providing diving services, and that it was for lawyers, not Masterson, to interpret the meaning of contracts.[32]

Flipping to page 12, MacNeill drew Masterson's attention to clause 14 which stipulated that Infabco was not allowed to subcontract out any work unless BNOC was specifically informed about it and had given prior approval.[33] When MacNeill asked Masterson if BNOC had been informed of his private arrangement with Offshore Co-ordinators, he replied that he didn't think so.

There were several obvious reasons for this subterfuge. First, from Masterson's point of view, he had to preserve BNOC's faith that his company, as the diving contractor, had the power to hire, fire, and control its staff of divers, because if the oil company discovered otherwise—that the divers were really a group of free agents and answerable to no one but themselves—then it would have summarily voided its contract with Infabco. For liability reasons, BNOC could not allow a gang of uninsured divers to tamper with its multi-million dollar tanker loading buoy. Infabco Diving Services was the signatory on the contract, and it was Infabco, not Offshore Co-ordinators, that was required to procure and maintain liability insurance "in an amount not less than one million pounds sterling (£1,000,000)" to cover liability for its negligence or that of its subcontractors.[34]

Second, BNOC was required by law to provide an "employer of divers" for any diving operation carried on in connection with its pipeline/SALM, and to ensure that the employer complied with the regulations, including the duty to employ divers under a "contract of service."[35] A contract of service essentially means that the divers were under operational control of their employer.[36] Had BNOC not complied with that provision, then the law states that the duties and responsibilities of the employer of divers falls back onto *the owner of the pipeline*,[37] and this was not a position that BNOC wanted to be in, because engaging in activities that lie outside its realm of expertise could potentially expose it to paying out civil damages in unwanted lawsuits.

While Masterson kept BNOC in the dark regarding his arrangement with Offshore Co-ordinators, informing BNOC about subcontracting the *Wildrake*, was a different matter. Earlier that week, MacNeill led evidence showing that Masterson had in fact received authorization from BNOC to subcontract the vessel work to Wilhelmsen Marine Services. This was stated in the contract. But nowhere was it written that Infabco had been granted permission to subcontract out the supply of divers.[38] MacNeill's point was clear: Masterson had shown performance of duty regarding the *Wildrake*, but not the divers. This meant he understood his legal obligations and deliberately circumvented the conditions of the agreement.

Continuing his examination, MacNeill asked if BNOC had a contract with Offshore Co-ordinators?

Not to the best of his knowledge, Masterson replied.

Then how was it possible for Offshore Co-ordinators, a company that had no contract with BNOC, to receive payment for the services of its divers? MacNeill asked.

Masterson explained that BNOC paid Offshore Co-ordinators through Infabco. As the SALM project reached a particular milestone, Infabco would bill BNOC for payment. It was Infabco that knew the details of the diving operation such as who deserved sat bonus, who made a surface dive for "depth pay," when the bell went into the water and so forth, because Infabco's contract with BNOC specified different rates of pay for different types of diving.[39]

After collecting this information from company time sheets, Infabco would release the information to Jersey so the people in the bank would know how much to invoice Infabco in order to pay the divers. It was a merry-go-round payment scheme and MacNeill demonstrated that, if Offshore Co-ordinators was in fact a legitimate business, much less the employer of Richard and Skip, it clearly took a backseat to Infabco.

MacNeill showed Masterson letters on Infabco stationery and signed by company officials appointing Don Cox and Brian Reid to supervisory positions, which Masterson explained away by saying the documents only confirmed their appointment in order to comply with the regulations. Offshore Co-ordinators was responsible for Cox and Reid, not Infabco.[40]

MacNeill produced Skip's application for employment on Infabco letterhead, which Masterson said he could not attest to but conceded it did suggest that Skip had applied for work with Infabco. MacNeill then turned the jury's attention to Skip's personal dive log. By law, divers were required to register each dive in their logbook and it was customary for the employer to stamp the dive to certify it. Inspector Giles had retrieved Skip's book and subsequently turned it over to MacNeill where it was entered into evidence as Crown Production No. 19.[41] The logbook revealed that, for more than a year, all of Skip's dives were stamped with the Infabco company seal,[42] the same seal that appeared on Infabco Diving Services' letterhead. When MacNeill asked Masterson to explain this he claimed there were actually two companies with the name Infabco Diving Services Ltd., one of which was registered in Ireland. The stamp in Skip's logbook must be from that company, Masterson ventured. In any case, the Irish company had nothing to do with his company and he didn't know who had stamped Skip's book.[43]

MacNeill's line of questioning was on point: what were the odds of an unaffiliated Irish company producing the same exact logo as the company Masterson owned in Britain if there was no connection between them?

Leaving the jurors to ponder those odds, MacNeill moved on to ask Masterson an even more puzzling question—if Offshore Co-ordinators was indeed Richard and Skip's employer, then why did he tell Inspector Roy Giles and Grampian Police Officer Bill Adams that he was the employer of the dead men?

Masterson couldn't give an answer to this question, and when MacNeill continued to press him on the matter, he said if he had made the remark it was untrue.[44]

Wasn't his response to Inspector Giles consistent with the other evidence before the Court? MacNeill asked—that the reason why Cox and Lane and the other divers approached Infabco to get their jobs, that company officials were appointing diving supervisors, that Infabco's letterhead was printed on Skip's application, that his logbook was stamped with the company seal—wasn't all of this evidence pointing to the fact that Infabco was the true employer of the dead men? To this, Masterson just reiterated that it was his company policy not to hire divers.[45]

When MacNeill asked why Mr. Kensey of BNOC sought out the details of Richard and Skip's employment from him in order to complete his accident report, Masterson flatly denied that Kensey ever approached him for the information.[46]

On cross-examination, Law introduced the Offshore Co-ordinators/Infabco contract into evidence, ostensibly to give credence to the legitimacy of the relationship between the two companies. When the QC sat down, MacNeill pointed out that the vaguely written document curiously made no reference to divers at all, only "consultants" which were to be supplied for the purposes of oil and gas exploration.

Signed by Masterson himself, the contract expressed no terms of payment, no indication as to how Offshore Co-ordinators would assess its fees or commissions, nor was it possible for MacNeill to discern the nature of the services that Offshore Co-ordinators were to provide.

When the fiscal intimated that the contract was merely a clever ruse to give Offshore Co-ordinators the appearance of being something more than a paying agent, Masterson waffled badly: one minute insisting that the divers were controlled by Offshore Co-ordinators, the next, painting the picture that, as independent consultants, they largely controlled themselves. When MacNeill asked how this could be achieved—either the divers were employed or worked independently—Masterson said he didn't know.[47]

But MacNeill couldn't let the ambiguity of that answer remain in the minds of the jurors. His entire case hinged on proving that Richard and Skip were employed by Infabco. This was the threshold requirement he had to meet, and, because of the way in which he had prefaced each of the charges with the key phrase "being the employer of," he knew that if one charge fell, they would all fall.[48]

His earlier attempts to refer to documents indicating that Infabco was the employer had been blocked by Law. He also knew that employment, as a legal concept, was not established by a piece of paper, even a written contract of employment. It was determined by the nature of the relationship between the two parties, and if he could prove that Infabco had given Richard and Skip their work orders, told them how to perform their jobs, supplied them with the tools and equipment to do so, and had the power to dismiss them, regardless of how they received their payment, then Infabco would be defined as the employer. So he began to inquire into the chain of command that had sent Richard and Skip to the bottom of the sea.

How did the divers receive their instructions? he asked Masterson.

Through Superintendent Peter Holmes, Masterson replied, carefully making the distinction that he never instructed Holmes on the specific tasks the divers would perform but merely discussed the contract program with him. Holmes directed the supervisors who then instructed the divers on their daily tasks—and everyone, Masterson insisted, from Holmes on down to the lowly tender, was employed by Offshore Co-ordinators.[49] The only Infabco employee on board the *Wildrake* was himself.[50] There were even phases of the work schedule, he said, in which he was not present.[51]

MacNeill asked if any officer of Offshore Co-ordinators was ever on the job.

Masterson said no.

Finally, MacNeill asked Masterson why he had failed to tell him about the role of Offshore Co-ordinators for more than a year. To this, Masterson replied that if he'd been asked these questions earlier then he would have given the information he was now providing to the Court. Dissatisfied with that answer, MacNeill persisted, wanting to know when he had informed his legal advisors that Infabco was not the employer of the dead men.

Masterson said he didn't prepare his case and couldn't recall if he'd ever given those instructions to his lawyers.

Everyone in court that day knew the truth, that Infabco was Richard and Skip's employer. But Masterson continued to bewilder the Court with the notion that Offshore Co-ordinators, a company which had no contract with BNOC and no manager on board the *Wildrake*, that never interviewed Richard and Skip for work, had contracted their services then jobbed them out like a couple of plumbers to re-install BNOC's SALM.

In the end, perhaps out of a sense of frustration, MacNeill issued a parting shot at his quarry by asking him if he was surprised to know that it wasn't until two weeks ago that he'd learned Infabco was not the employer of Richard and Skip.

Not surprisingly, Masterson's counsel lodged an objection to the complaint which Sheriff Stewart sustained, citing that the defendant was not obligated to make any response until the Indictment had been served.[52]

By the time all the evidence had been heard, 18 witnesses had testified for the Crown, while Mr. Law merely called Masterson's accountant for the defense. MacNeill then addressed the jury by summing up the evidence, feeling he had proved his case and that the jury would issue a conviction.

It was now Mr. Law's turn to make his closing remarks, but instead of doing so, he asked to speak to the Court on a point of law and requested that the jury be taken out. Sheriff Stewart complied with the request and then Law made a motion for the judge to find the accused not guilty on the grounds of insufficient evidence.

This was a common tactic used in courtrooms throughout Scotland, and Sheriff Stewart, due to the lateness of the hour, decided to adjourn the case, wishing to ponder the submission overnight.

Diving tragedy: Jury decide today

—— *Press and Journal*, December 19, 1980

All that week the weather in the City of Granite had been rotten: drizzling rain with temperatures hovering just above freezing. When court reconvened the next day the benches were crowded and Chris Roper was sitting in the audience, confused and angry at having wasted two days in the witness room without introducing so much as a word into the record.

In the absence of the jury, Sheriff Stewart entered the courtroom, took his seat, and then stunned the crowd with his opening statement:

> I have carefully considered the submissions made to me yesterday by Mr. Law and also I have reviewed the evidence in the case, and I have come to the conclusion that I must sustain Mr. Law's first submission. In my view, there is not sufficient evidence to entitle the Jury to find that the Accused were the employers of the divers mentioned in the Indictment. I propose to direct the Jury accordingly.

Sheriff Stewart's words took a few seconds to seep in, but it soon became apparent to the onlookers that he was letting Masterson's company off the hook. The judge then brought the jury into the courtroom to hear his directive. When all fifteen members were seated, he turned and faced them:

> Ladies and gentleman, as a result of submissions made to me yesterday in your absence, I am going to be giving you certain directions in law which may rather take you by surprise, which will have the effect of considerably shortening your task. I thought, in fairness to you, I would take the rather unusual step of bringing you back into Court in order to hear the reasons for these directions I shall be giving you.
>
> I have carefully considered the submissions made to me by Mr. Law, the contents of the Procurator Fiscal's address to the Jury, my own notes of the evidence and the authorities to which the Procurator Fiscal referred me. Mr. Law's primary submission was that the Crown had failed to lead sufficient evidence to entitle the Jury to find that the Accused company had been the employer of the deceased divers.

Sheriff Stewart then summarized the charges, pointing out that each alleged breach of the regulations was a duty imposed upon the employer of divers. The accused, however, had given proper notice that it was not the employer of Richard and Skip.[a] "It therefore became incumbent upon the Crown to prove this fact as an essential part of their case," Stewart said.

And then the sheriff launched into the legal requirements necessary to establish the existence of a contract of service between two parties. Referring to a case that had come before an English court in 1967,[53] he cited three conditions that must be fulfilled before a contract of service could exist: 1) that the servant agrees to perform work for his master in exchange for payment or compensation, 2) that he agrees, "either expressly or impliedly" to be "subject to the other's control in a sufficient degree to make that other master," and 3) that the other provisions of the contract are consistent with a contract of service. In other words, there are no conditions of the contract that would qualify the agreement as being something other than a contract of service.

In arriving at his decision, Sheriff Stewart said MacNeill had produced "no direct evidence" of what and by whom Richard and Skip were paid for their services, no evidence of who instructed them on the day of their last dive, and no evidence of any invalid provisions in the contract. MacNeill had presented evidence from some of the divers (namely Tim Newson, Randy Gunter, and John Brook), that they considered themselves to be employed by Infabco, but the sheriff discounted their testimony because in his judgment they had not been properly examined by MacNeill. In any event, Stewart said the divers' views were of "limited value" because "employment is a legal concept," and how Don Cox or Brian Reid framed their working relationship was irrelevant. The only

a Section 67 of the Criminal Procedure (Scotland) Act 1975.

thing that mattered was whether MacNeill had proved that Richard and Skip's relationship with Infabco fell within the meaning of a contract of service.

For some of the people in court that day, Sheriff Stewart's decision seemed outrageous, even faulty. But the judge's hands were effectively tied by the case that MacNeill had presented to him. He could make his judgment only and exclusively on the evidence presented. Whatever his personal feelings were, he could not set the law aside and say, "Well, I think Richard and Skip must have been employees of Infabco because Don Cox and Brian Reid must have been telling Richard and Skip what to do on their last dive." He could not invent the evidence nor base his decision upon gut feeling. The burden of proof rested with the Crown prosecutor, and in Sheriff Stewart's view, MacNeill had failed to present sufficient evidence to meet the test the law required.

Years later, *Press and Journal* reporter Norman Harper would remark that Sheriff Stewart was "not a man given to great emotion," and yet Harper saw the Scottish judge do something extraordinary that day. In his closing remarks, he expressed his irritation at MacNeill and delivered a sharp rebuke to a regulatory regime that permitted a man, responsible for organizing a diving operation, to escape the consequences of the law:

> In these circumstances, I have come, regrettably and reluctantly to the conclusion that the Crown has failed to lead evidence sufficient in law, to entitle the Jury to find that the Accused, at the relevant time, were the employer of either diver. My regret and reluctance arises from two considerations. One is that I am acutely aware that my decision cannot, as the law presently stands, be appealed against by the Crown, so that if I am wrong, the Prosecution has no redress. The second is that, given the serious nature of the charges here, and the fact that the Jury has patiently listened to four days of evidence, it is extremely unfortunate that they are to be deprived of the opportunity to adjudicate on the merits of the charges.
>
> However, despite these considerations, I am clearly of the view that the Crown has failed to prove this fundamental part of their case and I accordingly must formally direct the Jury to return a verdict of not guilty.
>
> I should add that I hope that the Department of Energy will take account of the evidence given in this case about the employment position of divers in the North Sea and will consider, as a matter of urgency, whether regulations require amendment to widen the definition of 'Employer of diver' or in some other way. It is clearly undesirable that persons responsible for providing diving equipment should be able to escape the consequences of any breach of the regulations applying to that equipment if such breaches occur. Whether they occurred in the present case is not, of course, for me to say.

The judge then turned and ordered the jury to find Infabco Diving Services not guilty. The jurors, having been so instructed, were asked by the clerk of court if they unanimously found the accused not guilty.

The foreman of the Jury replied, "Yes."

While Sheriff Stewart was speaking, Harper noted disappointment on the faces of the jurors who appeared to be fighting hard to conceal how displeased they were with the outcome. They had struggled to understand the technical aspects of deep-sea diving, had listened to one of the most shameless schemes ever conceived to escape responsibility, had begun to grasp the importance of what was being presented to them, and now were being denied an obvious decision.

When Harper glanced over at Masterson, he saw a man who appeared to be neither affected one way or the other by the verdict. Outwardly, his expression was inscrutable—there had been no flash of relief on his face, no display of emotion—but inwardly, Harper guessed that, having come through four days of intense pressure, the company owner "must have been in turmoil."[54a]

Sitting nearby, reporter Graeme Smith was equally appalled by the verdict. The outcome seemed "extremely unfair" to him, and today, when he looks back on the trial and recalls what happened to the two divers, it still seems just as unfair. "I thought it was terrible that [Infabco] escaped that criminal responsibility because of, I suppose, the law."[55]

But MacNeill's defeat was not the result of any deficiency in the law. He had Masterson cornered by indisputable violations, and the law defining employment was unambiguous. Several lawyers close to the case thought the fiscal simply committed an enormous blunder: one attorney commenting that MacNeill approached the issue of employment "without really considering what an employee meant," while another remarked that "a really careful prosecutor" would have moved to prosecute Infabco, not on the basis that it was the employer of Richard and Skip, but under statutory and common-law duties that obligated Infabco to provide a safe dive whether Richard and Skip were employees or not.[b]

Coming after the trial, these comments sorely resonate like typical hindsight rhetoric. But part of MacNeill's defeat may have been due to his being lulled into thinking that Masterson would plead guilty to the Indictment and take his medicine from the Court. At one point before the trial, the company director was apparently willing to plead guilty to the majority of the charges.[56]

According to Commander Warner, Masterson issued a statement two months before the trial acknowledging that he was in fact the employer of the deceased:

---

a Years later, Harper would study his paper's file containing oil industry related accidents from 1978 through 1981 (both diving and non-diving related), and discover "that virtually two-thirds to three-quarters" of the news accounts concerned the *Wildrake* tragedy.

b For example, section 3(1) of the Health and Safety at Work etc. Act 1974 provides such coverage: "It shall be the duty of every employer to conduct his undertakings in such a way as to ensure, so far as is reasonably practicable, that persons *not in his employment* who may be affected thereby are not thereby exposed to risks to their health or safety." (Italics added.)

> I am the owner of the company known as Infabco Diving Services Ltd., of 24, Carden Place, Aberdeen. The deceased Victor Guiel has been employed by the company since July 3, 1978. The deceased Richard Walker has been employed by the company since April 27, 1979...[57a]

But whatever intentions Masterson had in the autumn of 1980, they quickly evaporated at the preliminary hearing when he decided to disown Richard and Skip.

Caught off guard by this maneuver, it was only two weeks until the trial when MacNeill hurried off to St. Machar House to recover a copy of the BNOC/Infabco contract to see where it placed the responsibility for diving. But BNOC wouldn't give up the contract, forcing MacNeill to waste time applying for a search warrant.[58]

MacNeill's lack of preparation was also demonstrated in Brian Reid's surprise announcement that Infabco was not his employer. With the power to examine witnesses and obtain pre-trial statements called "precognitions," the fiscal would have discovered the existence of Offshore Co-ordinators long before Reid came under oath, instead of allowing his examination to deteriorate into a kind of fishing expedition in front of a jury.

Nor did he determine just who Offshore Co-ordinators really was and the role it played as the alleged employer of Richard and Skip. Only much later would MacNeill write to the bank in Jersey and receive a signed affidavit from one of its officials shedding light on the true nature of the Infabco/Offshore Co-ordinators relationship.

MacNeill did not subpoena anyone from Infabco to refute Masterson's assertions that Infabco was anything but the employer of divers. General manager Bob Lloyd, for example, who witnessed the signing of the Offshore Co-ordinators contract and who executed company policy, would have been in the unique position of knowing how that policy impacted upon the *Wildrake* divers.

Diving Superintendents John Nadeau and Peter Holmes and *Wildrake* Project Manager Bill Hemmings—more crucial links in the command structure—strangely never appeared in court.

Supervisors Don Cox and Brian Reid were not asked about giving instructions to Richard and Skip prior to the fatal dive. And MacNeill never called Lacey Lahren, Roger Palmer, Dave Mitchell, or Dave Boyd—the four divers in sat with Richard and Skip—who would have heard those instructions firsthand over the chamber speaker.

In the absence of these material witnesses, the job of connecting the dots from Infabco to Richard and Skip was incomplete, turning what should have been a slam-dunk victory into one of the most widely publicized legal fiascoes in North Sea history.

---

a This is incorrect. Richard joined Infabco in November 1978.

But MacNeill's impromptu performance is even more baffling in light of the fact that, for years, he was intimately aware that contractors such as Infabco were "laterally" employing divers through puppet companies.

Long before the Infabco trial, in March 1977, the fiscal was conducting an inquiry into the death of two Oceaneering divers when the subject of their employment status cropped up. A company manager gave evidence that the deceased were not employed by the UK company, Oceaneering International *Services Limited*, but by the Zurich, Switzerland company, Oceaneering International *A.G.* The UK company, the manager said, was only responsible for organizing the services of divers in compliance with contracts for diving work.[59]

Three months later, MacNeill brushed against this employment issue again at another inquiry into the death of Sub Sea diver John Howell. Howell had died under mysterious circumstances during a bounce dive the previous year, and on the witness stand the managing director of Sub Sea declared that the deceased was not employed by Sub Sea International Limited but by an "associate company" called Sub Sea Diving Services located on the island of Jersey.[60] (A month after this inquiry, Sub Sea would shift paying agents to Subaqua Services International Inc. and begin issuing paychecks from Panama.)[61]

Less than a month later, MacNeill found himself back in court looking into the death of Comex diver Clay Ellis. The young Texan had died when his bell accidentally surfaced in the Argyll Field, leaving his bell partner, Derek Bannister, paralyzed from the chest down.[a]

In what would be the harbinger of things to come, Comex officials took the witness stand and repeatedly framed Ellis's employment status in terms that only a diving contractor could appreciate. Again, it was the same old story—the dead diver was "employed" by a decoy company located outside British jurisdiction while the live one had taken orders from the UK company. And so it was with Clay Ellis. The Service Manager of Comex testified that the dead man belonged to an outfit called Contract Manpower International Limited whose place of business was, once again, located in the Channel Islands (Guernsey this time). And while Ellis was said to qualify as a Comex employee "for insurance purposes," it was asserted that he was "not technically" an employee, even though it was admitted that if one were to take the chain of command higher than the level of diver, it would be shown that the diving operation was actually controlled by Comex.[62]

At one point during the inquiry, the atmosphere in the courtroom started to heat up when Mr. Osborne, the attorney representing Ellis and Bannister, asked the Comex Safety Officer who he thought Ellis's employer was in relation to the regulations.

"Comex Diving," he replied.

Osborne put it to his witness that, if Comex was the employer, then the company must have been engaging its divers under a contract of service as the regulation required.

---

a See Chapter 6.

That logic brought the Comex attorney immediately to his feet, strenuously objecting and arguing that the witness was not competent to express his opinion on the matter, that the divers were self-employed, and that the regulations did not apply to Comex.[63]

This was, of course, a scene destined to be reenacted at the Infabco trial, but with a different cast of characters. And throughout the debate, as MacNeill listened to the Comex attorney, there was someone else paying close attention and asking questions too—the presiding judge, Sheriff James Law, the same man who would, in three years time, split MacNeill's case down the middle and implement Brian Masterson's escape route.

There were other cases that year for MacNeill with a similar pattern. The fatal accidents of divers Richard Dupuy, Michael Moore, Howard Spensley and Charles Meehan all sparked the familiar refrain that the deceased didn't work for the diving companies that sent them offshore to dive.

At the very least, these cases in 1977 should have forewarned MacNeill that Masterson was a prime candidate to try the same dodge, and that he should have avoided prosecuting Infabco under a statutory duty that had been in contention for years. Sadly, the terrible injury to Derek Bannister and the demise of eight divers was not enough to make it so.

In court, lawyers tend to split legal hairs *ad nauseam*, and to reduce arguments to such things as the "irreducible minimum," until the moral underpinning of the law seems to lose its meaning in relation to the tragedy. In the Ellis case, probably the most telling testimony came when accident survivor Derek Bannister, now confined to a wheelchair, told the Court that his transition from Comex to Contract Manpower International entailed signing some forms. When Bannister asked Comex why he no longer worked for Comex, he was told, "[This] is the way it is all done, this is the way you work for us."[64]

To Bannister, changing job locations and signing company documents in the process, seemed like "a straightforward, regular procedure."[65] And after he signed the forms, there was never any change in the way Comex controlled his movements: they still organized his trips offshore, told him what helicopter to take, what rig to work on, when to dive, what tasks to complete, and when to go home. Unwittingly, however, the man fated to lose the use of his legs had provided his employer with the means to escape responsibility in the event that something should ever happen to him.

Bannister's experience—that he was confused about the ramifications of his employment situation—was precisely the way the diving companies wanted it. The less the diver understood, the better for them. It was an arrangement they'd been cultivating for many years.

The theory that divers were "self-employed consultants" dates back to 1970 when the Inland Revenue began assessing divers as self-employed workers to encourage them to work in the North Sea. They could be paid in gross and claim certain expenses against their income (work gear, travel, etc.). But that classification was for *tax purposes only*. It had nothing to do with the legal status of the diver in relation to his employment.[a]

Meanwhile, the UK diving companies began insulating themselves from the legal and moral consequences of their actions by setting up "associate companies" offshore with several priorities in mind, one of which was to save money. According to an investigation conducted by the *Glasgow Herald*, paying divers through companies based outside the United Kingdom provided tax benefits to everyone involved, in particular the UK diving company and its offshore paying agent. Any profits accumulated by the offshore company would be taxed at a lower rate (or none at all) by the host country, and any payments the UK diving company made to the offshore company could be claimed as a deduction against UK corporate taxes.[66]

And there was another cost-saving benefit as well. If Infabco had been paying its divers through normal channels in Britain, it would have been required to contribute National Insurance co-payments to the government.[b] By paying the divers through Offshore Co-ordinators, Masterson avoided that cost. Former diving company owner Malcolm Williams says this arrangement was "just a mechanism."[67]

And that "mechanism" had the potential to play a long-range tactical role as well: to provide a smoke screen in an attempt to escape criminal and civil liability when the time came.

Sometime in the 1970s, the diving companies decided to appropriate the Inland Revenue's tax classification to include the divers' employment status as well. They realized that, because the diving regulations related *only to employees*, the diving contractor could not be found guilty of an offense in relation to anyone who was *not an employee*. In theory, the diving operation could be as unsafe as necessary to get the job done, and if the diver affected by the lack of safety was not an employee, then there could be no conviction. In the event of a fatality, the

---

a This point was re-confirmed in 1995 in a court decision handed down at an Industrial Tribunal between Diving Superintendent Ken MacIntyre and his employer, Stena Offshore Limited. At the tribunal, MacIntyre claimed that he had been unfairly dismissed by Stena Offshore. Stena, still clinging to the "Infabco defense" some fifteen years later, claimed that there was no case to hear because MacIntyre was "not an employee." Within the body of the tribunal's decision, the chairman wrote: "In my view it is also relevant to bear in mind that divers are highly skilled specialists in an arduous and sometimes dangerous occupation. For that reason, tax concessions were made to them in the Income Tax Act 1988, which incidentally mean that divers are not to be taken as self-employed simply because they are taxed under Schedule D." The chairman ultimately ruled that MacIntyre was an employee of Stena Offshore Limited. *Source*: The Industrial Tribunals (Scotland), Case No. S/4419/94, p. 6.

b Similar to Social Security payments in the US.

contractor would simply point the finger at the dummy company offshore and say, "There's the culprit."

Panama, Channel Islands, Switzerland—it was the perfect foil in the disguise of a tax fiddle as MacNeill correctly intimated in court. If an investigator actually traveled to Panama or Jersey to inquire about deep-sea divers, he would have found a secretary in a simple office typing up paychecks. Certainly no diver wanting a job in the North Sea would fly to these locations to seek work in the North Sea. Job instructions, promotion, and solutions to pay problems all emanated from the offices of the UK diving companies in Aberdeen. When Richard needed a pay advance in May 1979, he received it from Infabco General Manager Bob Lloyd in Aberdeen;[68] he didn't hop on a plane to sort things out in Jersey. And when details of the *Wildrake* accident were released to reporters on August 8, 1979, again, it was Bob Lloyd in Aberdeen who answered questions at the press conference, not Offshore Co-ordinators.

It was no accident that the diving contractors chose these distant lands to set up their bogus "diving" companies. They knew that jurisdictional issues and secrecy laws of the offshore financial institutions would make it virtually impossible for prosecutors like MacNeill to "pierce the veil" of incorporation in order to make the case that these so-called recruitment companies, with dubious names like Offshore Co-ordinators and Contract Manpower International (which do not appear in any diving company directory), were nothing more than agents of the UK diving industry. It was all part of a clever plan to remain free of responsibility in the event they were ever charged with committing negligent acts. Of course, the scheme was just theory until Brian Masterson, having manipulated the system with the skill and precision of a Machiavellian prince, proved that it actually worked.

But just because he took the witness stand and testified under oath that Richard and Skip were self-employed didn't make it so. And just because he diverted their pay through Jersey didn't mean that Offshore Co-ordinators was their employer. As Sheriff Stewart pointed out, how one member of an agreement frames its employment relationship is irrelevant; the relationship is defined by law and by practice. From the moment Richard and Skip stepped on board the *Wildrake*, they were told what to do, how to do it, and when to do it per the engineering plan devised jointly by BNOC and Infabco. Richard and Skip did not come on board announcing to everyone, "Hey, this is how you do it!" Nor did the diving equipment they use belong to them or to Offshore Co-ordinators; it belonged to Infabco. Whether Masterson admitted to it or not, Richard and Skip were not independent contractors; they were employees following orders, and they followed Infabco's orders until the day they died.

To diving company owner Malcolm Williams, "the recruitment company who hired divers and then contracted them to the diving company was, in fact, effectively an employee," and the outcome of the trial no less surprising to him than the rest of the spectators in that December courtroom.

"If the issue had been investigated and pushed, I think any reasonable legal interpretation would be that they [Infabco] were effectively the employer, without any doubt at all....I quite clearly remember I was puzzled at one of the two inquires why the issue wasn't more forcibly [argued]—saying, 'Well, come on it's only a mechanism, isn't it?' And 'You are effectively the employer,' because you were! In every sense of the word, legally."[69]

So the diving company shoved forms in the diver's face and told him where to sign (which he did to keep his job). The diver in turn carried on with work believing that the people in the office were doing the right thing by him. His paychecks now came from another part of the world and he had no reason to suspect anything improper about the arrangement. It all seemed completely straightforward and meshed neatly with the attitude of the typical diver who didn't care where his pay came from, just so long as he was paid for the time he spent on the bottom of the sea. But the truth was, while he was offshore working hard to enrich his company, the people in the office were committing treachery by preparing to sell him down the river if they were ever caught violating the law.

And this is what happened to Richard and Skip.

The scheme was perfectly legal and existed for years because the Crown never formally challenged the diving companies on the employment issue until MacNeill took Infabco to court. That's why the trial became a milestone case, because it represented an opportunity for the State to correct a long-term practice relating to master-servant relationships.

For the common diver, like Randy Gunter and Chris Roper, the Infabco prosecution represented a chance to see something positive emerge from the deaths of Richard and Skip. That's why they saved the welding wire and took it to the authorities and went to court to testify. They wanted the law to sit up and take notice with regard to the working conditions of divers offshore. As Gunter said after the accident, "All we are trying to do is prevent [this] from happening again, because this game is going to go on for years and years to come."[70]

But the trial turned out to be a farce and Roper didn't get to testify. After spending two days in the witness room, the not guilty verdict, on top of the manner in which Richard and Skip died, left him with a bitter memory. "I was really disgusted with the whole British way," he says looking back. "It was kind of like the system was not working right."[71]

As he left the courtroom, the young Canadian passed beneath a large plaque mounted on the wall, high above the sheriff's bench. It was the Royal Arms of Scotland. It was there to affirm that this was a Royal Court. The shield, encircled by the Order of the Thistle, was quartered with symbols representing England, Ireland, and Scotland. To its left was a large unicorn rampant, a mythical heraldic beast with lion's tail and goat's beard supporting St. Andrew's Cross, the flag of Scotland.

Opposite the unicorn was a lion rampant, emblem of the King of Scotland, holding St. George's Cross, flag of England. Both creatures were standing in a bed of roses, shamrocks, and thistle. The six-barred helmet sitting atop the shield, and the closed crown and the royal crest of Scotland above that, all signified a monarch's rank.[72] At the base of the coat of arms was the Latin inscription: *Nemo me impune lacessit.* Roughly translated it meant—no one attacks me and gets away with it. But as Brian Masterson stepped from the courthouse into a cold sleeting winter's day, Chris Roper knew that his former boss had just walked away with justice in his back pocket.

Ten days later, on the Isle of Man, Offshore Co-ordinators mysteriously vanished from the register of trading companies.[73]

Chapter Twenty-three

How Can a Mother Explain the Worth of a Son?

Children shouldn't have to die before their parents.

—— Shirley Guiel

Word of Infabco's victory traveled fast. Chris Roper telephoned Jeanne and Sheila to give them the verdict. Several days later he followed up with a letter describing the judge's dismayed demeanor during his ruling and the general feeling that "the jury would have found Infabco guilty."

Still dazed by the outcome, he tried to give the families some hope for justice:

> As far as the next step is concerned I think that Infabco will be taken to court in England. Met Mr. Colin Sanderson the day the trial ended. We spent a lot of time going over the whole trip and he feels that some justice can yet come out of it all.[1]

But Infabco would not be taken to court in England. Jurisdiction belonged in Scotland, and as Sheriff Stewart stated in his closing remarks, his decision could not be appealed by the Crown (a safeguard against double jeopardy). The judicial process had been served. Infabco was found not guilty, and there would be no new charges emanating from the fiscal's office, so the criminal threat against Masterson's company had reached its climax.

Jeanne's reaction to the verdict was double-edged, and understandably cynical: on the one hand she was "shocked" that Infabco had "used such a stupid, easy way out, [and] that somebody would be so careless as to not have covered those bases." But on the other hand, she wasn't surprised either because Infabco represented an irresponsible industry that basically did what it pleased, then justified its misbehavior later by abusing the legal system.[2]

She also had the sick feeling that MacNeill's case may have suffered because of information she had in her possession which she offered to MacNeill prior to the trial, but he had turned down.[3] Masterson had denied any direct link between Infabco and Richard, and yet Jeanne had Richard's pay slips from November and December 1978 which proved he was being paid directly by Infabco Diving Services, not Offshore Co-ordinators. In addition, she found a "Leave to Enter" letter among Richard's papers which stated that he was "engaged as a diver by Infabco Diving Services" and "would be employed for at least another year." It was signed by Craig Roberts on Infabco stationery.[4]

In Massachusetts, Sheila found a similar letter among Skip's effects. Dated six weeks before his death it read: "This letter confirms that Victor Guiel is employed on our contract as a diver on board the *Gulnare*. Mr. Guiel will continue to be employed in the North Sea for a further year." It was signed by Infabco's Operations Manager, Colin Duncan.[5]

Mr. Walker "would be employed for at least another year." Victor Guiel "is employed on our contract as a diver." Wasn't this language that only an employer would use? Jeanne thought so. Whether these documents would have survived the objections of Mr. Law to keep them out of evidence is probably unlikely. Still, Jeanne viewed the pay slips and Leave to Enter letter as a virtual admission by Infabco that it was in fact the employer of Richard.

Despite the turn of the events, Jeanne was determined more than ever not to give up the fight. Richard's last words were still branded deeply in her mind, and she refused to believe that the Scottish legal system had let her down. Perhaps because of Chris Roper's comment that a new case could be built in England, she began writing to Richard's co-workers in desperation, begging them to send her copies of anything that would prove they were employed by Brian Masterson's company.[6]

For Sheila, Masterson's victory caused her to blame herself for not being at the trial to defend her brother's memory. On January 14, 1981, she fired off a letter to MacNeill commenting on the verdict and the mystery company called Offshore Co-ordinators:

> Mr. Masterson came to the homes of Richard's widow and my own house and sat and we discussed the accident many hours. We never received any form of communication from 'Offshore Coordinators' and if they were the employers, we would have heard direct from them that <u>their</u> divers were in trouble. I received a call on Aug. 8th from Bob Lloyd, Infabco's office, of the trouble.[7]

Six days later, MacNeill responded by saying he was planning to hold a public inquiry into the accident and that he shared Sheila's disappointment with the result of the prosecution. He thought Sheriff Stewart had made a mistake:

> I personally was satisfied that at the Trial I had produced adequate evidence to establish that Infabco were the employers of the divers and feel strongly that Sheriff Stewart should have taken a different course and enabled his finding to have been taken on Appeal to the High Court.[8]

In January 1981, the North Sea was quiet. Since the deaths of Richard and Skip, there had been no diving related fatalities in the UK sector. It was an unprecedented interval. There had been no changes to the safety regulations, and the numbers of divers and construction projects remained at very high levels. There was, however, a strong undercurrent running within the diving community that there should never be another *Wildrake* accident.

And then, like a dormant volcano, Thistle came alive once again.

It was January 21, and the Diving Support Vessel *Stena Seaspread* was conducting routine diving operations on the Thistle SALM, where Richard and Skip had died. Wharton & Williams was now the primary contractor in the Thistle Field, and the *Stena Welder*, used in the *Wildrake* rescue, had been replaced by the *Stena Seaspread.* Wharton & Williams divers Philip Robinson and Jim Tucker had been sent down that morning to maintenance the SALM near the seabed. With a strong tide running that day the bell umbilical fouled on the loading buoy. At some point—presumably during the recovery operation—the lifeline was torn completely in half.

Once again, two men were hundreds of feet below the surface with no communications, no topside gas, and no hot water supply. This time, however, the crew kept its head and proceeded slowly. They established wireless through-water communications with Robinson and Tucker and determined they were okay. Unlike the *Wildrake* bell, which was lost in the dark on the seabed, the *Seaspread* bell was still suspended in mid-water by its lift wire. The main winch was functioning so the crew began hoisting the bell slowly to the surface, checking periodically with the divers to see if everything was okay.

At 165 feet from the surface, Robinson and Tucker ordered topside to stop the lift. The bell was losing pressure. They didn't know why or where the loss was emanating from, but topside had no choice but to return the bell to its working depth of 430 feet. And so they did. At that point, 2W declared an emergency and began notifying nearby diving support vessels to come to its aid. Once again, there was a race against time before the effects of hypothermia added two more names to the death list.

Twenty miles to the south, the *Uncle John* was stationed beside the Brent Bravo platform just as it was during the *Wildrake* accident. It was as though Mother Nature was conducting some bizarre déjà vu experiment to prove that Richard and Skip could have been saved after all. When the emergency call came in at 1158, Comex was monitoring a subsea pumping operation with an ROV (Remotely Operated Vehicle).[9] The crew immediately recovered the ROV while the captain of the *Uncle John* began the process of dewatering the massive legs of the vessel.[10] Normally it would have taken three to four hours to deballast the semi-submersible to raise it to transit depth, but the captain was accelerating the process by executing an emergency deballast, pumping water from the legs through huge fire cannons mounted on the aft end. As the *Uncle John* headed north on its rescue mission, huge rooster tails of cascading seawater trailed behind.

Three hours later the *Uncle John* arrived alongside the *Seaspread* to begin the rescue.[11] Four hundred and thirty feet below the surface, Robinson and Tucker were keeping warm in very thick mummy bags and using breathing masks which not only scrubbed carbon dioxide from their respirations but recycled their expired body heat. After the *Wildrake* accident, the diving industry recognized the futility of using space blankets as protection against hypothermia. Norway's Underwater Institute began conducting thermal studies to test survival systems for stranded divers,[12] and Wharton & Williams, in conjunction with a local technology institute, developed a new survival suit with high heat-retention properties. Robinson and Tucker were now using those suits.

Unlike the *Wildrake* accident, where the Infabco and 2W superintendents did not communicate with each other, the diving superintendents of the *Uncle John* and *Seaspread* put a plan of action together. And unlike in the *Wildrake* accident, where Infabco dictated the rescue strategy, the people onboard the *Uncle John* took charge of the *Seaspread* rescue, because, as 2W company owner Ric Wharton later pointed out, "they were doing the rescue."[13]

And the plan that Comex Supervisor George Head decided to execute was the one that Supervisors Don Cox and Mike Allen advocated on board the *Stena Welder* in 1979—perform a wet transfer—take Robinson and Tucker out of the damaged bell and put them into the safety of the rescue bell.

And so they did.

Smoking one cigarette after another, Head lowered the rescue bell with divers Joe Puttnam, Richard Taylor, and Ken Iversen inside. In the gloom and darkness of 430 feet, they established a swim line between the two bells, took a hot water hose to the stranded divers, warmed them up, then made the transfer by leading Robinson over first, then Tucker. For their part in the rescue operation, Iverson, Puttnam, and Taylor received the Queen's Commendation for Brave Conduct.

It was the world's first through-water transfer and it cost Wharton & Williams thousands of dollars to call in not only the *Uncle John* but also the *Tender Comet* as a backup vessel. But it was money well spent and today Malcolm Williams looks back upon the episode with nothing but pride for the role Comex played. "Brilliant on the part of Comex. Absolutely superb."[14a]

The new survival suits, the cooperation between superintendents, the need to move quickly and decisively with all available resources—this is why Robinson and Tucker were saved.

Meanwhile, in San Francisco, Gerald Sterns received a visitor from the UK. Although Sterns felt a US court would rule it had jurisdiction over Infabco, Masterson's insurance company disagreed. And it sent one of its solicitors to

---

a Two months later, on March 28, another successful wet transfer was performed in the Bay of Campeche, Mexico that saved the lives of two Taylor divers. *Source*: *Roustabout*, April 1981, volume 117, p. 19.

Sterns's office to deliver the message in person. "We're not going to defend this case," the man said. "We're going to rely on English Law."[15]

Masterson's insurers were gambling that, even with a default judgment against them, Sterns would never be able to recover because Infabco had no assets in the United States. Moreover, they believed that if Sterns attempted to have the US judgment enforced by a British court, he would lose, which would then force him to file a new case in Britain where the courts had a more restrictive view on compensation.

By late January, Infabco made good on its forewarning and defaulted on the lawsuit when it failed to appear in court. This meant the liability phase was over. Accordingly, Sterns filed a motion to take default against Infabco. All that remained now was for him to prove his clients' damages at an evidentiary hearing which was scheduled in April.

On the support issue, Jeanne's claim was more straightforward than the Guiel's. She and Marisa had relied upon Richard's salary which meant the law would ultimately look at compensating them in a different way. Intrinsically, Skip's life was worth no less than Richard's, but under the law, for the purposes of compensating dependents, it was.

In March, Sterns asked the Guiels to prepare statements about their relationship with Skip. He wanted to demonstrate to the Court the pain and suffering Infabco had inflicted upon his clients. So one by one, they sat down with pen and paper and described the unspeakable pain of having lost the person who'd held them together as a family.

In beautifully precise handwriting, Sheila wrote page after page in agonizing detail about Skip's life, how he had protected her when she was young, stood up to a father whose emotions sometimes exploded over a misunderstood word, and how he had dedicated his time, money and love to the people around him. Lastly, Sheila wrote that the hardest thing she ever had to do in her life was to find Roger and tell him that Skip was dead.

For Roger, losing Skip had destroyed his direction in life. All through childhood he had worshipped his older brother, emulated him, and lived in his shadow. Skip had taught him all the things older brothers teach younger brothers, but above all else, Skip had kept Roger on track morally, reminding him of his responsibilities whenever he veered off course. And Roger listened to Skip because he had an inner strength that commanded his love and respect. Even under pressure, Skip was always calm, never riled, and that impressed Roger tremendously.

But when Skip died, Roger's world began to cave in on him. He quit his job and drifted from place to place. His plans for the future seemed meaningless:

> Everything had changed. Vic was no more. I had no partner. Sheila took care of all his affairs after his funeral. We put everything in both our names. She gave me back the account and let me live just like Vic was still alive. We have a trust between the

> three of us and it can't be broken for any or all money because we've depended on each other since we were born. We were always there for each other and we knew that. Now Victor isn't there. I hope I never forget anything about him.[16]

On any summer's afternoon, if you passed by Shirley Guiel's house on Lowell Street, you might catch her watering her flower beds, sitting on the front porch enjoying a cigarette, or talking with a neighbor over the fence. Gone were the days of children building army forts in the tall grass, and the sound of hockey sticks clacking against each other on makeshift winter ice fields. There were no more kids to catch under her front porch smoking her Pall Malls, no more Mr. Wizard experiments exploding in her cellar. Life in Al's Corner was considerably quieter now; the Dayriders and the Nightriders had all grown up and moved away.

Lowell Street was home to Shirley. She'd been born here and it's where she wanted to die. After 52 years she was used to the sound of the railroad cars crashing through her neighborhood at two in the morning. She was a woman who'd worked hard all her life in a factory job to give her children a better life. But losing Skip had nearly destroyed her. After the funeral, Skip's grave became her shrine, a place to visit three times a week to water the flowers, pull a weed or two, and sit quietly beside his headstone to confide in him. Three times a week, she would look out over the rolling hillside to ask for his reassurance and guidance. Three times a week she struggled with the calamity of his death. Not until Sterns asked her to put her thoughts down on paper had Shirley voiced her feelings about what it meant to lose her eldest son. She came from a generation uncomfortable with self-expression, and she found it hard to bring her feelings to the surface after a hard life.

"I am writing this letter for Victor," she began in flawless script. "It's hard and painful. How can a mother explain the worth of a son?"

Shirley spoke of the day Skip sat her down to explain that his work was dangerous. He didn't want her to worry, but he wanted her to know that what he was doing was something he'd been preparing for all his life. Shirley understood because she adored him and respected his wisdom, and his courage. From the day the old men on the shoreline in Maine cried out, "Look, look, it's Lloyd Bridges! It's Lloyd Bridges," Shirley knew that Skip's heart belonged to the sea. But nothing could have prepared her for his loss:

> When I found out Vic was dead, I had no will to live. [It] seemed like my life was just over. I was getting a divorce, my health was poor. I never realized how much I or my other children depended on him for support in everything. If we needed advice on anything, he was there. Losing him is like losing ourselves. My faith in God has helped me accept it. I know very little about the details of his death because to accept losing him is hard enough. All of our lives have been changed. My other two children are still lost.

For more than ten pages, Shirley's anguish poured out on paper. In the final lines her thoughts were for him because she was his mother and she had been asked to explain the unfathomable:

> It is hard, but for him, we will do it. We can't let him down. We are all lost but because of what he stood for and believed we have to carry on.[17]

In early April 1981, the Guiels flew to California to testify before Judge Manuel Real of the US District Court in Los Angeles, and to meet, for the first time, the wife of the man who shared death with Skip. The two families were very different, with different backgrounds and lifestyles, but they would remain united in grief as time led them through their ordeal.

There was no discussion about the accident or the events leading up to it. None of the issues concerning wrongdoing were addressed because liability was no longer the issue. All the Court was interested in was Richard and Skip's history of earnings, their life expectancies, and whether they were supporting the plaintiffs. It was just a straightforward calculation of the damages.

A month later, Judge Real handed down his decision, ruling that the California Courts had the power to enter judgment against Infabco Diving Services and A/S Møllerodden (the manufacturer of the *Wildrake* diving system) on the grounds that they were doing business in California.[18] Jeanne and Marisa were awarded compensatory damages of $475,000 and the Guiels $75,000.[19] Sterns next move was to try to have the California judgment enforced by a UK court.

By this time, Malcolm MacNeill had notified Jeanne and Sheila that he had fixed the date of the Fatal Accident Inquiry for May 11, 1981.[20] Where the focus of the criminal trial had been to prove the charges against Infabco, the inquiry would reduce the accident to anatomical detail. It would be the longest inquiry ever conducted into the death of two North Sea divers, and this time, witnesses that might have altered the course of the trial would give testimony. This time the Dynamic Duo would arrive in Scotland with an armload of documents. This time, after struggling through the horrors of their imaginations, Richard and Skip's relatives would hear, in one shocking revelation after another, how their loved ones came to be trapped on the bottom of the sea.

This time, most of the truth about the *Wildrake* accident would come out.

Chapter Twenty-four

The Inquiry

I think in terms of neglect for the safety of divers who are in a bell under compression however deep—300, 400 feet down—it was extremely bad....I mean, it was absolutely crystal clear that there were so many appalling deficiencies in the equipment and the handling of the equipment upon which these guys' safety and their lives depended.

—— Reg Christie
Attorney for the relatives of the deceased[1]

Sitting in the front row of Courtroom No. 1, Roger Guiel had vengeance streaming through his veins. Time had not cured him of the hatred he felt towards the people who ended his brother's life, nor the system that failed to exact retribution for it. As far as he was concerned, the people in this country had killed his brother, orchestrated a sham trial and were now putting on another minstrel show for appearance's sake. *Pomp and pageantry, isn't that what the Brits were good at?* All these people scurrying about wearing white hair and black gowns had magnified Roger's feelings of stored anger and cynicism. He wanted to scream out, *"Give me a break, take the white wigs off!"* Instead, he sat quietly beside his brother-in-law, holding back his rage and waiting for the show to begin.

In the middle of this torrent, Roger turned his thoughts to Sheila who was with Jeanne somewhere in the building preparing to testify. He was intensely proud of his younger sister. She had been the one to organize the funeral arrangements and to write to the authorities and to gather the tax records and talk to the attorneys. He knew she was destroyed inside, too, but somehow she'd been able to drag herself through the paperwork while he remained paralyzed with anger. During those first few months, all he wanted to do was hop on a plane, fly to Britain, and "make it right." But like the victim of a hit-and-run accident, he didn't know whom to kill.

Above the judge's bench, Roger noticed the colorfully ornate plaque, unaware that in this very room Masterson had made a mockery of its motto. He tilted his head back and inspected the ceiling. *Christ, it's high enough to fly a kite in here!* The upper gallery behind him was empty. For that matter the lower benches were practically deserted as well. Already, the *Wildrake* accident was beginning to fade into obscurity.

Photo by Author

Reg Christie in 1996.

Roger stared over the banister separating him from a long wooden table where half a dozen black-robed solicitors were shuffling papers, among them a young Dudley Moore look-alike by the name of Reg Christie. Through his contacts in London, Gerald Sterns had found the Aberdeen solicitor and offered him the job of representing his clients.[2] Christie readily agreed to take on the case, and as luck would have it, no better choice could have been made.

With superb legal skills, Reginald G. Christie was a master at cross-examination and a virtual barracuda against even the most intransigent witness. He was also no stranger to inquests involving North Sea divers. In 1977, he aided the investigation of a bizarre accident that claimed the lives of two British divers working for Oceaneering International. Two years later, another dual fatality brought him to court to represent the relatives of two men killed in the 1978 *Star Canopus* accident.

Now Christie was making his third appearance on behalf of the relatives of divers. He understood the technology of deep-sea diving, and after studying the trial transcript Sterns had sent to him,[3] Christie knew the *Wildrake* case was fraught with negligence and likely to expose a lot of questionable diving practices that had been around for years. Only days earlier, Roger and the others had met with Christie and took an instant liking to him. He was patient when explaining Scots Law; he was completely committed to their cause, and he told them what to expect from the inquiry process. This would not be a trial, he warned them; there would be no winners or losers, no consequences for the guilty, only the search for truth.

Also at counsels' table was James Tierney, Masterson's solicitor, Mr. Strath MacKenzie, solicitor for Wilhelmsen Marine Services (the *Wildrake* owners), a man named James Steel representing British National Oil Corporation, Mr. Allan Crawford for Wharton & Williams, and Malcolm T. MacNeill for the Crown.

In time, battle lines would be drawn with MacNeill, Christie, and Crawford generally falling on one side of the field, peeling away layer after layer of violations and bad judgment, while Tierney and MacKenzie would attempt to deflect blame and prevent evidence prejudicial to their clients from entering the record. Steel, for BNOC, would speak little as the finger pointing was generally directed away from his client.

Photo by Author

Sheriff Douglas James Risk, 1996.

Suddenly, a tall, wiry, patrician-looking man stepped into the courtroom from behind the judge's bench. At the command of the court officer, everyone stood up as the tall figure, draped in black, exchanged bows with the group of lawyers then took his throne.

If ever there was a man who looked the part of a legal referee, Sheriff Douglas James Risk fit the bill. At age 40, his hair was already showing signs of turning steel gray above ears that tended to stick out a bit too far. His face had a chiseled look, as though an Aberdeen mason had produced it.

Glasgow born and raised, the son of James Risk and Isobel Katherine Taylor Dow didn't grow up dreaming about becoming a judge. He wanted to be a lawyer like his father, and in 1963, earned his BA from Cambridge. Two years later a law degree from Glasgow University followed. By that time, Risk had decided he wanted to earn his living as an advocate rather than a solicitor. A year later, at the age of twenty-five, he was admitted to the Faculty of Advocates. In 1967, he was back in the halls of Cambridge earning his master's degree, and for the next ten years practiced law at the Scottish bar.

Then, some time in the mid-1970s, the man destined to oversee the *Wildrake* inquiry decided to move to the "other side." He obtained a commission as a temporary sheriff and found the job suited him. In 1977 he applied for a full-time commission and was appointed Sheriff of Lothian and Borders at Edinburgh. Two years later he transferred to Aberdeen.

Sitting in his high-backed chair on the morning of May 11, Sheriff Risk had presided over dozens of fatal accident inquiries but none involving the offshore diving industry. The *Wildrake* accident would be his first case.

Not until much later would Roger and the others appreciate the measure of the man sitting behind the raised bench that day. For had the inquiry been overseen by someone with a less diligent nature, it's quite possible the details of Richard and Skip's deaths would have remained a mystery forever.

The mandate under which Sheriff Risk operated first arrived on the judicial scene in the waning years of Queen Victoria's rein. Spurred by deplorable working conditions during the Industrial Revolution, Parliament passed the Fatal Accidents Inquiry (Scotland) Act 1895. It charged the procurator fiscal with the duty of holding a public inquiry into all fatal accidents which occurred in the workplace. Verdicts were issued by a panel of seven jurors who were required to determine where and when the accident occurred and the cause of death.

In 1906 the Act was amended. The jury's responsibilities were expanded to ascertain the persons whose fault or negligence the accident was attributable, the precautions by which the accident could have been avoided, and any defects in the system of work which contributed to the accident. The scope of the accident was also broadened to include any "sudden and suspicious deaths," not merely deaths occurring at the workplace. In 1976, Parliament removed the jury system and charged the sheriff with delivering the findings.

In his book, *Sudden Deaths and Fatal Accident Inquiries*, Scottish Advocate Ian Carmichael writes that, in some instances before 1976, inquiries tended to be "mere formalities,"[4] with cases sometimes stacked up like cord wood and processed in assembly-line fashion. Particularly in the case of industrial accidents, the fiscal rarely addressed the jury after presenting evidence, and the sheriff "more often than not had ready a draft verdict," usually restricted to the date, time and place of death. Moreover, the sheriff frequently advised the jury not to assign blame although it had the power to do so:

> Thereafter the jury usually signified their assent without retiring, the verdict was recorded and the next inquiry called. When the matters under investigation were gone into with such speed, one wonders if those legitimately interested, such as the relatives of the deceased, received much in the way of enlightenment as to the causes of accident or death.[5]

But even after 1976, some sheriffs continued to ignore the 1906 mandate by issuing pathetically meager findings on the very day the inquiry closed.[a]

Fortunately for the relatives of Richard and Skip, Sheriff Risk was not cut from this cloth. He took his job seriously and beneath his sober exterior, he was a thoughtful and compassionate man with an impressive capacity to distill simplicity from the complex. He also knew a bald-faced lie when he heard one and suffered no fools. In fairness to the process, he would often interrupt an examination to give a wayward witness an opportunity to row back to truthful ground, and those that did not were frequently left with a case of self-inflicted humiliation. Nor was he the least bit timid when it came to assigning blame. Before the inquiry, Jeanne had heard rumors about Risk. "People said he was a hanging judge," she recalls. "We said, 'Good!'"[6]

For the relatives, the inquiry was their last hope of finding out what actually happened to Richard and Skip. They knew it would never achieve the kind of justice they wanted, but they hoped it would raise enough public awareness to prevent another *Wildrake*-style accident. They also needed the truth to free themselves from all the months of grief and sadness. They needed to know why.

That quest for truth translated into another lengthy battle over the thorny issue of Richard and Skip's employment status. At the Infabco trial, Sheriff Stewart complained that MacNeill had failed to prove Richard and Skip were employed by Infabco Diving Services under a contract of service. Most inquiries avoided this contentious subject, but Risk's interest had been piqued, despite strenuous and repeated objections from Masterson's attorney that it was not the purpose of an inquiry "to find out who was or who was not the employer of the divers," nor to "set the stage for putative defendants or defenders in civil proceedings." After lengthy arguments, Risk ultimately ruled that the contractual relationship between the deceased and Infabco was relevant and should be heard. As a result, MacNeill and Christie spent the better part of four days methodically and painstakingly introducing evidence to settle this dispute once and for all.

They began by putting Sheila on the witness stand, followed by Jeanne. Between the two women, documents were introduced that clearly showed how much Richard and Skip were paid and by whom.[7] When Christie asked Sheila why she was familiar with Skip's employment history, she said she kept her

---

a For example, at the conclusion of the Clay Don Ellis Fatal Accident Inquiry, which began on June 29, 1977, Sheriff James Law wrote twenty-two lines of text, half of which were dedicated to the location of the accident, the time of day, the names of the victims, and where they lived. The remainder described the vessel, what Ellis was trying to do, and lastly how the accident occurred. The Determinations were issued on the same day the inquiry came to a close, Friday July 1, 1977. Other examples: John Howell FAI, June 6, 1977; Michael Moore FAI, September 12, 1977; Howard Spensley/Charles Meehan FAI, January 30, 1978; Craig M. Hoffman FAI, June 26, 1978.

brother's books, did his taxes every year, and became the executor of his estate after he died. When she discovered a discrepancy in his earnings for the 1979 tax year and wrote to Infabco for clarification, a company official responded by acknowledging the mistake and correcting it. Infabco never suggested to her that they were not the ones that dealt with Skip's pay.

"You were not referred to anyone else such as Offshore Co-ordinators?" Christie asked.

"No mention at all."[8]

Towards the end of the first day, MacNeill launched an attack against Masterson's credibility when he put Infabco General Manager Bob Lloyd on the stand. At the trial Masterson testified it was his policy not to hire divers, and that the divers would apply to Offshore Co-ordinators when looking for work. It was Offshore Co-ordinators who controlled the divers, not Infabco.[9]

But Lloyd contradicted this. He said divers came into Infabco's Aberdeen office seeking work "on a regular basis," were interviewed, and if found suitable for engagement, their rate of pay was negotiated and they were hired. The employment information was then transferred to Jersey to set up the appropriate payroll account. Jersey had nothing to do with these negotiations, Lloyd said. In fact, "they wouldn't have spoken to the divers."[10]

When Christie took over for MacNeill, he revealed that Offshore Co-ordinators had been incorporated on the Isle of Man on March 7, 1979, and by the following day had managed to form a contractual agreement with Infabco.[11] This suggested the new company had been brought into existence for the sole purpose of paying the divers. When Christie asked Lloyd (and later Project Manager Bill Hemmings), in what other way, besides paying the divers, did Offshore Co-ordinators act as an employer of divers, both men said there was none.[12]

Christie then moved on to the crucial issue of who controlled the divers. The legal formula for distinguishing between employee and independent contractor takes into consideration several factors, among them: who controls the details of the job and determines how it should be done, who supplied the tools and equipment, who had the power to dismiss, and was the diver integral to the business or was he an accessory to it?[13]

Lloyd testified that Infabco supplied the tools and equipment, and that this provision was actually written into the SALM contract. When Christie asked if Infabco really operated in the field of diving, Lloyd answered yes.

"So that the divers were in fact a very integral part of the whole business of the company?"

"Oh, yes, in any diving operation."[14]

Lloyd went on to say that the chain of command began with Infabco and that operational instructions were passed through the diving superintendent, to the diving supervisor, and then to the diver. And the "divers would operate always under the orders and instructions of the supervisor," Lloyd said. The diving supervisor had the power to dismiss the diver. To make certain Sheriff Risk understood that control over the divers traced back to the office of the general manager, Christie inquired if the diving superintendent was responsible to the general manager of Infabco?

"That is correct," Lloyd replied.

"And that is you?"

"Yes."[15]

At this point, Sheriff Risk wanted to know if Lloyd was on a self-employed basis with Infabco or a full-time salaried employee?

"I was a salaried employee of Infabco Diving Services," Lloyd answered.

When Christie tried to pursue the issue even further by referring to a document over the objections of Masterson's attorney, Sheriff Risk chimed in, "I think, Mr. Christie, you have pretty well made your point with this witness."

"I am glad to hear that, My Lord," and he moved on to another subject.[16]

Lloyd's testimony had a significant impact upon the Court, but Malcolm MacNeill eviscerated any further argument that Offshore Co-ordinators was the employer of Richard and Skip when he introduced a sworn statement from a lady named Mrs. Brenda Rhensius, who acted on behalf of the enigmatic Offshore Co-ordinators Limited on the island of Jersey.[17] Dated three days before the inquiry, Rhensius stated she was employed by an organization which acted as Company Secretary to Offshore Co-ordinators Limited ("Offshore"). Her job was to invoice Infabco and pay the divers on a monthly basis "in accordance with instructions received from Infabco." She said she was familiar with the names of Victor Francis Guiel and Richard Arthur Walker and said they "were not employees of Offshore but were self-employed Consultants who were made available on a non-exclusive basis to Infabco."

Rhensius went on to reveal that Richard and Skip had been sent "consultancy agreements" by Offshore Co-ordinators for their signature, but they were never signed nor returned, yet Infabco continued to treat Richard and Skip as though they were engaged by Offshore Co-ordinators regardless:[13]

> Guiel and Walker were introduced by Infabco to Offshore. They were not interviewed nor to my knowledge were they ever seen by any member or employee of Offshore. No member or employee of Offshore assessed their qualifications or medical condition for their work with Infabco, had any knowledge of the work for which they were engaged, or instructed or supervised them in their work. No member or employee of Offshore was employed at Aberdeen or on any of the diving ships engaged by Infabco.

Here indeed was Masterson's paymaster, an abstraction of the industry that turned out to be a lady in a bank who wouldn't have recognized Richard and Skip if she had passed them on Union Street, much less be able to evaluate them as divers.

By mid-morning of the third day Brian Masterson entered the courtroom, faced Sheriff Risk and held up his right hand. Risk asked the witness to repeat the oath after him and then Masterson took the stand.

Jeanne and the others hadn't seen Masterson since August 1979. They were not at the trial to hear him deny employment of Richard and Skip, but over the next several days they would listen to him give conflicting, sometimes ludicrous, evidence that severely tested the obliging disposition of Sheriff Risk.

After introducing the smoking-gun affidavit of Brenda Rhensius, MacNeill wasted little time asking Masterson if it was not the case that Offshore Co-ordinators "had absolutely no contact with divers" and did not provide a single man or person as a diver for the SALM contract.

Masterson parried this challenge by saying that that was not his understanding of how the divers were engaged by his company. And then he proceeded to describe the system of acquiring divers as though he were trading in commodities, like corn or pork bellies. Infabco would make a request to Offshore Co-ordinators, and they in turn—because they had the "option" on the diver—would match the request with the name of the best candidate for the job.[19] Masterson said he couldn't see the system being "workable otherwise."[20]

But MacNeill was not buying that scenario. "Would it not be easier to see the unworkableness of a company having literally nothing to do with diving except payoffs, nothing to do with any divers that were sent to you. Isn't that the unworkable side of it?"

Calmly, Masterson replied, "You must get the right class of diver for the job and you must pay them the rate for the job."[21]

Beneath his mild-mannered exterior, MacNeill's composure was beginning to crack. He'd been plagued by Masterson's prevarications at the trial, and now his journey into what seemed like the Infabco Twilight Zone was starting all over again. Realizing he was getting nowhere with this line of questioning, he decided to confront his nemesis directly by asking if he knew Mrs. Brenda Rhensius.

Masterson acknowledged that he did.

And how was Mrs. Rhensius to know the divers to engage for Infabco, MacNeill asked.

"Through the organization that they had," Masterson replied. "I would expect them to be capable of contacting the divers for the provision for this contract, and other contracts."

"Don't let's beat about the bush, what organization did they have that would enable them to know the divers, enable them to say who was working and who was not, what organization are you talking about?"

"I really don't know."

"You don't know?"

"No."

"Was Mrs. Rhensius working for Lazard Brothers?"

"Yes."

"They are not a diving company?"

"No."

"They are bankers, are they not?"

"Yes."

MacNeill read portions of Rhensius's statement and put it to Masterson that Infabco hired divers in the usual way, but all he received for his trouble were denials and glib responses that went nowhere.

This struggle to get Masterson to yield to the facts continued for some time, then took a decisive turn when the Infabco owner made the startling admission that Offshore Co-ordinators played absolutely no part in the creation of the agreement to supply divers to Infabco.[22] MacNeill then embarked on a campaign to discover the motive behind setting up Offshore Co-ordinators in the first place, but Masterson began bobbing and weaving again with a lot of slippery, "I don't know that myself" responses, claiming the people below him must have been responsible, not him.[23]

This got MacNeill's back up again, accusing the witness of giving the false impression that he was unaware of how the rest of his company was operating when, in fact, that was not the case.

"Absolutely not," Masterson agreed.

"Well, that is the impression I am getting," MacNeill retorted. "I don't know what impression anybody else is getting. You are saying that anything I have asked you about was something that was by a person in a junior grade to you and you were not knowing about it?"

"I was fully familiar with the way of it, we had meetings about such things."

"Well, why not give your evidence in a way which indicates a bit more clearly that you knew what was going on and what was not going on," he boomed. "Is the fact of the matter not this, that the divers were engaged in the way I have described, and that the purpose of the payments was reported, or your office reported it to Offshore Co-ordinators the name and address of the chaps who wanted to be paid through Offshore Co-ordinators?"

"I don't know the answer to that."[24]

Masterson went on to say he had discussions with Lloyd and Hemmings about these things: the contract with Offshore Co-ordinators was in fact his policy being carrying out. But when MacNeill suggested that the purpose of the arrangement was to use Lazard Brothers simply as paying agents, Masterson continued his maddening responses, "That was not the purpose of Offshore Co-ordinators from my point of view."[25]

Fed up and exasperated, MacNeill rattled back, "Can you say clearly what your position was concerning Offshore Co-ordinators, what was your purpose?"

"To make certain that when the particular contract was awarded to us, that we had sufficient personnel with the correct qualifications to execute the contract."

At this point, Sheriff Risk came to MacNeill's rescue.

"And were you satisfied that Offshore Co-ordinators, on the date when you signed the contract, had sufficient expertise and organization to supply your needs?" Risk asked.

"From the recommendations that were made to me, yes."

"You had recommendations that this company was good and efficient in supplying personnel at the date when you signed the contract?"

"That is perfectly correct," Masterson replied, straight-faced.

"Would it surprise you to know that that was the day after the company was incorporated?"

"I would not be surprised about that."

Risk looked at Masterson sharply. "So they had built up their expertise and organization in 24 hours, is that what you say?"

"In a similar way as we do at times, yes. It depends on the personnel you have in the company."[26]

That incredibly fatuous response was an insult to the Court's intelligence and if Masterson was not careful, he could have found himself in contempt of court. There was a common-law power vested in all sheriffs which enabled them to punish a witness summarily if he continued to prevaricate under oath.[27] But Sheriff Risk restrained himself that day, giving MacNeill the opportunity to move on to greener pastures.

On the fourth day, Christie put the whole employment matter to bed when he focused on the statement of Brenda Rhensius, particularly the part about Offshore Co-ordinators never interviewing Richard and Skip or having knowledge about their qualifications or medical history. Masterson told the Court that Infabco assessed the qualifications of the divers and knew the work they were to be assigned. When Christie asked if he knew anyone at Offshore Co-ordinators who knew "the first thing about diving," Masterson said, "No."

When Christie asked who employed the diving supervisors and superintendents, Masterson answered, "Offshore Co-ordinators."

"And they are appointed by Infabco, are they not?"

"That is correct."

"To [their] respective positions?"

"That is correct."

"Infabco reserve the right to select any diver for any particular position?"

"Oh, yes."

"It was only Infabco who could do that, because Offshore don't know the first thing about it, do they?"

"This is correct."[28]

Had MacNeill been able to lead the same evidence at the trial, one can only speculate whether Sheriff Stewart would have allowed the jury to deliberate a verdict or not. But from the evidence at the inquiry, Sheriff Risk ultimately found that, "apart from the actual delivery of wages, Infabco continued to exercise the normal functions of an employer in relation to divers."[29] Senior officials had actively interviewed divers, offered them jobs, negotiated their salary, and then passed that information off to Jersey. But it was Infabco Diving Services, he said, that "exercised operational control" over the divers and had the power to dismiss them if need be. In his concluding remarks, he said:

> It is perfectly obvious, notwithstanding Mr. Masterson's contrary assertions, that the function of Offshore Co-ordinators Limited had nothing to do with diving, but was simply to have its registered office in the Isle of Man and its place of business in Jersey, so as to confer certain fiscal advantages upon those who utilized its services.[30]

It didn't happen by design, but by the fourth day of testimony, veteran reporters were convinced that the *Wildrake* inquiry was like no other. Sheriff Risk had already granted MacNeill and Christie an unprecedented amount of time to debunk a mythical Jersey-based diving contractor. But the inquiry covered other matters those first few days which produced episodes of heart-wrenching drama that catapulted the *Wildrake* tragedy far beyond anything Aberdeen had ever seen.

Until now—though the events of the accident and subsequent trial were captivating and shocking—nothing had happened to suggest this was anything but just another sad story about two divers who went offshore never to see their families again. The press had no pictures of Richard and Skip, no background information on them, so they remained as anonymous as the day they died at Thistle 21 months ago. But all that changed on the first day of the inquiry.

After Sheila's testimony, MacNeill called his second witness. A moment later, a side door of the courtroom opened and a tall slender woman dressed in black walked across the floor and climbed the hooded witness box. When the fiscal asked her to identify herself, she replied, "I am the widow of Richard Arthur Walker."[31]

After confirming the identity of her husband and his occupation, Jeanne explained how she was notified of the accident, who was with her at the time, and who Richard's employer was. This was the typical format, and at other inquiries relatives scarcely took up more than a few minutes of the Court's time. But once again the *Wildrake* inquiry did not follow the norm. Towards the noonday break, Christie took over and sparked a media sensation by asking Jeanne to read Richard's last diary entry. With tears welling up, she read her husband's stunning last words into the public record.

No one in court that day knew much about the hardships of saturation diving, or what caused Richard to confess his misgivings about the job just hours before his death. No one understood the basis behind the worrisome phrases, "Poor topside management," and "I'm no longer impressed." But as Jeanne read on, it became perfectly obvious the author of the diary feared for his life. No one had to be a diver to understand what he meant by: "Dear god, I want out," and, "Oh god, please help me to exercise my talent and will to pull out of it." Everyone understood the desperation of: "I don't even know if I'm gonna get out of here alive."

If it had been anyone else reading these words, the diary might have been dismissed as a fake or a vulgar prank. But to see Jeanne dressed in black, clutching the blue notebook, and hear her voice quivering as she spoke the words—there was no doubt in anyone's mind that they were listening to the man she loved on the eve of his death. In utter disbelief, the spectators remained transfixed as the voice of a ghost echoed across the room confessing his weaknesses, his addiction to the big paycheck, and how Marisa would never understand that he had fallen prey to an occupation that was dangerous and unhealthy.

When Jeanne finished, she looked up, her eyes glistening with tears.

In the press box, reporter Graeme Smith sat there mesmerized by the spectacle of a young, very striking woman wearing funeral attire, reading the last words of the man she obviously loved. With Richard's final thoughts now in the public domain, Smith couldn't help but think this tragedy was taking on a life of its own. And if he thought the revelations were going to end there, he would have been mistaken. In the days that followed, Jeanne and Sheila would give interviews to the press and supply pictures of Richard and Skip which appeared on newsstands from London to Aberdeen under tragic headlines adapted from the diary. Suddenly, the two men who died at Thistle were no longer faceless divers anymore; they had become the symbol of all the fathers and sons and brothers who had died on the rigs.

Heartache and suffering extended well into the afternoon that first day when MacNeill put the man who'd performed Richard and Skip's autopsies on the witness stand. Dr. Hendry was the head of the Department of Forensic Medicine at Aberdeen University, and in cold clinical terms, he cleared up any residual doubts that Richard and Skip might have died from something other than hypothermia.

Hendry's dark words were emotionally draining to the relatives, but the anguish they caused was nothing compared to the shameless attempts by some of the attorneys to substitute reality with pure fiction at the expense of two men who could no longer speak for themselves.

It was at the very end of the doctor's testimony when MacKenzie, the attorney for Wilhelmsen, asked a few perfunctory questions about Richard's clothing, then suddenly shifted to the real topic on his mind.

"Did your examination of the remains of these two unfortunate divers show whether or not they were under the influence of drugs or stimulants to any extent?

"No," replied Hendry. "Analysis is not done for drugs in general unless there is something pointing to them."

"You do, in fact, exclude alcohol and carbon monoxide poisoning?"

"Yes."

"But you do not specifically exclude drugs or stimulants?"

"No."

"And this is because you did not test for them?"

"Because I did not ask for them to be tested, yes," Hendry countered.

Unfazed by the doctor's rearrangement of his question, MacKenzie pressed for the answer he wanted to hear. "So there might have been?"

"There might have been, yes, but there is no evidence pointing to it that I am aware of."[32]

And later, there was a similar episode when Masterson's attorney, over the strenuous objections of Christie, succeeded in having a second passage of Richard's diary read to the Court. The entry was dated July 19, when Richard was at the airport preparing to return to the North Sea, and in one of his cheerless moods because he was leaving Jeanne and Marisa and his dog, Jozie:

> LAX again. Broke down and sobbed last night. Flight to London in ten minutes. I don't feel good about leaving my fine two loves—three loves. I'm really broken hearted. How I love them and my eyes fill with tears as my mother's eyes fill with tears and maybe I'm wrong about so many things, and, well, I will survive. I feel so wasted...

Notwithstanding that three weeks separated this entry from the one Richard wrote shortly before his death, Tierney thought there was "an interesting parallel" between the two entries that disclosed "a continuous state of mind" which might have played some part in Richard's ability to function at the time of the accident.[33]

Tierney and MacKenzie were almost certainly trolling for evidence of contributory negligence that might lighten the liability load for their clients in future court proceedings. But despite these preposterous insinuations, Sheriff Risk would later conclude that the state of mind of Richard and Skip, "as evidenced by letters and diaries," had no bearing on the events which occurred. He also said that, upon reflection, he regretted having allowed himself to be persuaded to admit the diary entry into evidence, which he said, "served no useful purpose and which must have occasioned distress to the relatives of the deceased."[34]

Risk was right about that. The mere suggestion that Richard and Skip had been diving on drugs or were mentally unstable at the time of the accident was outrageously offensive to the relatives who would never forget, nor forgive, the men who tried to smear their loved ones' memory. Sadly, this was the first attack on their memory, but it would not be the last.

During the first week of the inquiry, Jeanne and Marisa, who was now three years old, stayed with a friend of Chris Roper's in a small village overlooking the North Sea. Every morning, Roper would drive out, pick Jeanne up, and take her to court. When a room became available during the second week, mother and daughter moved to Roper's where the Guiels had been staying. On some days the foursome would walk around the city or visit with Christie to talk about what happened in court. Testimony they didn't understand, he would explain. Questions they wanted asked, he'd find a way to work into his cross-examinations. Having Christie on their side, Jeanne said, "made all the difference in the world."

Around town the two women stood out and reporters constantly approached them for interviews. Roger and Rick didn't interfere. They stood back and watched in admiration as these two very strong women took control of family decisions. But there were times when the emotional strain on Sheila nearly overwhelmed her, when she wanted to dig a hole and crawl into it and die. It was during these times that she'd turn around and hear Roger and Rick chanting like a couple of high school cheerleaders, "You're doing okay. You're doing okay." And so she forced herself through the next day of dread, hoping it would be her last.

The remainder of the inquiry dealt with the events of the failed rescue attempt. BNOC official Ron Wilson verified the log entries and identified Masterson as the man in charge of the rescue mission. Tim Newson explained how the bell umbilical was destroyed using the ship's crane. Randy Gunter recounted the aborted surface dive to 500 feet, the lack of coordination on deck, his conflict with Superintendent Holmes over rigging procedures, and the threat to charge him with mutiny if he tried to take the rescue out of Masterson's hands. Supervisors Don Cox and Mike Allen related the events from the rescue ship: the frantic efforts to put the *Stena Welder* diving system back together, Masterson issuing instructions to them from the *Wildrake* bridge, and the rescue divers working under life-threatening conditions with virtually no hot water, no comms, and no bell lights. At one stage of the inquiry, Lacey Lahren described to the Court what he saw after breaking into the bell—recalling the look on Skip's face, his glazed eyes, the gash on his forehead and the froth running down his face. Momentarily pulled back into the past, the diver who vowed never to go back into sat, broke down on the witness stand.[35]

Day after day, the relatives listened as the unfolding drama metastasized into an unbelievable disaster. MacNeill had managed to subpoena just enough witnesses with different vantage points to provide a clear picture of what had happened that August night. Not until the final stages of the inquiry, after Holmes and Reid had taken the stand, did Jeanne and the others begin to realize, with frightening clarity, that the seeds of the tragedy had been planted long before Richard and Skip stepped on board the *Wildrake*. As the second week of the inquiry progressed, they heard evidence that the swivel and clump weight had been discarded, that the tire in the power block, which was supposed to have had a solid interior, was in fact empty at the time of the accident,[36] and that the motion compensation system and the moonpool aeration system were not working. Even the communications link between the bell van and crane cab was out of order.[37] They also heard Holmes acknowledge that it would have been safer for the bell to have a stage, yet did nothing to ensure that it have one before it entered the water.[38] The growing suspicion in the minds of the relatives was that safety had become an annoying burden to the job, a suspicion that turned to absolute certainty when the inquiry began addressing the subject of the drop weights.

Under cross-examination by MacKenzie, Masterson admitted that his divers had tampered with the system. And he acknowledged that if officials of Det Norske Veritas had ever found out about it, they could have revoked the certification.[39] When MacKenzie asked if this "fairly important" alteration had been reported to Veritas, Masterson said that "all of these minor modifications" had been reported, through Peter Holmes, to Møllerodden (the designers) while the ship was still being fitted out in Norway.

But MacKenzie was skeptical of this, because it implied that Veritas and Møllerodden knew about, and sanctioned, the lashing of the drop weights, and his next question put the whole matter into proper perspective.

"It seems somewhat odd to me, that either the manufacturers of the design of the actual weights system, or Veritas would approve a simple method of destroying the secondary means of recovery of the bell by attaching it to a length of rope. If the weight system had been found to be inefficient or not working in any of their trials, then they would have modified in a proper engineering fashion and not merely lashed with rope. Do you see that?"[40]

As an engineer, Masterson had to agree with this view, but then he turned right around and discredited himself by making the astonishing disclosure that it was his "company policy" to lash drop weights.

Flabbergasted by this preposterous remark, MacKenzie blurted back, "This is the policy of your company?"

"Yes," Masterson answered.[41]

Once again the Court was asked to surrender reality to fantasy, notwithstanding that such a policy, if it even existed at all, conflicted with the law. To test the veracity of Masterson's remark, MacNeill put Randy Gunter on the stand and asked him if it was standard practice to have the drop weights lashed on diving bells. Not only was it "not a normal practice,"[42] Gunter replied, but the people with whom he'd previously worked with and instructed, did not consider it a normal practice either.[a] Later, he was asked if it was possible for the bell to go into the water without the supervisor or superintendent knowing about the ropes. Gunter said no.[43]

In the audience, Sheila occupied her time by taking notes, while Roger glared at the witnesses and focused on Sheriff Risk's reactions to the evidence. Rick

---

a The bell on board the *Gulnare* did not have its drop weight lashed, and Gunter's former employer, Sub Sea International, never made its drop weight systems inoperable in any way.

scribbled questions down when the testimony didn't make sense to him, and Jeanne did the same. At the end of the day, her British Airways notepad was usually full of troubling questions. At one point, she wrote: "Why were the drop weights secured to bell via 1/2" nylon?"

In answer to that question, Masterson said the ropes were justified in order to protect the divers in the bell. This rationale was consistently put forward throughout the inquiry, but it was during Supervisor Brian Reid's testimony when Reg Christie stumbled over evidence that belied that claim.

Asked if he was present during a drop weight test conducted *after the accident* by DOE Inspector Roy Giles, Reid said no, but that he had been present at the one *before the Wildrake set sail* on its maiden voyage.[a]

Unaware that there had been such a test, Christie asked, "And what did you find?"

"You couldn't release the weight if the bell was sitting on the seabed."

"Without lashings?"

"Yes."

"Why was that?"

Reid explained that, without a stage or some similar device to stand the bell off the deck, the weights had no place to fall to and therefore remained stationary within their cradles.[44]

Christie characterized this as "a double catch-22 situation," where Richard and Skip were prevented from getting out of the bell to cut the lashings, but even if they had been able to, the bell still wouldn't have risen to the surface. Reid agreed with this, and then provoked an attack from Christie when he claimed he never knew why the ropes were attached in the first place.

"You were supervisor on a series of dives at this time when the bell went down with a drop weight system which was plainly inoperable, isn't that right?"

"Yes."

"Was it not your responsibility to see that that was sorted out?"

"Yes."

"You didn't do so?"

"I did, I told my superiors about it. I told him I didn't like the rope lashed on there at all. When I asked why they had a rope on, they said it was a safety device, because they didn't like the way the drop weights worked, because if it [the bell] was just dropped down, you would lose the weights straight away, and the bell would come to the surface. So, it was like—they said it was a safety factor. So, I said if you are going to be that extreme, why don't you put a wire strop round as well in case the rope breaks—being sarcastic."[45]

Of all the testimony at the inquiry, this was some of the most distressing for the relatives, because it meant that long before Richard and Skip boarded the

---

a No other witness was asked about this test at the inquiry, but in 1996, in an interview with the author, diver Tim Newson confirmed Reid's version of the events.

Wildrake, the people supervising the job could have taken precautions to avoid a known risk but chose not to do so. By virtue of the initial drop weight test, they had hard evidence—proof beyond a shadow of a doubt—that those weights would not work under the conditions with which they were prepared to dive the bell. And this is why the quality of negligence on board the *Wildrake* was so notoriously unique and criminally appalling—because it had the stench of premeditation to it.

Sitting in the front row that day, Jeanne, Sheila, Roger and Rick accepted the proposition that deep-sea diving comes with certain risks, risks that might lead to death or injury. But undermining a diver's safety net has no defense. Drop weight systems, like bailout bottles, exist for those rare occasions in which a diver must redeem his life from the brink of disaster. However one feels about their efficacy, these life-saving devices should be treated with respect and maintained, not arrogantly abandoned in a fit of machismo.

On board the *Wildrake*, safety received a lot of lip service, but when it began to interfere with work schedules and profit margins, it was the first consideration to be tossed aside. Those ropes were not a safety precaution. They may have created the illusion of being a precaution, but they were really just an excuse to keep diving.

At the inquiry, one of the *Wildrake* supervisors, under pressure from MacNeill, testified that, "Once the system has been certified and approved by the certifying authority, you can't just change what you want."[46] Unfortunately, Richard and Skip died before he experienced that epiphany.

Reg Christie poignantly summed up this issue when he asked DOE Inspector Roy Giles if the drop weight system was regarded as a last-resort means of recovering the bell?

"Yes," Giles answered.

"But from the diver's point of view, when all else seems to have failed, do you agree that it would be a last resort to which they should be entitled?"

"That is correct."[47]

The dismaying realization that Richard and Skip had become trapped on the bottom of the North Sea by an attitude of depraved indifference understandably made the relatives angry. But it was the behavior and poor judgment of certain crew members during the rescue attempt that made them positively furious.

To Roger, the reckless decision to raise the bell by its umbilical was probably the most difficult to stomach. "Damn, I wouldn't have tied off the whole umbilical. I would have separated it at the fucking stress line and tied it off to the stress line. You know that plastic stuff on the outside is going to stretch until it breaks. That's probably what made me the maddest is because I could understand it! I felt I could have done a better job, and I've never even been on a boat."[48]

And there were more incomprehensible acts: removing the transponder from the bell before the fatal dive; the fact that the superintendents of the *Wildrake* and the *Stena Welder* failed to confer with each other, Holmes and Roberts surrendering control of the operation to a businessman with no diving experience—all of these actions defied not just common sense, but good sense. And then, after the rescue diver spent hours struggling to attach a new cable at great personal risk, there were more mind-boggling decisions: starting the bell recovery without vertically aligning the *Wildrake*, failing to have the rescue diver clear the load off the bottom, dragging the bell into the SALM base and then trying to dislodge it by whipping the crane boom back and forth. When Christie suggested that this is what caused the wire sling to break, Masterson refused to accept that explanation.[49]

And probably the most perplexing act of all was the stubborn adherence to an unorthodox, overly complicated rescue strategy that was plainly taking too long, and the rejection of offers of help from nearby ships. In Jeanne's mind, this only made sense if the rescue was being influenced by extraneous considerations. On her notepad, she framed her suspicions with another penetrating question: "Was the reason for the ist [*sic*] lift by Infabco political? i.e. not to bring other comp. [companies] onto scene." It was a question that would be answered in time, but not by the inquiry.

In the end, no one claimed responsibility for the worst diving accident in North Sea history. While Craig Roberts told the Court he felt "those in charge were exercising proper care and skill and competence in what they were doing,"[50] Masterson never fully accepted the view that some of his decisions were "ill-advised," as one of the attorneys suggested.[51] When Christie suggested he should have taken no part in the rescue operation whatsoever, but left it to those with experience and skill of diving systems, Masterson replied, "I don't agree with that at all."[52]

As for Peter Holmes, on several occasions he ended up giving evidence so poor that, it not only damaged his credibility as a witness, it raised serious doubts about his competence as a superintendent. In his judgment, Richard and Skip were not freezing to death at all during the first ten hours of the accident, but were still receiving hot water until around midday when the crew began to recover the bell.[53] Christie found this difficult to comprehend because the dive log contradicted this view, and because Holmes was unable to speak to Richard and Skip during this period due to the loss of bell communications at the onset of the accident. When Allan Crawford, attorney for Wharton & Williams, tried to determine what evidence Holmes had to support this belief, he conceded he was only "assuming" Richard and Skip were receiving hot water that morning.[54] That disclosure produced the following exchange.

"So you really didn't have too much concern about the question of hot water?"

"Yes."

"It proved to be ill-advised, that lack of concern, because the divers had hypothermia, did they not?"

"Yes."[55]

At the end of the day, the *Wildrake* superintendent appeared to be all out of answers when Christie put this last question to him.

"Finally, what we have here is this, is it not, that first of all, the clevis pin comes out at the socket, then you find the power block is unable to pull up the umbilical, and then the umbilical comes apart, and parts of it break, you then find the crane does not work properly, the heave compensator is not working, the through-water communications don't work, the clump weight system is not used and is inoperable anyway, the drop weights can't be used. Isn't that an alarming catalogue of defects to have on a vessel where you are diving?"

"Yes."[56]

When MacNeill had his turn he charged Holmes with running a job on a ship containing a diving system which was "less than the minimum in which to go diving."

"If you put it that way, yes," Holmes replied.

Irritated by the tone of that answer, MacNeill drove home his point. "You were just able to put the bell down and you would be able to get the bell back all right as long as the hoist wire stayed in one piece?"

"Yes."

"But that thing did not happen, the hoist wire did not stay attached, you could not get the bell up again. That is how—that is what I would describe as less than minimum for going to do the dive?"

"Yes."[57]

Listening to the appalling evidence, Roger finally came to the conclusion that Richard and Skip were simply fated to die because the whole operation reeked of failure. "It was, like, inevitable that this was going to happen," he says.

As for Jeanne, the inquiry left her feeling empty and dissatisfied. Upon leaving Aberdeen she told reporters she still didn't know what had killed her husband. The inquiry had provided some answers, but there were two glaring questions that gnawed at her night and day. Were there political reasons for not bringing in more help? And why did it take so long to call in the *Seaway Swan*? It didn't make any sense. In her mind, there had to be more to the story behind Richard's death.

On May 22, after two weeks and nearly 2,000 pages of testimony, the investigation into the death of Richard Walker and Skip Guiel came to an end. A month later, Sheriff Risk listened to summations by the attorneys. For Reg Christie, this was not a difficult case to argue. Negligence permeated every aspect of the tragedy, and so it was no surprise to anyone when he stood up and began counting off the alarming list of failures which turned a fairly straightforward recovery into what he coined "a succession of disasters."[58] Overall, he concluded there were so many defects in the system, "that this vessel should not have been diving at all at this particular time."[59]

"Saturation divers, who put their lives at risk in working at sea depths of 500 feet or more, are entitled to expect that every reasonable step is taken to maintain their safety and the safety of the system."[60] The fact that the crew did not do this, he said, raised "very serious doubts as to the skill of those on board, their knowledge of the system and their general competence."[61]

MacNeill also spoke and said that the *Wildrake* was in an "incomplete state" at the time diving operations were being carried out.[62] In his view, the recovery could have been made in time to save Richard and Skip, and the fact that they were not saved, demonstrated "the want of reasonable precautions and defects in the system."[63]

When James Tierney took his turn, he tried to mitigate Infabco's misconduct by saying the company had acquired the most modern diving ship available and operated the system in the way it received it, "and the way it was certified."[64]

It was late June and the relatives had long since returned to the United States to await Sheriff Risk's Determinations. Several months passed, and then several more. No sheriff had ever taken so much time to publish his findings about a diving fatality. By the time Sheila received word that Risk had completed his deliberations on November 13, crimson maple leaves were falling in West Springfield.

To those who read his findings, it was abundantly clear that Sheriff Risk had spent a great deal of time wrestling with the evidence in order to assign responsibility. His 22-page long narrative was eloquently written. It covered every aspect of the tragedy from the contractual relationship between the respective parties, to the equipment and working practices, to the initiation of the accident, to the attempted rescue, and finally to his recommendations on how to improve safety for divers working in the North Sea.

Risk pointed out in his opening remarks that a fatality offshore poses unique problems because it occurs far out at sea and in an industry with customs and practices generally unknown to the investigator. Certain witnesses who could have shed more light on key events and documents did not testify. And because the inquiry was held nearly two years after the accident, memories had obviously been "dimmed by the passage of time."

In most cases, Scottish judges seldom include any finding of fault.[65] One school of thought believes that sheriffs are better off not spending valuable time wandering into the uncertain waters of determining defects in systems, precautions, and matters of culpability, which are issues often "imperfectly focused and, of necessity, inadequately explored."[66] Indeed, this was James

Tierney's position during his summing up when he urged the Court not to render its decision based upon hindsight. "Questions of fault and appropriation of blame," he said, "would be better left to another tribunal."[67]

But others, such as author Ian Carmichael, argue that the relatives of the deceased "are entitled to more than just a repetition of the death certificate if the evidence presented justifies it:"[68]

> Where the evidence is sufficiently compelling, the responsibility of exposing and finding fault should be accepted. The whole object of impartial public inquiry is to get at the truth, to expose fault where fault is proven to exist, and in all cases to see to it so far as is humanly possible that the same mistake, whether it arises through fault or other reason, is not made in the future. The public interest, in whose name inquiries are held, requires and deserves no less.[69]

Sheriff Risk was of this same mind, and in unusually harsh language, proved he had no stomach for those who came into his courtroom and lied under oath:

> With one exception, I did not suspect any witness of deliberately trying to mislead the Court, but there were instances where I suspected that witnesses had persuaded themselves of the truth of assertions which, on the basis of other preferable evidence, simply could not be correct.
>
> The one exception was Mr. Brian Masterson, Director and controlling shareholder of Infabco, who, when describing the arrangements which existed involving Infabco, a company call Offshore Co-ordinators Limited and various divers, gave evidence which was so obviously false, that, if a Fatal Accident Inquiry were not a serious process, it would have been laughable.[70]

On the decision not to use the clump weight and guide wire system, Risk said that it "constituted an important safety factor in the whole diving system." And if it had been in use during the time of the accident "it would have been relatively simple to recover the bell when it parted from the main hoist wire:"

> The decision to carry on diving without it rested with the diving contractors, namely Infabco, and was a serious error which made a material contribution to the deaths of the deceased.[71]

On the importance of a bell stage, he said that it was recognized "by all hands," that the bell should not come into direct contact with the seabed and that a bell stage could have prevented this from happening:

> It cannot be said with certainty that this contributed to the deaths of the deceased, but it does suggest that the diving contractors were more concerned with speed than with safety, in that they were prepared to run a known risk for which there was a known remedy, rather than wait until the remedy (i.e. the underframe) was available. Such an attitude would be entirely consistent with the earlier decision to dive without guide wires and clump weight.[72]

On the material effect of not field testing equipment prior to diving operations, Risk regarded it as "rash" that diving operations had begun without first determining that the only means of secondary recovery would be effective:

> As it turned out those means failed on the first occasion on which they were tested which, unfortunately, was while the deceased were under water.[73]

And with respect to Masterson's decision to raise the bell, Risk said, from the evidence presented, it was clear that the bell had been dragged against the SALM base:

> The sling itself caught on some projecting part of the base and what caused it to break was the pull of the "Wildrake's" crane against the unyielding base.
>
> It is surprising that such a possibility does not seem to have occurred to Mr. Masterson who, although not a diver, was an engineer. He knew that the bell was not far from the SALM base since that is where the divers had originally been working. The angle of the crane wire and the strain, which was many times greater than could have been imposed by the weight of the bell, must have indicated that something substantial was impeding the lift. If in fact the impediment was the SALM base the lift had no prospect of success unless the crane wire was freed. It seems to me that in instructing the crane to be slewed and the lift to resume without first investigating the cause of the obstruction Mr. Masterson committed a grave error. Mr. Masterson said that he did have such a check carried out by the "Stena Welder" divers but I do not believe him. He was not corroborated by any other witness and there is no entry relating to such a check in the log book (Production No. 6). Had such a check been carried out it might well have been possible to free the sling from the corner of the SALM base and to enable the lift to continue successfully.[74]

But for that action, Sheriff Risk said, "the lives of the deceased could probably have been saved."[75]

Risk made recommendations for securing clevis pins when used for human payloads; that diving bells be equipped with transponders as a matter of regular practice; and because of the lack of command structure which occurred on board the *Wildrake*, he said every diving crew should have "one individual expressly designated to take charge of any emergency recovery operation, whose identity would be known to the other members of the crew and made known to any outside parties who come to assist in the operation."[76] And, just as Sheriff Stewart commented at the trial, Sheriff Risk recommended that "if regulations for the safety of divers are to be effective they must be framed in such a way as to place responsibility upon those responsible for the organization and conduct of the operation who may not necessarily be 'employers' of divers."[77]

Lastly, he commended the "great skill" of Captain Mellor in keeping the *Stena Welder* on location for 17 hours and the heroism of the 2W rescue divers "who worked to the limits of their physical endurance in circumstances under which they would not normally have been expected to dive at all."[78]

Chapter Twenty-five

The Dark Side of Humanity

I don't say they necessarily went out to kill people, but if the choice was between jeopardizing the contract and letting somebody die, they'd let them die.

—— Bryan Gould
TV Eye Investigator[1]

By the time the *Wildrake* accident came to the attention of Bryan Gould, he'd lost the sharp edge of his New Zealand accent, had spent time in Her Majesty's Diplomatic Service and served five years in the House of Commons as a Labour MP representing the district of Southampton. In 1979, when Margaret Thatcher came to power, Gould lost his seat in Parliament and went to work for a weekly current affairs program called TV Eye at Thames Television.

When Gould took up his new assignment, North Sea accidents had been producing banner headlines for more than five years. But unlike other tragedies, revelations about the *Wildrake* accident continued to appear in the press with increasing intensity and momentum. The horrifying news of the disaster in August 1979 had been followed by the Crown's 1980 judicial blunder. As appalling as these events were, it was the firestorm of publicity surrounding the Fatal Accident Inquiry that catapulted the issue of offshore safety to a national level, and brought the deaths of Richard and Skip to the attention of the former lawmaker.

Initially, Gould thought the *Wildrake* accident would be an interesting story to explore because of its connection to offshore safety and because the subject of oil, and what was being done with the revenues, was still very much on the national political agenda. In the autumn of 1981, Gould moved his small team of investigators to Aberdeen and began asking questions about the diving industry in general and the death of Walker and Guiel in particular. He read the inquiry transcript, talked to members of the diving industry, and traveled to Edinburgh to confer with an expert on safety practices in the North Sea who felt that safety was being compromised for oil.

While Gould found people willing to provide information, there were times when his inquiries were constantly frustrated. He learned that BNOC's security division and the local police had investigated the rumor that Brian Masterson had bribed BNOC officials in order to win his contracts in the Thistle Field. When Gould asked the police about the results of that investigation, they would not reveal the outcome. The more questions he asked, the more convinced he became that there was a darker side to the scandalous diving accident, and that it involved "skullduggery" and "nefarious dealings" that no one knew about.[2]

In December 1981, Gould televised his findings in a documentary which added another piece of the puzzle to the *Wildrake* tragedy. The program, entitled *The Last Dive*, began with an opening statement:

> Working 500 feet below the North Sea requires a gambler's instinct. But when Richard Walker and Vic Guiel died, were the odds unfairly stacked against them? And if so, why?

For the next thirty minutes, the former MP recounted the accident, the failed rescue, Infabco's escape from justice and Sheriff Risk's criticisms of the operation. Gould interviewed several *Wildrake* divers on camera as well as Captain Mellor of the *Stena Welder* who still showed signs of anger over the event. Jeanne also appeared in the film and read Richard's last diary entry. After learning Infabco had not yet compensated Jeanne for her loss of support, Gould took his television crew to Infabco's office and confronted Masterson directly with the question of compensation. Startled by all the camera lights, Masterson darted into his office and closed the door.

In terms of negligence and deficiencies in the diving system, the TV Eye broadcast clearly illustrated that the odds were stacked against Richard and Skip. But midway through the program, Gould's narrative took an unexpected turn when he began to focus on how Brian Masterson had won his initial contract with British National Oil Corporation during the summer of 1978. This was the contract which supplied diving services from the *Gulnare*, parked alongside the Thistle platform. One of the directors of Infabco described the award as "a once in a lifetime deal," but Gould found out that it had been rigged.

Gould discovered that during a secret pre-bid meeting at BNOC headquarters in Aberdeen four members of Infabco were allowed to examine the bid details of the competition. Armed with this information, Masterson was able to manipulate his tender to make it appear that he had the winning bid. Gould identified the man responsible for organizing the secret meeting as Joe Singletary.

As BNOC's Offshore Construction Manager, Singletary held one of the most powerful positions within St. Machar House. If a major decision had to be made on the construction side, Singletary was involved. He oversaw procurement, movement of men and materials, hookup, budget, scheduling, and quality assurance. He also had the power to influence contract awards. One former BNOC official familiar with the process says that, although awards were determined by committee, they could be "leaned" one way or the other through the use of subjective commentary by indicating that this contractor or that had better equipment and better expertise.

By the time Gould learned about the corruption, Singletary was no longer working for BNOC, but the former MP managed to track him down by telephone on the island of Curaçao as Singletary was arranging for entry papers into Venezuela.[3] When Gould questioned Singletary about the pre-bid meeting, he denied the allegation. When Gould indicated he was prepared to reveal the names of the people who attended the secret meeting, Singletary interrupted Gould in an angry voice and told him he was "talking about a bunch of crap" that had already been investigated by the authorities. "It's all a matter of record," he said. "You start trying to ask me if I remember something here in the middle of this, and I don't even know what you're talking about, so it's just a waste of time."[4]

But Gould wasn't talking crap. He had hard evidence from two very reliable and unshakable sources who verified that Singletary had been in league with Masterson. He never revealed those sources on camera but they were in fact Brian Masterson's former business partners, Ken Duell and Ed Goltz.

In the spring of 1978, Infabco Diving Services was the brainchild of Duell and Masterson. Duell was working for Sub Sea International and Masterson owned a firm providing engineering services to BNOC. Both men had been involved with the Thistle Field for years and when they learned that BNOC was not going to renew its diving contract with Sub Sea International, they quickly formed Infabco Diving Services hoping to fill the void left behind by Sub Sea. Duell then coaxed Sub Sea Superintendents Ed Goltz, John Nadeau and Craig Roberts to join Infabco with a promise of partnership. They bought a small diving system, set up shop in the city of Montrose, and submitted a bid to provide diving services to BNOC.

And then one Saturday morning, Duell and Goltz received a telephone call from Singletary.[5] The BNOC manager wanted to see them. So the two divers drove the 35 miles to St. Machar House, parked the car, took the elevator to the seventh floor, and entered one of the offices several doors down from Singletary's. The Offshore Construction Manager was alone in the room, and spread out on one of the tables like a smorgasbord were all the bids of the competing companies for the new contract. Goltz says there was no mistaking what was on the table. "You could obviously see what they were." And then he says Singletary walked out of the room without saying a word, "just like you see at the movies."

At that point, Ken and Ed looked at each other wondering, *What the hell's going on here?* And for a moment, Ken says they stood there, dazed and confused, "like two virgins standing in the middle of a whore house."[6]

And then Brian Masterson walked into the room.

Duell says Masterson had a spreadsheet with him and used it as a comparator to adjust some of the rates in his bid. The meeting didn't last long, but to Ken and Ed it seemed like an eternity. When they were dismissed, Ken says they drove back to Montrose feeling "pretty shitty" about what had just happened and angry with Masterson for putting them into a situation that was clearly illegal. "In our naiveté in the whole thing," Duell says reflecting back, "it was Masterson's way, probably, of tying us in."[7]

Feelings of guilt and self-recrimination were on the horizon, but Duell says they didn't have much time to feel sorry for themselves. "We weren't in the office more than ten or fifteen minutes, and the goddamn telex machine came on awarding us the contract! And our first reaction was, 'Holy fuck, hope nobody finds out.'"[8]

And no one did. By the time the shock had worn off, Duell and Goltz were well into the frenzy of outfitting the sat system, and the incident at BNOC headquarters was largely forgotten. "It was a mistake in my life," Duell says, reflecting back. "Goltz and I were as green as the goddamn grass."

At this stage, Duell hoped he had seen the last of Masterson's shenanigans. But the corruption did not end there. Once Goltz installed the diving system on the *Gulnare*, the Teletype machine began spitting out telexes from BNOC requesting extra equipment, which was normal after a contract began, but what caught Duell's eye was the level of pricing Masterson was charging for the items which the oil company accepted without question. Duell was familiar with the costing of equipment and says Infabco's prices were clearly out of the ballpark.

"An air diving spread, $5,000 a day, a welding machine was like $500 a day. It was fucking unbelievable! So, what turned out to be what was an economical contract to start with, i.e. a four man sat complex—grew within a month's time of being a very significant contract. Prices that you just couldn't believe, and there was no negotiation. Masterson would just simply call us up and say, 'This is the price we can get for it. Go ahead and put it in.'"[9]

So Duell submitted telexes to BNOC, confident that the figures would be rejected. But to his astonishment, authorization came back almost as fast as he sent the numbers off. And as fast as the money rolled in, Duell says a lot of it went out the back door in the form of "expensive gifts" to several members of BNOC's staff.[10]

It was during this initial period of invoicing when Duell learned that the original *Gulnare* contract had been issued, not to Infabco Diving Services as expected, but to Masterson's parent company. Today, Duell attributes this oversight to his lack of experience.

Because Masterson spent much of his time out of the Montrose office, day-to-day operations fell to Duell. With bills piling up and no way to access the money in Masterson's company account, he began charging costs against his personal credit card to keep diving operations running.

Meanwhile, the incident with Joe Singletary left Goltz feeling vulnerable. He'd been promised shares in the company which he had yet to lay eyes on, and now Ken was having difficulty obtaining reimbursement from Masterson for his credit card expenses. He wanted out. When he finally told Ken he was leaving, Duell understood. The corruption had been eating away at his conscience as well. Several weeks later, in a showdown with Masterson, Duell said he was tired of the corruption and demanded that it stop, but he says Masterson's response was, "Well, that's just business."[11]

"Well, it's not," Ken shot back, and the following day he and his wife signed over their shares and left the company. Later, Ken heard that the *Gulnare* contract continued to grow until it became a virtual "money machine."[a]

By that time, Richard and Skip were diving for Infabco. During January 1979, the SALM buoy broke free and was subsequently rescued. Two months later, BNOC made the decision to repair and re-install the SALM. Then, on June 1, Singletary sent out a telex inviting the diving community to bid on the re-installation of the SALM. The invitation was issued at 5:55 on a Friday afternoon when offices were shutting down for the weekend. Wharton & Williams owner Malcolm Williams says normally the oil company gives the contractor "something like three to four weeks" to respond to the terms and conditions of the contract. Then it would take, on average, another three to four weeks to evaluate all the bids before making a decision.[12]

But on June 1, 1979, Singletary wanted an answer to his bid package "not later than 12 noon on Wednesday 6th June 1979."[13] That left prospective respondents a mere two and one half business days to assemble their tenders if they wanted the job.

While the competition was scrambling to react to Singletary's telex, Masterson was already out of the blocks. A week before the telex, he'd sent his general manager to Norway to obtain an exclusive right to the *Wildrake*.[14] Industry insiders say there's nothing unusual about this type of preemptive arrangement, but what struck Bryan Gould odd was the way in which BNOC's vessel requirements (in Singletary's telex) so closely matched the *Wildrake's*. The exclusive right to the *Wildrake* was Masterson's "trump card," Gould pointed out, because it was clear that anyone offering the *Wildrake* had the best chance for winning the contract.

So Masterson submitted his bid before the deadline, then sent his diving experts to Norway to begin the mobilization of the *Wildrake* weeks before he received the award. The travel expenses and the salaries of these men were costs that could not be recovered from BNOC.[15] But in the summer of 1979, Masterson was a confident man and word spread throughout the company that the SALM contract was in the bag. Even Richard, on board the *Gulnare* at the time, was under this impression. On June 15, he made a note in his diary to that effect: "On the chopper. The last one. Craig's gonna want me to hang around

---

a At the time of the *Wildrake* accident, Masterson had two complete sat systems on board the *Gulnare* (the original Scan Dive system and a rented Oceaneering system).

for sea trials on this new diving boat..."[16] Three days later, BNOC awarded the contract to Masterson.

In Aberdeen, BNOC's growing affection for Infabco had not gone unnoticed by 2W owners Ric Wharton and Malcolm Williams. They'd been doing business in the Thistle Field for several years and considered themselves to be the dominant contractor. When Masterson, a man with no bona fides in the diving industry, suddenly appeared on the scene muscling in on "their field," they were mystified. Then, a disturbing rumor began to circulate that Masterson and Singletary had formed a secret alliance. Williams took note of this alarming talk and wondered if it was true. When Masterson won the SALM contract, Williams's suspicions intensified, and he began to ponder what Singletary was receiving in return for his services.[17] He had no proof of any collusion, but he says it was "painfully obvious" to him that Singletary was biased in favor of Masterson which meant that all of his company's contracts with BNOC were in jeopardy.

Bryan Gould, however, did not have to speculate about the link between Singletary and Masterson. He had eyewitness accounts from two former Infabco partners who suddenly abandoned the company for the sole reason that they couldn't stomach the corruption and were extremely reluctant to tell anyone about it. When TV Eye investigators repeatedly hounded Goltz in California for an interview, he told them, "Look, I've got nothing to say. That was years in the past."[18]

But against a production deadline Gould and his team turned up the heat by enlisting the support of a reporter from Goltz's hometown newspaper who contacted him and suggested he come down to his office to confirm what the British journalists had uncovered. Wanting to get them off his back, Goltz finally relented and verified what they had.[19]

Gould determined that the odds were unfairly stacked against Richard and Skip. But the former MP believed that those unfavorable odds extended far beyond matters of defective equipment and negligence.[20] The revelation that Singletary was tilting the scales in favor of Masterson raised a multitude of questions, the most troubling of which was whether the Singletary-Masterson alliance had prejudiced the outcome of the rescue. Listening to the evidence at the inquiry, Jeanne suspected that politics might have interfered with the decision not to bring in more rescue ships. With all the problems the *Stena Welder* encountered, why did BNOC take eight hours to call in the *Seaway Swan*? Surely the oil company had a vested interest in preventing loss of life in its field. Or was Masterson's handling of the rescue mission typical? The answer to that question resides with history.

On August 29, 1973, Roger Mallinson and Roger Chapman surfaced in a small mini-sub called the *Pisces III* 150 miles off the south coast of Ireland. The time was 9:18 in the morning and the two Rogers were tired and hungry after spending eight hours burying part of a transatlantic telephone cable.

In the distance, the mothership, *Vickers Voyager*, radioed the two men and then initiated a recovery. Within minutes, divers in a Zodiac pulled alongside the sub and attached a towline from the *Voyager*. But in a strange twist of fate, the tow rope fouled on a cover plate to an unoccupied aft compartment, knocking it free. As seawater filled the compartment, the stern heeled back sharply and the submarine began to sink. A few minutes later the towline snapped and the vessel plunged 1,575 feet to the bottom of the Atlantic Ocean. Incredibly, the two men survived the impact. Mallinson and Chapman quickly assessed their life-support situation and radioed the *Voyager* for help. They had 79 hours before their air supply would run out.[21]

When Sir Leonard Redshaw, Chairman of Vickers Shipbuilding, heard of the accident, he told General Manager Peter Messervy, "Get everything you may possibly need—everything. Assume that the odds are against you."[22]

And Messervy did just that. Thirty-five minutes after receiving the emergency call, he launched one of the largest international rescue efforts of its kind by ordering the transport of the *Pisces II* (working in the North Sea) to Teesport, England where it was airlifted to Cork, Ireland. Eleven minutes later, he contacted International Hydrodynamics in Canada, who offered the assistance of the *Pisces V*, which was then flown by the Royal Canadian Air Force from Halifax, Nova Scotia to Ireland. Four hours later, when the US Navy offered to fly a team of technicians and their unmanned Controlled Underwater Recovery Vehicle, *CURV III*,[a] from California to Ireland, Messervy accepted the offer.[23]

Over the next 76 hours, an armada of ships assembled at the accident site, including the research vessel *HMS Hecate*; the US salvage ship *USS Aeolus*, and the *USS John Cabot*. At the same time, the captain of the *Voyager* sent out an SOS which was picked up by the *HMS Sir Tristram* and *British Kiwi*, owned by British Petroleum.[24] Both ships immediately began sailing to the accident scene. The Royal Navy ordered two submarines in (*HMS Valiant* and *HMS Andrew*), and in Aberdeen, Dyce Airport was opened in the dead of night to fly a British diving expert to the scene of the accident.[25] J. Robert Starr, the senior pilot of one of the rescue subs was later quoted as saying, "The most sophisticated equipment in the world will be used. We are leaving nothing to chance and we are optimistic about the outcome."[26]

That optimism and Messervy's leadership saved the lives of Mallinson and Chapman. Just before their air ran out, *Pisces II* and *CURV III* attached two recovery lines to the stranded sub. Chapman would go on to write a book about his ordeal and said it was only after he reached the surface to see all the ships and planes overhead that he realized the magnitude of the rescue operation. To this day, the recovery of the *Pisces III* remains the deepest underwater rescue ever conducted.

---

a In 1966, predecessor *CURV I* recovered a nuclear bomb which sank in 2,550 feet of water off the coast of Spain. *Source*: Sontag and Drew, *Blind Man's Bluff*, p. 58-60.

Five years later, on March 9, 1978, a small *Perry PC-9* observation submarine was launched from the deck of the *Subsea 2* into waters licensed to British Petroleum near the Frigg Field in the North Sea. Submariners Jamie Jones and Bill Cornwall were ordered to clear debris around an abandoned wellhead and set charges to cut the guide base free.[27] Four hours later, the submarine became trapped on the seabed when a polypropylene rope snared the submarine's propeller. The owners of the sub, P&O Ferries, immediately began marshaling means of support. They ordered the sistership, *Subsea 1* with a diver lockout submarine (*Mermaid III*), to sail the 240 miles from Montrose to the accident scene; and they contacted the Diving Support Vessel *Seaway Eagle* and the *Intersub 3*, also with a diver lockout submarine. Three vessels from three competing companies began converging on the scene. A spokesman for P&O Ferries said they were "pulling out all the stops" to save the two men.[28]

When the *Intersub 3* arrived on site, South African diver Andre Van Heerden and German diver Eric Arndt were selected to go into sat. Van Heerden says there was a sense of urgency to the operation "because with the mini-subs you didn't have much heating, hypothermia being the biggest problem."[29]

At 0744 on the morning of the 10th, the 20-hour crisis came to a happy ending when Van Heerden exited the sub, cleared the rope from the propeller and cut the explosive charge away. After the rescue, Ivor Davis, the deputy general manager of P&O Ferries, was quoted as saying, "This is the camaraderie of the sea. When something goes wrong, we all muck in to help each other."[30]

Later that same year, when the *Star Canopus* lost her bell in the Beryl Field, Martin Dane, the company official representing Star Offshore, mobilized three vessels to the accident scene: the *Intersub 4* with a manned lockout submarine, the *DSV Tender Carrier,* and the semi-submersible *Uncle John.* In the middle of a gale-force storm, the rescue ended unsuccessfully, but the diving company spared no expense in trying to save the two men.

Each of these accidents had its own set of circumstances, and yet the men managing the rescues shared a common philosophy about the sanctity of human life. They took immediate action and didn't hold back; they went for overkill and called in multiple ships, coordinated with the rescue teams, and if the situation demanded it, they changed tactics accordingly. In essence, what they did reinforced the unwritten law of the sea, that, in times of disaster, you pull out all the stops and do everything possible to save the people who need your help. In contrast, the attempt to save Richard and Skip bears no kinship to these rescues and continues to stand alone in history as an aberration that has never been repeated.

One hour and fifteen minutes into the accident, Brian Masterson realized he had to ask for assistance in order to save Richard and Skip. His crew had blown the main umbilical and could not raise the bell. Worse yet, with no hot water getting to Richard and Skip, they would soon freeze to death. If anyone had a reason to call in the cavalry, it was Brian Masterson. But Masterson did not call in the cavalry. Instead, he engaged in conduct that defies the historical record.

At the time of the *Wildrake* accident, four ships were at hand. The *Stena Welder* was a few miles away while the *Uncle John*, the *Seaway Swan*, and the *Talisman* were 20 miles to the south in the Brent Field. Masterson did not issue an SOS, nor did he make immediate phone calls to these vessels to ascertain whether they were operational or not. Instead, he says he compared the respective capabilities of all four vessels, wanting the one with "the best sea keeping qualities"[31] that could operate in deep water and stay on location without anchors even in deteriorating weather.[32] And yet he selected the *Stena Welder* which he knew had no dynamic positioning system.[33]

He said the *Stena Welder* was asked to make the rescue because it was the closest vessel to the scene and that it "could do the job very easily."[34] And yet it took the 2W crew three precious hours to reassemble the diving system and launch the bell (0257-0607). He says, "At no time during the operation was he told that the *Stena Welder* was unfit to dive,"[35a] and yet he personally contacted the *Welder* "on successive occasions" to inquire about the delay in getting the rescue bell into the water and was notified that there were problems with the hot water machine and that Captain Mellor was having trouble holding his ship on station.[36] At the inquiry Superintendent Peter Holmes testified that, once Don Cox transferred to the *Stena Welder*, "then we were informed of the situation."[37] That was just before 4 o' clock in the morning, still plenty of time to call for more help to save Richard and Skip.

Even after the rescue bell had gone into the water, Masterson knew that problems had developed with the communication and hot water systems to the rescue diver because he continued to speak to Don Cox "throughout the day" to ask why it was taking the rescue diver so long to complete his assigned tasks.[38]

"Everybody was wanting it to go quicker," he said at the inquiry.[39]

Masterson says he decided not to use one of the other vessels because he felt it would have taken a lot longer to put a request through Shell's marine department to get one of the Brent vessels.[40] But with mounting delays and the *Stena Welder's* poor performance becoming increasingly obvious, why not put that request through? If there was the slightest chance that one of the Brent vessels could save Richard and Skip, wouldn't a prudent man make that call? Brent was 20 miles away; it was not Halifax, Nova Scotia.

Masterson was in the middle of a full-blown diving emergency. Had he asked Shell for one of its vessels, the oil company would have complied without hesitation because, at the time of the emergency, the UK Offshore Operators' Association had in place a "Code of Practice and Plan" which divided the North Sea into Sector Clubs to respond effectively to an offshore emergency and

a Supervisor Don Cox, *Stena Welder* Superintendent Jim Hurney, and *Stena Welder* Captain Mellor strongly dispute this.

provide a pool of equipment which an operator could call upon "to enhance his own resources."[41] Thistle and Brent were part of the Red Sector and Shell Oil in the Brent Field was the designated emergency coordinator for that area.[42] And as history has shown, in a time of crisis when one operator has a vessel the other needs, it says, "Right, off you go."

But Masterson didn't have to go looking for help: it came looking for him. Not long into the emergency, Captain Chauvelin of the *Talisman* called the *Wildrake* and offered his assistance.[43] Of the three diving vessels working for Shell, the *Talisman* was the best equipped to save Richard and Skip because it was not a slow moving semi-submersible; it was a reasonably fast, very mobile diving support vessel similar to the *Wildrake* with the capability of dynamic positioning. Captain Chauvelin estimates he could have made the crossing to Thistle in two hours, including the recovery of the DP taut wire and the time to reposition over the accident site.[44] But Masterson turned his offer down.[45]

This reluctance to allow another vessel into the operation manifested itself again when Masterson's own diving experts advised him the rescue was taking too long and that he should call Brent for backup, specifically the *Seaway Swan*.[46] Masterson says he eventually called the *Swan*, but instead of requesting it to come to the rescue, he told it to standby[a]—that he would "get back to them"—but he never did.[47]

Captain Mellor of the *Stena Welder* ran up against this virtual blockade as well. As he struggled to keep his ship on location, he requested, through his BNOC representative, that Masterson bring in the *Uncle John*.[b] But the Infabco boss replied that "it would not be necessary at this time."[48] Randy Gunter says Masterson had taken the position that additional ships were unnecessary because he thought the situation was not really as serious as it seemed to be.[49] And yet one can hardly imagine a circumstance more life-threatening than being marooned on the bottom of the North Sea without hot water and communications.

Unlike Peter Messervy in the *Pisces* rescue, Masterson did not have to organize a submarine airlift from Canada or contact the Royal Navy for assistance. He merely had to say to the *Talisman*, "Yes! We need you," and Captain Chauvelin would have begun steaming to Thistle immediately. He could have said to the *Seaway Swan*, "Don't wait, come now," and cancelled his request if he managed to recover the *Wildrake* bell during the interim. At any point during the rescue Masterson could have marshaled additional support, and yet he chose not to do so. Instead, he continued to stall for time hoping the *Welder's* problems would

---

a There is no ring of truth to this. Calling the *Seaway Swan* to tell them to standby 20 miles away makes no sense. It is more likely that the *Swan* called Masterson to offer assistance.

b The *Uncle John* remained on station at Brent Bravo while Comex divers locked out on bell dives No. 262, 263, and 264 to remove a 12-inch crossover valve. *Source*: Comex Construction Superintendent Don Higham's personal diary.

go away. But they did not go away. In fact, they deteriorated to the extent that "friction" developed between the two ships over the delays in executing the rescue.[50] Even when the wire sling broke at 1220 (during the initial lift), it was BNOC, not Masterson, that finally delivered the call to the *Seaway Swan*.[51]

At the inquiry, when asked why he made this crucial decision or that, Masterson referred to his paymasters 11 times.[52] He said BNOC was involved in the discussions about the attributes of the various rescue vessels.[53] He said they were the ones who alerted the *Seaway Swan* and the *Uncle John*,[54] and were "putting suggestions forward" on how to effect the recovery.[55] And when the *Stena Welder* supervisors suggested that they should recover the *Wildrake* bell, he said, "that particular suggestion...went into Aberdeen."[56] Even the decision not to use the Brent vessels, in his view, traced back to BNOC. According to Masterson, who claimed to be "fairly conversant with semi-submersibles,"[57] St. Machar House told him it would take eight to nine hours for the Brent vessels to de-ballast and transit the 20 miles to Thistle.[58] But history has since proven that semi-submersibles travel at speeds much faster than what the experts at BNOC believed.[a]

But whatever credence Masterson attached to BNOC's information, one cannot escape the question: what reason could there be for keeping the *Talisman* at bay? Both BNOC and Masterson knew it was not a semi-submersible and it had offered its services. The *Talisman*, the *Seaway Swan* and the *Uncle John* were operational and willing to provide assistance if only Masterson would give the word. Those vessels had the capabilities he desired. But there was one thing the Shell boats had that the *Stena Welder* lacked. They had a price tag.

Under charter to Wharton & Williams, the *Stena Welder* was already working for BNOC and so the issue of compensation would not have raised its head. But the *Uncle John* was a different matter. It was under contract to Shell, and, at an estimated cost of £30,000 per day,[59] it was one of the most expensive diving vessels of that era. Exactly who would reimburse Shell for its use would largely depend upon who requested it. Under the terms of the UK Offshore Operators' Association agreement,[60] if BNOC asked Shell to bail Infabco out of its predicament with the *Uncle John,* Shell could have charged BNOC for the use of the vessel. BNOC could then try to back-charge Infabco for the cost. But a former managing director of BNOC says Masterson probably could have foiled that attempt by taking the position, "Look, I would have been all right. I was moving forward with the rescue and would have gotten the job done, but you decided to step in with the *Uncle John* which increased the cost of the operation, and now you're trying to charge me? No way!"

But if Masterson had explicitly urged BNOC (or Shell) for the services of the *Uncle John*, then the oil company would have had a strong case for making those charges stick. On the other hand, if he kept silent and continued to hang on

a During the 1981 *Stena Seapread* accident the *Uncle John* made the transit in just over three hours (1158-1505).

using the *Stena Welder*, it would be unlikely that BNOC would charge Masterson for a ship already under its control.

Once the *Stena Welder* became the designated rescue boat, Masterson was able to influence the direction of the operation in a manner in which he almost certainly would not have been able to if he were dealing with one of the Brent vessels. After taking charge of the operation, he sent Supervisor Don Cox to the *Stena Welder* to act as his "personal authority."[61] There was no reason to take this unusual step; Masterson could have easily spoken to the 2W supervisor by radio himself. But Cox had been a former 2W supervisor and was highly respected by its crew, and by having him on board the *Stena Welder,* Masterson was able to advance, and exercise control over, his interests with little or no objection. At the inquiry, he admitted that all of his communications went through Don Cox.[62] And those communications began with an overly complicated, unorthodox plan, to recover the *Wildrake* bell with the *Wildrake* crane. And when that plan began to sour, Masterson rejected other strategies that would have required him to surrender control of the operation to his arch rival Wharton & Williams. When the *Stena Welder* supervisors requested they recover the stranded bell and hand it off to the *Wildrake*, that suggestion was rejected. When they suggested a wet transfer of the divers, again, the answer was no. With time running out Masterson clearly had disaster looming in his face, and yet he would not give up the idea of getting his bell back using his crane on his ship. TV Eye investigators believed this behavior was motivated by financial considerations.

At the time of the accident, a third BNOC contract was up for bid. According to Captain Mellor, it was "the daily conversation" on board the *Stena Welder*, and naturally it would have been the target of Masterson's ambition. But the accident forced Masterson to seek help from his competitor to save Richard and Skip. Had Wharton & Williams taken control of the rescue and upstaged Masterson's rescue effort, then it would have left a real black mark on his reputation which might have ruined his chances to win future contracts, with or without Singletary's help.[a] But if Masterson could carry on and actually save Richard and Skip, then 2W's role in the rescue could be marginalized later by spinning a report that it was really Infabco's strategy and Infabco's equipment that got Richard and Skip back alive.

In the *Pisces* rescue, Peter Messervy assembled everything he could lay his hands on to rescue Chapman and Mallinson. But, far out at sea, Masterson was not pulling out all the stops, or changing tactics, or going for overkill to save Richard and Skip. Instead, he was trying to keep his costs down, turning down offers of help, and rejecting advice from his experts. This behavior should have raised eyebrows at St. Machar House, not just because time was drifting alarmingly by, but because BNOC officials were fully aware that the *Stena Welder* was ill-equipped to handle the rescue long before the emergency call came in from Thistle Alpha. BNOC had its own representative on board the *Welder* who

---

a According to the TV Eye documentary, Masterson was not awarded the contract.

knew the diving system was under repair, was witness to it, and was required to telex St. Machar House about it in the form of Daily Progress Reports[63] because the oil company had a graduated system of payment based upon whether the ship was working or on standby.[64]

But if the DPRs were not sufficient evidence for BNOC to request backup, then Malcolm Williams and Ric Wharton's visit to St. Machar House on the morning of the accident had to be. Concerned that the *Welder* might fail to save Richard and Skip, Wharton says he and Malcolm urged BNOC to bring in the *Uncle John* as a backup for the *Stena Welder*. "Our attitude was...you've got to go for overkill when people's lives are down there. You don't piss about."[65]

They had good reason to feel this way because they knew the antiquated diving system on the *Welder* was under repair and had a temperamental history. Moreover, one of their chartered ships, the *Tender Carrier,* had been involved in a rescue attempt eight months earlier to save two divers in the Beryl Field. The *Carrier* was one of three diving support vessels called to the scene. And yet, even with the additional support, the two men died. So when Richard and Skip became stranded in the Thistle Field, the *Canopus* accident was still fresh in Ric and Malcolm's minds and they did not want the *Welder* to be left all alone in the event it couldn't handle the job.

"We were the closest," Wharton recalls, "but we were not the best equipped to do the job. And if we had waited until we had failed to bring in the *Uncle John*, it would be too late, and the *Uncle John* really needed to start steaming immediately to get there, and that was the point why we suggested it."[66]

But Wharton says BNOC refused his request, so he telephoned Ken Duell hoping Duell could persuade Masterson to call in the *Uncle John*. Duell was offshore at the time, but his wife, Yvonne, took the phone call.[67] As Wharton explained the crisis, Yvonne sensed an element of desperation in his voice. Hours later, when Ken arrived home to hear the bad news, he tried to contact the *Wildrake* but was unable to get through, so he placed a call to BNOC to speak to the diving manager, but Duell says the man wouldn't entertain a conversation at all.[68]

Why did BNOC refuse to call in the *Uncle John*? According to Ric Wharton, "They said no because of cost, because they would have to pay for it."[69]

In 1979, BNOC was in its fourth year of existence and positioned to become involved in 23 of the 24 fields currently in production or under development (including its role as co-licensee through participation agreements).[70] With access to the National Oil Account and a book value approaching £1 billion,[71] Britain's oil broker would have no reason to quibble about financing a rescue mission to save the lives of two trapped divers in one of its fields. Such an expense would pale against the types of financial impulses oil companies routinely engage in. During the winter of 1974, for example, British Petroleum spent £400,000 *per day* to remain on schedule in the Forties Field.[72] In 1976, BNOC's development and exploration costs alone amounted to £91 million.[73] And roughly seven months before the

Wildrake accident, it mounted an expensive rescue mission to save the SALM buoy when it broke loose from the riser. A 669-foot long crane barge estimated to cost $150,000 per day was rushed to the scene; explosive experts were quickly hired away from Shell Oil; a London-based project engineer was flown to the site within 24 hours, and special equipment was hurried offshore to raise the buoy out of the water. There was a sense urgency to the operation and money was no issue because the potential financial loss was considered to be significant.

But strangely, BNOC officials did not muster the same sense of urgency to save Richard and Skip. Still, the proposition that someone at St. Machar House refused to call in the *Uncle John* because of cost considerations not only sounds profoundly absurd, it defies historical precedent, and cuts across the views of industry workers who are adamant that, when it comes to saving lives, money is no object.

"Money is not an issue at all," one long-time oilman says. "There is no discussion about money, there's no discussion about what we do with cost....You call in whatever is available....Then after a couple of months, when the bills come in, then it becomes an issue. But we're not involved anymore at that time. And it's the accountants and the bean counters that take over."[74]

That attitude, of course, is what would be expected. And yet BNOC's own dive log is crystal clear on one indisputable point: the oil company stood silent for eight hours as preference was given to a ship it knew was in no condition to mount a rescue operation. Via the radio link to the *Wildrake*, the people at St. Machar House heard about the mangled umbilical, the loss of hot water, the delay in launching the rescue bell, the difficulties of finding Richard and Skip in the dark, and the reattachment of the crane wire. And they knew that, by allowing the *Stena Welder* to perform the rescue, Captain Mellor would be forced to live-boat the rescue divers over the accident scene. They knew this because Masterson and a member of the BNOC team were relaying progress reports to St. Machar House throughout the day of the accident.

At St. Machar House, BNOC's diving manager was an American by the name of Terry Slusher. Slusher was a former diver who had worked an office job with Comex Diving and then at Wharton & Williams before signing up as BNOC's resident expert.[75] His job at BNOC was to keep abreast of all matters relating to undersea projects;[76] and on August 8, 1979, it was Slusher who was communicating with Masterson on the other end of the radio link,[77] and who kept Joe Singletary apprised of the developing emergency.[78]

As diving manager, Slusher knew the *Stena Welder* was not a DP vessel, and via the Daily Progress Reports, he would have known the diving system was in the midst of an overhaul. Had he any doubts about the *Welder's* state of readiness, he could have called Captain Mellor to ascertain the severity of the situation, but Captain Mellor says Slusher never radioed him at all. Had he done so, he says he would have told the diving manager what he thought of the situation "in no uncertain terms."[79]

Meanwhile, over on the *Wildrake* bridge, ship-to-shore communications were flashing back and forth and painting a foreboding story. Project engineer Chuck Alexander specifically recalls talking to Slusher after the main umbilical was destroyed at 0400. "I called Slusher then and said, 'This is a serious problem. You better start working on some solutions because this ain't going to be easy.'"[80]

That sobering comment should have prompted Slusher to exercise his authority to bring in additional rescue vessels (or advise Singletary to do so), but he never did. Instead, he continued to give Masterson *carte blanche* treatment and gamble the success of the operation on a vessel that was ill-equipped to handle the job. Even if Masterson was relaying overly optimistic reports to Slusher, Slusher should have been horror-stricken by the passage of time. As a former diver, he had to know that, without hot water, the two trapped divers were in serious trouble, that helium accelerates the onslaught of hypothermia, and that, in waiting eight hours to send in the *Seaway Swan*, he might as well have waited eight years.

While Slusher's bizarre behavior remains an inscrutable mystery,[81] Joe Singletary says that, whatever failure there was in mobilizing additional vessels, it had nothing to do with him, even though he acknowledges that Slusher was passing progress reports in his direction. In fact, he says he wasn't involved in the emergency at all. "It was something that was happening offshore and other people were taking care of it. As I remember, I didn't get into the thing."[82]

In the midst of a lost-bell diving accident, that is a curious attitude for a senior manager to take. At the very least it would have a damping effect upon the need for urgency by sending a message to the people below him that the client is taking a back seat to the contractor, which is not how master-servant relationships work in the oil industry. Nonetheless, Singletary says he "didn't really have hardly anything to do with it," and that it was really Slusher who had his finger on the pulse of the accident and understood diving matters, not him. To illustrate his point, he offers this example:

> Let's say DeBakey works for me and he's doing heart surgery and there's a problem in there; he certainly can't ask me what to do because I really don't know. I mean, I wouldn't have a problem okaying calling in a ship if that's what Slusher recommended, but to know whether it ought to be called or not to be called, I'm not a saturation diving person.[83]

No, but Singletary had been in the oil business since 1957, and although he was not a diver, he was an engineer and understood the dangers of the North Sea in an equally dramatic and meaningful way. Thistle Alpha had been designed to withstand wind speeds of up to 150 mph and wave heights of nearly 100 feet. It was also common knowledge that the temperature of the water could incapacitate an unprotected man within 30 minutes, and for years, the local newspapers had been reporting in graphic detail the perils of this legendary sea. And whether one is off the coast of Scotland or Singapore, downed bells and submarines by nature have always been handled with a sense of urgency because running out of time is what ultimately kills the occupants. That anyone, especially the Offshore Construction Manager of British National Oil Corporation, could imagine that Richard and Skip were not in imminent danger is patently not credible.

Singletary was the client and he was positioned at the heart of BNOC's health and safety management system. He represented the company's vital interests which conferred upon him the responsibility of monitoring the situation closely to ensure that Masterson was not engaging in behavior that could possibly compromise Richard and Skip's well-being. It was his duty to make sure that everything that could have been done was being done. By disassociating himself from the recovery operation, not only made him remiss on a professional level, it put his actions directly at odds with company policy.

At the time of the accident, BNOC had a set of emergency procedures which were written in part by a man named Bill Philliber.[84] Philliber worked in BNOC's London office and held the position of Manager of Projects. But before he took over that position he had, in fact, set up BNOC's Offshore Construction Division in Aberdeen and was Joe Singletary's boss. Philliber says the emergency procedures he helped design were intended to relieve people from thinking in terms of money during a crisis and to make sure vital information got to the correct sources as quickly as possible and proper actions were taken. The emergency procedures were also a form of legal protection for BNOC, he says, because everyone needed to know what to do when the time came. "If you get into an emergency situation, and you haven't prepared yourself, and you don't know what to do, you'll make all kinds of mistakes."[85]

Above Philliber, and below the post of Chairman of BNOC, Managing Director Don Shimmon says "everybody" knew what the emergency procedures were and that there was no confusion about them. Was it possible for members of the senior staff not to know what to do during an emergency? "No, No," Shimmon says. "Anybody that had come up with that would have been gone. That's not a mature attitude."[86]

Philliber says Singletary was a competent guy and knew the business, and that "no manager or senior executive" within the company would have hesitated for a moment to overrule the diving contractor if there was any fear of loss of life, whatever the cost. BNOC would have simply ordered in extra resources and picked up the tab.[87] But Singletary insists he didn't get involved and adds that the cost of a rescue vessel had to be considered because the company was "way over budget" at the time.

"Those things cost $100,000 a day, or in that neighborhood, as I remember it, depending upon which one you're talking about...and they probably wouldn't come unless they got somebody to okay paying."[88]

Singletary is right about that: permission is required. But both Philliber and Shimmon say that verbal authorization is all that is necessary to get the ships moving during an emergency. "If I was in Aberdeen," Philliber says, "and Shell called me as Offshore Manager and said that they had an emergency—'We have an immediate need for a helicopter...or we need a boat.'—I would immediately authorize it. I would get the man's name on Shell so that we could follow-up later on, but you just move on those things....There would be back charges, but you wouldn't agree costs or anything in advance, you would go for it. If there is human life at stake, you accept the costs."[89]

But Singletary says that that was not his understanding of BNOC's emergency policy and today questions whether he even had the power to pick up the telephone and call for more support. As far as additional resources were concerned, he says "that would all be handled from the field, it wouldn't be handled from the office." And he adds that if Ric Wharton or Terry Slusher or Brian Masterson or anyone else had recommended he mobilize the *Uncle John*, he would have had to contact London to receive authorization from the Thistle consortium to pay for the cost of the rescue.

"I know anytime I wanted to get a ship out for rapid response or anything like this," Singletary recalls, "I went directly to Don Shimmon, and he called various other partners until he got 51 percent okay, then he called back to say, 'Go ahead,' because we were way over budget, and there was a matter of paying."[90]

"That's bullshit," Philliber says emphatically. "You just don't build in delays like that with an emergency. You just don't do it...certainly not when there's life at stake." And polling the owners of the Thistle Field is not done, he adds, because that would create an "extensive delay."[91]

Don Shimmon concurs with that view and maintains that Singletary not only had the authority to call in other vessels, it would have been "expected" of him to do so without majority approval because BNOC was the designated operator of Thistle and carried the responsibility for safety. And he adds that if Singletary or Slusher or even the man offshore had launched an expensive rescue mission to save life and limb, "their actions up to the point of advise and consent would have been backed up, period."[92]

But despite what Philliber and Shimmon say, Singletary maintains he was not in the command loop, which left the fate of Richard and Skip in the hands of an industry neophyte. "Brian was calling the shots out there," Singletary says, "and if he had told us to get the *Uncle John* over there, I'm sure that we would have done it."[93]

That hands-off attitude was exactly what worked in Brian Masterson's favor. It allowed him to pursue his immediate interests by controlling the circumstances of the rescue. It also reflected the nature of his relationship with Singletary, a relationship that poisoned an otherwise normal emergency response policy and produced an outcome that could have been prevented. When BNOC saved the SALM buoy seven months earlier, it responded in its usual way: expeditiously, and in accordance with company policy. There was no vacillation, no delay, and no argument about money, because, absent Brian Masterson's interests, there was no hidden agenda to prevent a quick response. BNOC simply went in with all due speed and protected its investment, as it should.

But in the *Wildrake* accident there was a telltale eight-hour delay in sending in a second vessel. This delay is the proverbial canary in the mineshaft that exposes the obvious deference shown to Masterson by those at BNOC who were more concerned about their comrade's predicament than with the lives of two divers. Only when it became clear to them that there was a real potential for loss of life did it become necessary to make a show of action by sending in the *Seaway Swan*, but by then it was too late for Richard and Skip.

Just months before Richard and Skip died, BNOC issued a statement in which it declared that its highest priority was to protect human life, "including contractor's staff," and that the safety organization within the company "must be properly delegated and monitored continuously to ensure that it is working properly, is always ready to deal with emergencies, and is fully respected at all levels."[94] Trouble is, when it came time to put those words into action, the people at BNOC, in deference to Masterson, chose to ignore them which cost Richard and Skip their lives.

In wrongful death cases, the courts frequently have to determine whether an act qualifies as negligence, which has been defined as "the omission to do something which a reasonable man...would do, or doing something which a prudent and reasonable man would not do."[95] After uncovering the link between Masterson and Singletary, Gould recognized that the failed rescue was not just a simple case of negligence, but that it was such "obvious negligence" that the people at BNOC had to know what they were doing as they stood by on the sidelines. "They knew they were risking the lives of these men," Gould says looking back, "and they were willing to do that, it seemed to me, and they had to have a very powerful reason for that."[96]

Greed, politics, personal pride—these are affairs that men should bar from their consciences when lives are at stake. They should not be asking themselves, "Well, hang on a minute, what edge can I gain here?" They should be following the example of men like Redshaw and Messervy; they should be trying to save the helpless as though they were trying to save themselves.

There is no legal obligation on a bystander to come to the aid of another. It may be morally reprehensible to stand back and watch a man die, but the law stands silent on things not done. In the meantime, those who live through events such as the *Wildrake* tragedy are left with emotional scars they have to bear for the rest of their lives. It leaves men like Captain Mellor forever asking himself, "What price is life?" knowing there is no price.

Following the accident, Mellor tried to go back to sea. But after a brief attempt, he gave up the life, too sickened and embittered by the experience to put himself in that position ever again. Today, when he looks back on the accident, he does so with a mixture of smoldering anger and pained resignation. Anger, because he's certain he could have saved Richard and Skip, and resignation because he knows that in those days "cost was a big factor." And that's what sticks in his throat, knowing that two decent men were "killed needlessly" because of politics and cold-blooded commerce.

People who know Brian Masterson are fairly consistent when they comment on his character. They say he's a very friendly, decent sort of chap. He's intelligent, and very very clever. He doesn't smoke, he doesn't drink, and he tackles work with missionary zeal. But they also say that the lust for money is bred into his bones. One man who had business dealings with him says Masterson is unable to "tread the straight and narrow," and that keeping the Shell boats at bay for financial reasons was well within his nature.

"That's the sort of thing he would do. He's quick enough to realize that he would have to pay, and if it was a question about whether to pay with Brian, he wouldn't."

He goes on to say that while most people wouldn't think twice about money in order to save two desperate divers, that isn't the way Brian Masterson thinks:

> If Brian looks at anything, if he looks at a business that he wishes to take over, he doesn't look at it from the point of view of making it a success, he looks at what he can do with it to gain the most financial benefit to himself, irrespective of who he stands on or who he has to knock to one side. He always looks to himself, and that would be the case with these divers. He would instantly think, OOPS, that's costing money. No, we'll do it my way.

Within weeks of the deaths of Richard and Skip, rumors about corruption at St. Machar House climbed the corporate ladder to Chairman Lord Kearton's office. An internal investigation (what Joe Singletary refers to as "the Big Stink") was launched by BNOC's chief security officer. Masterson, Singletary, and Ken Duell were hunted down and interviewed. With documents to back up his story, Duell told the detective everything. Several months later, Slusher was no longer working for BNOC. When the press got wind of the bribes probe, they began asking questions about the diving manager's sudden departure. The official word was his contract had expired, but the man who replaced him says Slusher "was told to leave." A few months later, in early 1980, Singletary pulled up stakes and returned to South America, back to where his career began and where reporter Bryan Gould would eventually track him down.

After being involved in the most notorious scandal in North Sea history, Masterson's reputation in the industry was irreparably damaged. Within several years of the TV Eye broadcast, his company went into disarray and he gave up the diving business and drifted away. No one seemed to know where the disgraced contractor had vanished to. Some people naturally assumed he learned his lesson and retired to a quite corner of Britain to stay out of the glare of the press. And that probably would have been the case had it not been for the untimely death of a British tycoon who was destined to topple off of his yacht into the dark waters of the Canary Islands.

Chapter Twenty-six

An Incestuous Relationship

The Minister should understand that some of us feel deeply that an improper relationship is involved. Indeed, it is an incestuous relationship. The sponsoring Department has no business getting involved with health and safety.

—— W. G. Carson
The Other Price of Britain's Oil, p. 174

A few months after Bryan Gould aired *The Last Dive,* a well documented book about the UK offshore oil industry was published entitled *The Other Price of Britain's Oil.* It was written by W. G. "Kit" Carson. Carson was not a man particularly attracted to the high drama of North Sea diving accidents. He wasn't a diver, a rig worker, or a journalist looking for his next day's headline. In fact, he'd never worked offshore at all, and yet he knew more about the North Sea oil industry in ways that even Richard and Skip would have found astonishing. Dr. Carson was a professor of criminology at the University of Edinburgh, and like Bryan Gould, became inextricably drawn to the *Wildrake* accident, but for entirely different reasons.

With a long-standing interest in British factory legislation, Carson wanted to investigate the offshore oil industry, its safety practices, and how the government regulatory system was working. With the sudden discovery of "unanticipated wealth" off the coast, Carson suspected that Britain's quest for oil "might turn out to be a rough and tough affair in which questions of safety were pushed aside in the pursuit of riches previously undreamed of."[1]

In 1976, he received a small government grant to determine whether such an investigation was even feasible. When his initial research indicated that it was, he was given a final grant to complete his work. In 1982, he published his landmark study which effectively linked Britain's policy of rapid exploitation to the North Sea's poor safety record and ultimately to the kinds of tragic consequences that fell upon men like Richard and Skip.

Carson discovered that the origin of this link could be traced back to the mid-1960s when the country's reliance on imported oil was putting pressure on the national Treasury, and at a time when seismic testing in the North Sea was suggesting that the Continental Shelf might contain significant quantities of oil and gas. Desperate to head off mounting economic woes, it didn't take government officials long to commit the nation to an ambitious program where rapid exploitation became the dominant theme towards curbing the balance-of-payments problem.[2] In the years that followed, speed underscored all developing events. It was a fundamental ingredient in the drive towards self-sufficiency and it explained why successive administrations allowed offshore operations to proceed without first ensuring that adequate safety provisions were in place.[3]

By the end of 1973, Britain's economy was in dire straights. War had broken out in the Middle East with OPEC nearly quadrupling the price of crude. The country's oil importation bill, which was once calculated in the hundreds of millions of pounds in the mid-1960s, was now on a collision course with the unheard of sum of nearly £4 billion by the end of 1976.[4]

On the streets, an angry citizenry was feeling the weight of the energy crisis and uncontrollable inflation. Food prices had risen dramatically, power restrictions were in force, and industry was on a three-day-work week. On the eve of a general election, the country was in a state of emergency and Prime Minister Edward Heath was under tremendous pressure to take the nation's energy policy out of the hands of the colossal Department of Trade and Industry (DTI) and elevate it to cabinet level. Not to do so was considered to be "a serious disadvantage,"[5] and the Prime Minister agreed.

On January 8, 1974, Heath created the Department of Energy and installed heavyweight Lord Carrington as energy czar. The mandate of the new agency was to assist the oil industry with such things as licensing and the development of energy policy. But its responsibilities also extended to the creation and enforcement of regulations that the oil companies had to comply with. In effect, the DOE wore two hats, and to some North Sea observers this spelled trouble for the offshore worker because it set up a competition between the government's exploitation policy and the role of safety. There was a real fear that Department officials, caught between conflicting priorities, might lack the necessary independence to impose the law upon, or even criticize, the very industry it was sponsoring. And, in a larger context, the question could be asked: would an administration really punish an industry for misconduct when it lays at its feet the resources which enhances its ability to stay in power? These fears gain further momentum by something else Carson discovered: not only had authorities incorporated the administration of safety into a framework where the rush for oil had become a top economic priority,[6] but they also "vertically structured" the chain of command so as to leave those responsible for safety "answerable to superiors whose brief included other, and not necessarily compatible, objectives."[7]

Unlike safety inspectors who worked onshore within the Health and Safety Executive, DOE inspectors were rarely rotated to new locations to prevent fraternization. Inevitably, in their struggle to discover what industry was learning, regulators developed a "close relationship" with the regulated.[8] The two bodies began exchanging personnel which helped to strengthen bonds and ensure a "common outlook" in terms of achieving economic goals.[9] To oil executives, there was nothing improper about enticing government regulators to come work for them, but to critics this practice smelled like a system of "deferred bribery."[10]

While authorities in Norway would remove the responsibility for health and safety from the industry's sponsoring department, British officials, in contrast, not only resisted this approach, but took extraordinary measures to ensure that the supervision of safety remained in the hands of an organization with a less than autonomous view of the industry it was regulating.

Ultimately, Carson says the oil industry's self-definition became "institutionalized" within the bureaucratic structure of the DOE.[11] Even Energy Secretary Tony Benn conceded that the relationship between regulators and regulated tended to get "a bit cosy."[12]

That cozy relationship later became an issue before the Burgoyne Committee when it began looking into the DOE's handling of offshore safety. In its final report, two dissenting committee members, Miller and Lyons, warned that the potential for conflict of interest was real and could not be overlooked. In their view, the whole approach to the administration and enforcement of offshore safety was "unacceptable" and adamantly stood on the principle that:

> a Government Department substantially responsible for the direction and control of an industry should not in any way be responsible for the standards and enforcement of occupational health and safety in that industry.[13]

Not surprisingly, the Department of Energy bristles at any implication that safety was compromised by the pressures of oil production. One junior Minister responsible for offshore safety is quoted by Carson as saying that everyone...

> in the Department of Energy, from the Secretary of State downwards, is dedicated to safety as a number one priority. We realize the value of the resources we have in the North Sea; we are anxious to have those reserves brought to the surface as quickly as possible; but we are not going to sacrifice one life in order to achieve this, if we can possibly help it.[14]

But Carson says "such bland assurances" were contradicted by a bias within a state structure that, on occasion, firmly weighted the "relative priorities" of production against the enforcement of safety by the DOE.[15] Whether that bias gained a foothold in the small branch that regulated diving is not something Carson specifically addressed in his study. But his research does at least raise the specter that the same kinds of pressures affecting the parent agency as a whole may have created a situation where the diving industry's reputation for danger became a convenient excuse to obscure a death toll that was unnecessarily high for other reasons. Moreover, when one looks into the past and examines the case histories of other fatalities, it becomes clear that the Diving Inspectorate's special relationship with industry fostered conditions where the *Wildrake* tragedy became the inevitable consequence of years of regulatory neglect and industry abuse. At least one *Wildrake* diver is convinced that it was and contends that, by 1979, the industry was poised for such an accident, and that when it finally occurred, it exposed all the tragic political and human consequences of a failed enforcement regime.

At the root cause of this failure lay the conflict that put safety inspectors in the untenable position of having to weigh the political and economic consequences of interrupting the flow of oil against the concern for safety. While DOE officials were keen to invest themselves with bumper-sticker slogans like "Safety First," they showed no such enthusiasm towards solving the internal problems of inadequate staffing, poor salaries, and lack of experience of diving inspectors. Nor was there an aggressive campaign to carry out one of the diving division's prime directives, that of preventing similar accidents from repeating themselves.

As written, the law in some critically significant areas was manifestly vague. That vagueness allowed contractors to circumvent the intent of the legislation for economic gain. And while they employed resource after resource to obfuscate the law or head off new legislation to close the gap on unsafe practices, divers continued to die. In one remarkably sad and unaccounted for period, inspectors deliberately ignored the obvious lethal consequences of one particular diving practice that accounted for one third of all fatalities within the first ten year period of the Diving Inspectorate's tenure.

While it had the power to shut down unsafe operations, the Inspectorate never exercised those powers, even after serious accidents. Instead, it avoided that responsibility and allowed a profit-motivated industry to police itself and, on occasion, to flagrantly ignore the law. Even in those rare instances when the Inspectorate found criminal proceedings to be unavoidable, its watchdogs not only displayed a lack of knowledge in preparing evidence for prosecutions,[16] they sometimes appeared in court defending the very practice that contributed to the fatality.

In short, instead of discovering evidence to support the premise that strict law enforcement was the standard of the day, what one finds is the final piece of the puzzle which led to the *Wildrake* tragedy—an agency that cultivated a questionable enforcement record, was all too willing to wear its sponsorship hat more often than its safety hat, and took little or no corrective action to curb the abuses that laid the foundation for what would happen to Richard and Skip.

Chapter Twenty-seven

The Watchdogs

It has been suggested to the TUC that the Department of Energy's Inspectors have insufficient experience of enforcement of health and safety regulations to be really effective. Unions have questioned their experience of investigative procedures, in particular in relation to the taking of statements.

—— Submission by the Trades Union Congress
to the Burgoyne Committee August 1979
Burgoyne Committee Report, p. 292.

In 1974, retired Royal Navy Commander "Jackie" Warner was asked to take over the diving division of the Department of Energy. Warner appeared to be the right man for the job. He had a kind of "I'll huff and I'll puff and I'll blow your house down" aura about him. He was a decorated World War II veteran, had served aboard the submarine *Tallyho*, and qualified as a clearance and bomb disposal diver. His resume also included assignments as Commander of *H. M. Diving Vessel* and the Far East Operational Diving Team in Hong Kong. Later, he took charge of the Royal Navy Diving Trials Team, Experimental Diving Team, and was the Deputy Superintendent of Diving for the Royal Navy. He was also associated with several US diving research programs including "Man in the Sea" and "Sealab Three."[1] In 1967, Warner was awarded the Member of the Order of the British Empire for his work in underwater explosive ordnance disposal.

As Chief Inspector of Diving, Warner says his "remit was safety," and found briefing ministers easy because "safety should be the whole truth and nothing but the truth, strictly a-political."[2] To arrive at that truth, he naturally had to know something about how the government's safety policy was being administered offshore and whether it was being adhered to. One way this could be achieved, he says, was to encourage his inspectors to spend as much time offshore as possible "on site—actually inspecting and advising on safety."[3] Not only was this "the most important part of an inspector's job, apart from accident investigation,"[4] but it fulfilled an ancillary function as well: deterrence. In Warner's words, sending inspectors offshore would be "like parking police cars every 10 miles on a motorway."[5]

But in actual practice no such goal was ever achieved. In fact, it was not unusual for a diver to work in the North Sea for years and rarely cross paths with an inspector.[6] When one long-time British supervisor was asked to comment on the presence of the DOE offshore, he remarked: "I don't know how many years I worked in the North Sea—15, 16 years—something like that, off and on—and I've only been on a work site twice in me life when the D of E actually came and visited to inspect."

This lack of oversight was due, in part, to staffing problems. For the first two years of his appointment, Warner worked alone as the only inspector responsible for overseeing all diving operations within the British sector of the North Sea, an area covering 102,177 square miles, or about half the size of France. Five years later, when the *Wildrake* accident occurred, his task force had only increased to four, which Warner says "was not achieved without some battles with the establishment."[7] Most likely, those battles concerned salary levels, which the Department concedes was part of the reason it had "difficulty in attracting the right kind of applicant to fill vacancies."[8]

Still, even when it did manage to employ an applicant, the Diving Inspectorate seemed to place a higher premium upon the candidate's title, than on his actual level of commercial diving expertise. When Richard and Skip died, three of the inspectors (including Warner) were ex-Royal Navy Commanders—which gave the agency a kind of "jobs-for-the-boys" image—while the fourth had gained his commercial experience outside the North Sea. Moreover, none of these inspectors had any practical saturation experience, a point not lost on one leading diving contractor who said of Warner's people: "There was nobody with any really serious background in that area."

That lack of experience was noted offshore. Although divers rarely came into contact with inspectors, those that did often came away with the impression that the inspections were not thorough, that they seemed to be conducted on the basis of "what caught the eye," and that the inspectors were not very knowledgeable about deep-sea diving on a technical level.

"It was just an exercise that they chose to go through as a bureaucratic function," one diver said about the inspection he witnessed, while another recalls that when an inspector visited his work site "you could have told him anything, literally...It was the general feeling that he didn't really know what was going on. He'd just come out there and you'd give him a run through, give him lunch, and put him on a helicopter, and send him back to the beach."

In addition to these subjective views, inspectors sometimes gave testimony at Fatal Accident Inquiries which gave one pause to wonder if they were learning any lessons at all from the accidents they investigated. At the *Wildrake* inquiry, for example, a Department inspector testified that the double shackle arrangement on top of the *Wildrake* bell was "not a bad practice" provided that the shackles were properly moused.[9] And later, when asked whether raising a bell which was not vertically plumb with the crane could create any problems, the inspector answered, "Not normally."[10] Yet both practices were clearly part of the cascade of failures that led to the deaths of Richard and Skip.

In 1979, this lack-of-experience issue became a bone of contention before the Burgoyne Committee when it was asserted by the trade unions that, overall, Department inspectors did not have the necessary experience to enforce the health and safety regulations, and that this deficiency was particularly apparent in the quality of their incident reports.[11]

When the Committee looked into this allegation, it was struck by how poorly some of the investigations had been conducted so as to make any conclusions about them invalid. Out of five accident reports filed by the Diving Inspectorate in 1974, four of them were written in such a way that the cause of the accident could not be established beyond doubt.[12] Commenting on the overall management of accident reporting, the Committee said "accident reports received by the Department are sometimes so brief—no more than 10 words or even less—that the cause of the accident remain totally obscure."[13] In response, the Department of Energy conceded that their system of reporting was "not entirely adequate," and that inspectors had not been gathering sufficient detail in order to make their reports.[14]

Today, with 20/20 hindsight, some of the most withering criticism of the Diving Inspectorate comes from British divers who comment that the agency was largely composed of people with "the wrong temperament," who were overseeing a profession for which they were not qualified, and who did not fully comprehend the scope of the work, nor the dynamics of the relationship between the diver and the company that employed him, as this North Sea supervisor relates:

> Very, very occasionally the Department of Energy would send people offshore onto a work site, and they would inspect and very briefly interview the divers that were on that work site. And I remember presenting my shift in the conference room on the *Stena Seaspread* to have their log books verified...and after they had finished looking at the troops and their log books...I gave them my log book and said, 'Can you authenticate mine,'...and they looked at me and they said, 'Why? You're a supervisor.' I said, 'Yeah, I am on this job, but I could work for another company next year and I could be back to being a diver.'
>
> That's how much they knew about commercial diving; they didn't understand the realities of it.

While Warner's office was staffed by men with sparkling military resumes, divers argue that Navy diving and commercial diving are worlds apart. "It's not the same as 'Civils' or bloody attack swimmers in the Royal Navy," one British diver insists, who adds that, if the government had really been serious about enforcing the law, then it would have employed ex-commercial divers who understood the industry and knew where the corners were being cut on safety:

> It's simple, you set a thief to catch a thief, or you employ a soldier to run soldiers; you don't use ex-Royal Navy torpedo officers from submarines to run commercial oil field divers. Come on—that's like getting your village blacksmith to run a bloody Apollo moon shot!

And it would be this lack of firsthand experience, coupled with the DOE's inability to separate its sponsorship duty from its enforcement duty that would remain a fundamental flaw within the structure of the Diving Inspectorate. It would impinge upon the inspector's assessment of what caused an accident, it would interfere with his ability to impose sanctions, and it would prevent him from learning the necessary lessons in order to prevent future accidents from occurring.

Among the case histories, probably the most glaring example that the Diving Inspectorate had the wrong people investigating accidents, in terms of their sleuthing capabilities, involves the mysterious death of 25-year-old Sub Sea diver John Howell from Liverpool. This case also illustrates how the judicial system, through no fault of its own, could be easily led astray by an inspector's erroneous conclusions, only to leave the surviving diver under a cloud of criminal suspicion.

Howell had been a Royal Navy diver since the age of 16. The military trained him as a mixed gas diver and even commended him for preventing a deep-sea diving accident.[15] In August 1975, he resigned from the service to seek his fortune in the North Sea, and soon after, was hired by Sub Sea International. On January 10, 1976, Sub Sea sent him to the *Western Pacesetter I*, an oil rig engaged in drilling an appraisal well about 40 miles southwest of Thistle.

The day after Howell landed on board the rig, Diving Superintendent Jack Highley called a bounce dive and instructed Howell and Jim Ramble to go down and attach two guide wires to the Blow Out Preventer (BOP) which was standing on the seabed in 514 feet of water.[16] The two men suited up, ran through bell checks, and at 2140 that evening they were lowered to 482 feet.[17] Fifty minutes later, Highley started the dive clock when Howell began blowing down the bell.[18]

In the dive van, Supervisor Mick Tooke stood beside Highley. While Highley ran the dive, Tooke was busy decompressing two men from a previous dive.

At 2243, Ramble locked out and swam down to the BOP about 15 feet away. The job was not particularly difficult; just stab the ends of the guide wires into the guideposts and hammer a few bayonet fittings in place.

About twenty-five minutes into the dive, Highley interrupted Ramble and ordered him back to the bell because the hot water machine was on the blink.[19]

Five minutes later, he radioed the bell and said he had fixed the problem. So Ramble locked out a second time, but soon after arriving at the BOP, he began to feel sick. Something was wrong; he didn't know what, but he began to feel "very uneasy."[20] At this point, he told topside he was going back to the bell, and Highley said, "Okay, get moving."[21]

In the bell, Ramble took off his gear and asked Howell if he would finish the job for him, and Howell agreed.[22] As is customary, the two divers traded positions, with Howell donning the same gear Ramble had been using. At 2340, Howell disappeared through the trunking with Ramble still feeling "a bit drowsy."

In the bell van, Highley was listening to the sound of Howell's breathing through a pair of headphones when he heard a dramatic change in his breathing pattern. Sensing that something was wrong, he asked Howell if he was okay.[23]

Howell, now swimming to the job site, managed to utter, "No, I'm in trouble. I'm returning to the bell." A moment later, Highley received Howell's last transmission: "I'm foul...[fouled]."[24]

Well over six feet tall, Jack Highley was a quick-thinking man who'd been around the diving industry a long time. In 1965, he was on board the first rig sent out into the North Sea,[25] and now, eleven years later, he surmised that Howell had probably passed out. Turning to his gas panel, he quickly dialed up the supply regulator to overload Howell's mask to prevent him from drowning.[26] Then he ordered Ramble to go out and rescue his bell partner.[27]

Five hundred feet below, Ramble had been peering through one of the portholes to try to catch a glimpse of his partner, when he noticed that Howell had unexpectedly returned to the bell and was struggling unsuccessfully to cram himself inside the trunking. When Ramble took a closer look he noticed that Howell's bailout bottle was caught under the lip of the trunking.[28]

Using his foot, he freed Howell, grabbed his harness, pulled him chest high into the trunking, then took off his free-flowing mask and tossed it aside. Howell was unconscious, but he was still breathing.[29] Ramble then reached up and unfurled the diver recovery hoist from the top of the bell and repeatedly tried to lift his partner inside, but, because he was not powerfully built and because he was still feeling sick, he failed to do so. On several occasions he attempted to float Howell inside by partially flooding the bell,[30] but that didn't work either because Howell had become so rigid in the cold water that his legs were bent underneath the lip of the trunking.[31]

Twenty-six minutes past midnight, the dive clock ran out. With Howell stuck in the trunking, Highley began to raise the open bell to its first decompression stop at 460 feet. Twelve minutes later, Howell stopped breathing.[32] At this point, Ramble started giving mouth-to-mouth resuscitation and cardiac massage and continued to do so for the next sixteen minutes.[33]

Meanwhile, as the bell was being raised through its water stops, Ramble's condition began to deteriorate. He started experiencing severe stomach cramps and had a strong desire to lie down and fall asleep.[34] Suspecting that the gas supply was contaminated, Highley switched to a new supply on the control panel and flushed the bell with fresh gas, but this had no effect upon Ramble's condition. "In fact," Highley said later at the inquiry, "he got steadily worse."[35]

Ramble was not able to revive Howell. And over the next three and one half hours, Highley had to continuously talk to his remaining diver to keep him awake and alive.[36]

At 0147 the bell was brought to the 370-foot stop. At 0200, Ramble was cast into darkness when the bell light suddenly burned out. At 0229, Highley discovered that Ramble had failed to change out the CO_2 canister as ordered to ninety minutes earlier.[37] The change out, which would have normally taken only a minute, dragged on for about thirty minutes because Ramble was still sick and racked by cramps.[38]

Just before 0300, Ramble tried again to hoist Howell inside, but was unsuccessful. Finally, at approximately 0425, Highley decided that if he didn't get Ramble on board as soon as possible, he might lose him.[39] He told Ramble to tie Howell's body to the makeshift ladder propped against the trunking. It was one of the most difficult decisions he ever had to make, and for Ramble, it was the worst day of his life.

At 0505, the bell was recovered onto the sat system.[40] Exhausted and feeling poisoned, Ramble had to be assisted out of the bell and put to bed by the two divers already in the system.[41] In the days that followed, it would be determined that John Howell had died as a result of drowning.

The next day, one of Warner's inspectors was flown out to the accident scene. Like Warner, he too was an ex-Royal Navy Commander, and during his initial investigation inside the bell, he discovered a gas valve, which was normally supposed to be open, was in fact closed. It was a secondary valve located between the penetrator valve behind the inside hatch and the manifold valves which fed the divers' umbilicals.[42] The fact that this valve was in the closed position struck the inspector odd, and two weeks later he returned to the rig to try to determine why it was closed and why Howell got into trouble in the first place.[43]

Based upon an experiment he conducted with Supervisor Mick Tooke, the inspector ultimately came to the conclusion that, when Howell was exiting the bell, he "inadvertently" bumped the handle of the valve closed with his bailout bottle,[44] then breathed down the remaining gas in his umbilical, which caused a pressure loss in his mask and allowed sea water to enter.

But according to Tooke, Howell could not have knocked the valve completely closed. Asked at the inquiry whether he was able to move the valve handle during the experiment, Tooke said, "You could move it a bit, to quarter shut or something like that, but that is about all."[45] The Commander, on the other hand, saw things differently, and later told the court he was "absolutely convinced" that he and Tooke had demonstrated it could be done.[46]

Still, the Commander knew his theory contained a fatal flaw. By his own calculation, Howell had "perhaps two or three breaths" in his gas hose before he locked out,[47] and if he had bumped the valve closed, then it would have been impossible for him to exit the bell, swim part way to the job site, alert topside he was in trouble, then clamber back to the bell all on these few breaths. Realizing that Howell could not have done this, the inspector backpedaled and admitted on the witness stand that his theory could not be correct.[48]

Such an admission might have ended all the guesswork then and there, but no sooner had the Commander's initial hypothesis buckled under the strains of reality than another was immediately produced like a rabbit out of a hat. The inspector was now convinced that the only way Howell could have drowned was if someone had closed the gas supply valve when he was out of the bell. Since it could not have been Howell, then it must have been Ramble. This was "the only possible conclusion" he could come up with, the Commander told the Court.[49]

But the inspector's interpretation of what happened was at odds with the evidence on a number of key points. First, Superintendent Highley heard Howell breathing through his ear phones well beyond the point when Howell realized he was in trouble,[50] something he could not have done if the crucial valve was closed. When one of the attorneys pointed this out to the inspector, he could offer no explanation for it.[51]

Second, the dive log clearly states that Howell was still breathing when Ramble took off his mask,[52] which directly contradicts the inspector's assertion that Howell had drowned in his mask before he was pulled into the trunking.

And third, when Ramble removed Howell's mask, he said he tossed it to one side and recalls hearing it free flowing in the water-filled trunking.[53] Again, this could only be true if the supply valve was still open. Furthermore, Ramble's testimony corroborates Highley's evidence on another important point and explains why the Commander found the gas valve closed during his initial inspection of the bell after the accident: namely, when Highley realized that his diver had passed out in the water, he immediately overloaded the supply regulator from 375 psi to nearly 550 psi,[54] which was more than enough pressure to overwhelm the demand valve in Howell's mask and cause it to free flow. As Highley correctly testified at the inquiry, "the only way [Ramble] could shut that gas supply off [to the mask] was to turn the valve in the bell because he could turn off all the valves in the mask, and the regulator would still be overloading and pumping gas into the mask."[55]

When the logic of this scenario was put to the inspector, once again, he began grasping at straws. "That could be," he replied. "This is a clue I don't understand. I can't understand. I wonder whether, when the mask was in the water, it wasn't just bubbles coming up from trapped air [in the mask]. I am not sure whether [Ramble] was fully concentrating on watching a mask when trying to give artificial respiration."[56]

That response was as inadequate as it was implausible and the lawyer for Sub Sea International quickly delivered a lethal blow to it. "How can there be trapped air in the mask if the man in the mask had drowned?"

"He would have taken the mask off when his head was out of the water, and any water that was in it would have dropped out," the inspector replied.[57]

Again, that answer did nothing to solve the mystery of how a man could drown in a free flowing mask. The Crown produced an expert witness to say the mask was in no way defective,[58] and that, only by a drop in pressure, could water enter it.[59] In a free flowing mask this would be impossible.

The inquiry did, however, examine one other possible explanation for the death of John Howell. Gas contamination. Ramble's sudden onset of drowsiness, his stomach cramps and his failure to respond appropriately to topside commands, led Jack Highley, Mick Tooke, and Jim Ramble to testify that the accident had all the earmarks of a corrupted gas supply.[60] If this was indeed the case, Ramble could have taken the initial hit, returned to the bell before it did any further damage, with Howell then locking out to receive the lethal brunt.

But the connection between Ramble's sudden illness and Howell passing out in the water seemed to have had no impact upon the Commander's powers of deductive reasoning. When one of the attorneys asked if the two contiguous events were "just an enormous coincidence," the inspector explained away the entire misfortune by speculating that Ramble's symptoms were probably due to an "emotional" rather than physical origin caused by either working too hard underwater or being affected by the cold when his hot water supply temporarily shut down.[61]

Here again, the inspector's conclusions fell far outside the line of evidence given by Ramble who stated that his assignment that night did not require much physical effort,[62] and that he had returned to the bell well before he became cold because Highley had forewarned him his hot water supply was about to shut down.[63]

Asked whether he had closed his mind to the possibility of gas contamination, the inspector answered: "I cannot close my mind to anything, but I think it is unlikely as a result of the analysis which has been carried out."[64]

When Highley was asked if he'd ever witnessed an accident caused by gas contamination before, he said, yes, that early in his career, a diver's gas supply had been compromised with a substance called trichloroethylene, a highly toxic dry cleaning agent that divers sometimes used to clean the interior of gas hoses.[65]

On board the *Pacesetter*, the crew were not using trichloroethylene, but Freon, a refrigerant, which the inspector pointed out was non-toxic. But it is a well known fact that certain agents which are non-toxic at surface pressure, can turn dangerously toxic at higher pressures. Breathing the normal amount of oxygen in air at surface pressure, for example, is not toxic, but feeding the diver the same mix at 500 feet is extremely dangerous. And experts point out that even if a substance is not actually toxic at low pressure, it may have a tendency to become narcotic at higher pressures. According to a British diving medical officer, Ramble's symptoms (stomach cramps, drowsiness, failure to pay attention to topside commands) are "very common effects of toxicity, reflecting an irritant effect on the bowel and narcotic effects on the brain."

After the accident, the thought occurred to Highley that someone in his crew might have left some Freon in the gas lines that supplied the bell van. He asked if anyone had used any chemicals to clean the gas manifolds and was assured that they hadn't.[66] When it was pointed out to him that analysis of the gas quads revealed no contamination, Highley replied, "I know they were found to be clean, but there is some doubt in my mind."[67] (A "refrigerant," however, was detected in samples from the gas quads, but according to a government expert, "it was well within the threshold limit value and may not have been toxic.")[68]

In the end, the more probable explanation of gas contamination seemed to carry little weight in the courtroom that day, the effect of which was to leave Jim Ramble at the mercy of the Commander's desultory theories which were apparently circulated among the attorneys and the sheriff prior to the proceedings. Indeed, even before Ramble uttered a single word on the witness stand, he was given "the caution" by the sheriff.

"You understand," he said to Ramble, "there may be a suggestion that perhaps not only that you made a mistake but that you committed some crime or offense." And then he proceeded to advise the witness that he need not answer any questions that might incriminate him.[69]

But Ramble had nothing to hide and answered all of the lawyers' questions forthrightly and to the best of his recollection. That he was suspected of wrongdoing demonstrated the tremendous power and influence of the inspector's opinion. The proceedings even exhibited overtones of a class-conscious struggle for controlling the truth. On one side of the contest was an ex-Army grunt with no prefix to his name and no authoritative cache who had been left with a mash of hazy memories about an experience that nearly took his life at 500 feet. On the other side was a man of distinction, a retired Naval Officer, who represented the personification of unquestionable integrity and expertise.

But the Commander was not a diving expert in any real sense of the word. This was a man who had no personal experience as a commercial diver and only his imagination to rely upon to explain a mystery that ultimately left Jim Ramble under a cloud of suspicion in a public venue. When his initial theory blew away like grains of sand in a windy desert, he quickly ginned up another—that Ramble, suffering from some presumed emotional impairment, unwittingly closed the crucial valve and caused the death of his partner. But it was all pure speculation based upon the discovery of a closed valve that was shut to stop Howell's mask from free flowing. When this possibility was put to the inspector, he conceded that if Highley had over pressurized the gas line, Ramble "could have turned it off."[70]

Unfortunately, this concession came too late for Ramble who had already been forced to endure the inspector's allegations through a series of rigorous, and sometimes hostile, cross-examinations. What damage this had upon his mind and career after the inquiry is unknown.

Without more information no one can say with certainty why John Howell died on the night of January 11, 1976. But one thing is clear; the man who barely escaped with his life did not "inadvertently" cause the death of his partner, and the man who lost his life did not cause his own demise. And yet, sadly, what survives today in government archives, as part of the official picture, is not the lamentable story the diving inspector finally settled upon at the inquiry, but the one he originally dreamed up—*the one that cannot be correct*—where Howell bumped his own gas valve closed.[71] That this is so demonstrates that, when the Diving Inspectorate decided to close the book on this investigation, it paid little or no attention to the evidence presented at the inquiry, and that, rather than admit it did not know how Howell had died, was willing to accept just about any explanation from its inspector no matter how inaccurate it was.

Chapter Twenty-eight

Benevolent Enforcement

Looking back, of course, it was education. Prior to the oil companies going into the North Sea, they were dealing with Arabs, people like that, and the management didn't really give a damn about life or death...so keen to get the oil pumping. In the North Sea, they soon realized, I think, that they had to conform to the law—and remember the law was only as we had written first in the world.... We were all learning and, yes, the oil companies eventually bent over backwards to do their best to be safe. Now, that took a little while.

—— Chief Inspector Commander 'Jackie' Warner[1]

In 1973, when Richard arrived in Scotland, there were no safety regulations covering diving operations in the British sector of the North Sea. When Commander Warner took office the following year, he says he was determined to change this, "to regulate, govern and guide the offshore industry." Eventually legislation was drafted which Warner says "was of a remarkably high standard."[2]

But, in contrast to the terrific pace drilling operations were taking place offshore, Carson found Parliament's legislative response to safety issues "laggardly" and "highly deferential" to the oil industry.[3] Moreover, government spokesmen were adverse to burdening oil companies with heavy-handed "inflexible regulations." Instead, they gave assurances that "the overall plan was to proceed by 'benevolent enforcement...generally advisory in nature'."[4]

In the end, legislators decided to write the law in general terms to obviate "the need for constant changes to keep pace with technological innovations."[5] Detailed prescriptive regulations, which addressed every contingency, were to be avoided. When legislators did get round to writing statutory provisions, more often than not, they turned out to contain anomalous gaps and overlaps that left the safety of thousands of men like Richard and Skip hostage to the uncertainties of jurisdictional confusion and misinterpretation of the statutes.

Even Commander Warner found many of the regulations "fragmented amongst a host of separate Acts and Statutory Instruments."[6] Furthermore, British legislation on many key issues never spelled out how a particular safety requirement had to be achieved, merely that the diving contractor "had to be safe."[7]

In retrospect, some former North Sea divers think the broad-brushed method was the wrong approach. In their view, deep-sea diving is an industry where the regulations should be detailed in order to prevent confusion, as pointed out by this English diver when comparing how Britain and Norway addressed the issue of access into and out of the bell:

> See in Norway—if I'm going to be a diver, I'm going to Norway. The regulations are laid out in black and white. [In] the Norwegian legislation, the bottom hatch on a bell must be a minimum of 800 mm...in our legislation it must be big enough for easy ingress and egress in the bell. What does that mean? Get a dwarf as a diver and just make it big enough for him?

As more legislation appeared on the scene, safety inspectors would come to realize that what they were dealing with was a tangle of piecemeal code stitched together in the language of compromise and peppered with vague and uncertain phrases. Consider, for example, the expressions "so far as is reasonably practicable," "novel structures," and "effective means of communication." What do these tortured phrases really mean? More importantly, how would they impact upon the lives of offshore workers? At the time of their creation no one really knew, not even the men who coined them.

The practical effect of a broad-brushed system, written in vague and uncertain terms, was to create a kind of legal "gray zone," leaving front-line inspectors feeling ambivalent and confused about how far they should venture when carrying out their duties. And in some instances, it was not the language of the law that presented headaches, it was the disconcerting revelation that the Inspectorate had failed to draft regulations to cover the type of diving taking place offshore.

In January 1976, just six days after the death of John Howell, the legal branch of the Department of Energy ran aground in this gray zone after the bell accident that killed Comex diver Clay Ellis and paralyzed Derek Bannister. Accordingly, Warner's office prepared to investigate the accident believing the Offshore Installations (Diving Operations) Regulations 1974 were in force. But Comex claimed the 1974 regulations did not apply because they related only to operations conducted from "rigid platforms."[8] Ellis had been diving from a converted supply ship, the *Smit Lloyd 112*. A set of regulations did exist pertaining to ships—the Merchant Shipping (Diving Operations) Regulations 1975—but to the dismay of the DOE, they were equally unhelpful because the *Smit Lloyd 112* was neither "a ship registered in the United Kingdom or a British ship of a specified description."[9]

As a result, the legal department of the DOE determined that "there was no legislation covering ships engaged in diving operations on pipelines at the time of the accident," and ordered Warner not to investigate Ellis's death.[10] Six months later, this gap in the law was plugged when the Submarine Pipe-lines (Diving Operations) Regulations 1976 were put into effect. Had they not been in place at the time of the *Wildrake* accident in 1979, then the investigation into the deaths of Richard and Skip might have suffered the same fate.

The hole in the law through which the Ellis fatality fell was not alone. There were other anomalies. In 1980, fifteen years after North Sea operations had begun, there were still no regulations for the safe construction and operation of pipelines. The laws pertaining to the interface between pipelines and subsea completions, single-point moorings, flare installations and floating production units contained "'grey' areas."[11] More troubling, the legal branch of the DOE did not believe that accommodation platforms, or "flotels," were offshore installations "in the accepted sense."[a] Thus, when Richard and Skip were working on board the *Gulnare* in 1978, they were apparently doing so outside the purview of Warner's office.

For all practical purposes, these legislative gaps translated into loopholes by which companies could potentially escape their responsibilities. They also reinforce Carson's point about how the law constantly lagged behind offshore developments, and that it was the constant focus on speed that underscored the government's primary objectives, not the establishment of adequate safety laws prior to the onset of operations.

In its defense, the DOE says it was keenly aware that some of its regulations were "inadequate and difficult to enforce."[12] But it felt confident that, over time, the offshore oil industry would ultimately come to understand the laws "largely through explanation and interpretation," and that any uncertainties about the language could be overcome by coupling the regulations with a set of detailed technical Guidance Notes designed to clarify the meaning of the legislation and "educate the industry."[13]

The trouble was, Guidance Notes were not law, did not affect the meaning of the law, and could not be directly enforced, an essential fact not lost on the oil companies. When regulators tried to rein in industry by referring to these Notes, they were constantly reminded that the Notes were not law and that compliance with them was strictly voluntary.[14]

As for the Diving Inspectorate, no Guidance Notes were ever drafted during the 1970s to clarify the intentions of the diving regulations. Instead, it issued "administrative advice" in the form of Diving Safety Memoranda which offered recommendations, and purported to counsel contractors on how the regulations should be interpreted. According to Commander Warner, the DSMs also carried

a The DOE would later reverse its view on this issue. *Source*: Burgoyne Committee, *Offshore Safety*, p. 228.

an implicit warning. "The memorandums I sent out were really to say, 'Watch it! Something has happened that indicates so and so and so and so, and don't do it.'"[15]

But as with Guidance Notes, these warnings lacked the force of law, and amounted to nothing more than old bromides which had little or no effect upon the behavior of the men who controlled the North Sea diving industry. Even following serious accidents, the language in them was often pedantic and irresolute. For example, when two Oceaneering divers died from hyperthermia in September 1975, the Inspectorate issued a telex with the main thrust of the message:

> Failure to maintain the diver in a safe thermal balance can be fatal. Hyperthermia is just as dangerous as hypothermia.[16]

There was no red flag about the problems the Oceaneering crew encountered, no general discussion about the lessons learned from the tragic event, and no recommendations for preventing similar accidents in the future.[a]

After Comex diver Clay Ellis died in 1976, Warner sent out another DSM. If he intended to send a stern warning to contractors against the dangerous practice of landing the bell on the seabed, it scarcely could be found in a memorandum laced with mind-numbing technical jargon about "drag forces," "thixotropic properties," and "hydrostatic pressures." It was only in the last two paragraphs when the Commander finally got down to business:

> No absolutely safe design for underslung ballast which would be safe in all types of sea-bed conditions has yet been designed.
>
> When using the underslung ballast technique with diving bells the bottom door of the diving bell should always be shut when the ballast weights are being broken out of the sea bed.[17]

It's difficult to imagine a hard-nosed diving contractor taking this conflicting message seriously. The memo mentions the inherent dangers of "sea bed suction," and "risk of hawser breakage." But more significantly, there is the sobering statement: "No absolutely safe design for underslung ballast which would be safe in all types of sea-bed conditions has yet been designed."

That being the case, the Inspectorate might have presented a more consistent image in its War On Unsafe Practices if it had punctuated this caveat with an outright ban on a technique that was clearly unsuitable for North Sea conditions. But far from going the full measure and conveying the message, "Watch it! Don't ever do that again," the memorandum suddenly flashes a green light to industry to go ahead and continue experimenting with the learning curve even though that experiment killed one diver and paralyzed the other.

---

a Although the Oceaneering system contained a defect which contributed to the accident, the DOE inspector did not stop bell operations from continuing. Instead, he determined that the system was safe and told the company that, "subject to the divers being happy to dive, they could carry on." *Source*: Holmes/Baldwin FAI, p. 316-317, 423, 431, 439-440.

In a broader context, this official notice reflects what Carson and others have described as the conflict of interest operating within the Department of Energy. In circulating this memo, the Inspectorate wanted to be seen at the forefront of the battle on unsafe practices in the North Sea, so it "warned" the diving community that the Comex technique was unsafe. At the same time, it did not wish to appear unduly oppressive either, so it donned its sponsorship hat and eviscerated the warning by allowing the practice to go forward. It was an unnecessary concession to industry, and fortunately for the divers at large, contractors recognized the Comex experiment for what it was worth and never repeated the mistake.

Although Diving Safety Memoranda were not law, there was a tendency by inspectors to interpret them as law or assign a special legal status to them. Contractors were expected to abide by them, and according to one Senior Diving Inspector, it was very much to their "peril" if they were ever caught disobeying the recommendations. But to industry, Department recommendations were just that—recommendations—not law; and introduced into this legal jumble was the added complication that contractors were permitted to mitigate the safety advice by other means. The overall rationale behind this thinking was to give companies the freedom to march in stride with technological advancements; if they could come up with a better idea for complying with the law, so much the better. Where the contractor put himself into jeopardy, in the government's view, was if he chose not to follow the recommendations, had an accident, was brought up on charges of breaching the regulations and could not demonstrate to the Court's satisfaction that the other method was equally as effective as the recommended method.[18] In such a situation, the prosecution's case would be enhanced and the Court might find him at fault.

But in practice this system had serious flaws. How the other method should be achieved was left entirely to the discretion of the diving company, which was under no legal obligation to inform the government that it planned to stray outside the boundaries of the recommendations. Moreover, because decisions to deviate from the norm were typically made hundreds of miles offshore (as in the case of modifying the *Wildrake* system), government inspectors never knew if the departure would, or even could, rise to the regulatory standard the law intended. And without a statutory requirement to evaluate or prove that the other method was safe before it was put into effect, operations always had the potential to lapse into highly questionable and, sometimes, illegal experiments using divers as guinea pigs. This, in fact, is what happened to Clay Ellis and Derek Bannister in 1976, and three years later it would happen to Richard and Skip.

But what if there was no accident, no breach of law, and the Inspectorate discovered that the contractor was not obeying the recommendations by any means? Would the Department still walk into court and argue that the Safety Memoranda actually had the power of law? The general impression given by

Commander Warner is that it would, although he quickly points out that no such challenge was ever tested.[19] Thus, in real terms, the actual peril to the contractor for breaching the recommendations was only marginal at best, and the campaign to root out unsafe practices in the North Sea, weighted in favor of solving problems *ex post facto*, after an injury, or as with most diving accidents, a fatality.

From London, the Department of Energy viewed its memoranda as an effective counterbalance against neglect and abuse, but from Aberdeen, the heads of the diving companies did not share this perspective. They saw London as lacking the political will to carry out its threats, and very early on, began to ignore the memoranda.

On August 6, 1975, the Diving Inspectorate issued a communiqué entitled "The Effect of Cold on the Diver." In typical bureaucratic parlance it reminded contractors that "the North Sea is a cold environment," and that, "ideally," the diver "should have" his breathing mixture pre-heated with a gas heater.[20] But because contractors knew the recommendation could not be imposed upon them, they saw no reason to spend the extra money to comply with it, and many did not. When Richard and Skip died in 1979, they were diving beyond 500 feet of water without gas heaters nearly four years after the Inspectorate issued the memo. Sub Sea International also routinely infringed on this recommendation, as it did not begin to incorporate gas heaters until the early 1980s.[a]

On February 3, 1976, a month after Clay Ellis was killed, the DOE sent out a memo listing nine reasons why supply vessels should not be used in deep-sea diving operations, the majority of which related to the unstable nature of this class of vessel. At the end of the text, the DOE voiced a very clear recommendation:

> The conversion of the normal North Sea supply and work boats into deep diving vessels is not recommended.[21]

Once again, this advice was ignored. Eight months after Ellis died, two Sub Sea divers nearly lost their lives in a bell accident in the Thistle Field while diving from the converted supply ship *Active Duke*, and three years later, the Swedish supply vessel, *Stena Welder*, would be involved in the rescue attempt to save Richard and Skip.

This practice of ignoring official recommendations would continue long into the next decade. Restrictions on the length of the diver's umbilical, advice on bell run durations, and procedures for pre-dive bell checks were routinely disregarded with impunity.[22] And if contractors could ignore advice on these fundamental safety issues, it could ignore advice on just about anything: oxygen safety, accident reporting, even the responsibilities of a diving supervisor.

---

a Government would not make gas heaters mandatory for dives in excess of 492 feet (150 meters) until 1981. The Diving Operations at Work Regulations 1981. Regulation 12 (h)(ii).

Early in 1976 Richard was in Santa Barbara and Skip was in dive school in Wilmington, California. The *Wildrake* accident was three years away and already contractors were taking advantage of an oversight agency that was staffed by the wrong people and unable to cope with problems created by imperfect laws. And while the Inspectorate professed a desire to "learn from the lessons of the past and not repeat the errors,"[23] it failed to ban a practice that accounted for more than one third of all fatalities within the first ten year period of its tenure because it was more interested in accommodating industry than raising standards of safety.

In some areas of the world, where sea conditions are less severe, the use of scuba in commercial operations is routine. But scuba is arguably the most dangerous form of diving there is, and the North Sea no place to be risking the lives of divers to test the soundness of that argument. The whole purpose for employing surface-supplied diving techniques is to give the diver the added advantage of topside support. Diving on scuba removes that advantage. The diver has a limited air supply, no communications, and typically no hot water. And without communications, the diver cannot talk to topside during an emergency, nor can the supervisor monitor the diver's breathing or receive information on his activities underwater. Complications can occur if the diver exhausts his air supply, and because he is not tethered to the vessel, he can fall prey to tidal flows which have been known to wash men away before help arrives. And without the use of a full-faced mask, if the diver becomes unconscious, he will drown if his regulator drops out of his mouth. In short, the number of things that can go wrong on scuba extends well beyond the argument that would justify its use. That scuba carried a host of increased risks was not unknown to the Diving Inspectorate. And yet, it stood strangely silent while a pattern of fatalities developed before its very eyes.

This pattern began on October 14, 1974, when Comex diver John Clarke was thrown against an anchor bolster of a drill rig, fracturing five of his ribs.[24] Dressed in scuba gear, he had no lifeline, and no communications with his supervisor.[25] His regulator subsequently dropped out of his mouth, and by the time he was recovered onto the deck of the rig, he was dead from drowning.

At the time of the accident, scuba was an unregulated activity in the North Sea. Commander Warner had only been appointed Chief Diving Inspector earlier that year and he was still adding finishing touches to a new set of regulations that were due out in a few months. Whether Clarke's death caused any alarm in his office is not known. But if it had, and even if Warner wanted to ban scuba outright, by law he was not allowed to do so without first consulting with the people who would be affected by the regulation which were the diving contractors, or in practical terms, their lobbying arm, the Association of Offshore Diving Contractors (AODC).[26] Indeed, as one former AODC Chairman recalls, regulations did not come into effect without heavy input from their organization:

> Nearly all the legislation—the initial drafting—was done within the AODC, and then Jackie's Department would modify slightly and put a stamp of approval on it, then it would be imposed on us.

And in late 1974, the AODC did not want to see scuba eliminated from the North Sea. According to the former Chairman, the pretext for this position was that there were always going to be some jobs where the use of surface-supplied diving gear presented a greater hazard to the diver than scuba:

> ...if Jackie had come along and said, 'We want to ban scuba totally,' we'd have been saying, 'Yeah, we agree, it's less safe than other methods, but there are certain specific jobs where the job cannot be done unless you use scuba; there is just no alternative. So if you ban scuba totally, then the oil company's got a major problem, and the diving industry has got a problem because it's our job to do the work underwater.'

But what jobs are these? Surface diving in the North Sea involves a wide range of tasks and challenges, but nothing that would absolutely demand the use of scuba. Clarke could have easily accomplished his mission by using surface gear. The truth is, underlying scuba's tenuous link to safety were other motivating factors that drove the AODC to push for its continued use. Scuba offered powerful economic and logistical advantages that surface gear lacked. It reduced the equipment requirement for topside support; it was easier to transport, and setting up to dive took less time—no humping 600-foot surface umbilicals around on deck. These incentives had nothing to do with safety, yet produced irresistible inducements to ambitious contractors, particularly when attempting to undercut competitors during contract negotiations. To critics, the AODC position was really just a "head fake" to allow its members to continue conducting diving operations on the cheap.

While the AODC saw "no alternative" to using scuba, there was in fact another option open to the Diving Inspectorate, namely, the dispensation. It could have banned scuba for safety reasons, then relaxed that ban on those rare occasions when the contractor justified its use and guaranteed additional safeguards to make up for the increased risk. This approach, however, was never adopted. In the end, the Inspectorate cut a deal with the AODC, and when the new regulations debuted at the beginning of 1975, the fruit of that deal was revealed.

On the plus side, free swimming was banned in the UK sector of the North Sea. The diver now had to be "securely connected to a lifeline" which was attached to either a person on the surface, or to another diver in the water who was connected to a person on the surface.[27] In addition, he had to be provided with "an effective means of communication,"[28] and could not go underwater unless he had a means of communicating "with the diving supervisor."[29]

At first glance, these provisions appear to have shut the door on the use of scuba. But on closer examination, under the sections headed by "Prohibited Diving Operations" and "Restrictions on Diving," scuba was not listed. What's more, the legislation left unclear other key issues. For example, it did not specify the nature of the diver's lifeline, which might be interpreted as nothing more than a piece of rope; nor did the uncertain phrase "effective means of communication" necessarily translate to radio communications in the minds of some diving contractors. Again, the DOE left these definitions open to shrewd interpretation, and as a consequence, Clarke's death would have a number of unfortunate sequels to it.

Of the 25 divers who died on the UK Continental Shelf between 1974 and 1983 (excluding territorial waters), seven, or 28 percent, were using scuba. And if one includes the names of the divers killed while working on oil and gas projects inside territorial waters (that the Diving Inspectorate either co-investigated or was aware of), that figure jumps to 36.6 percent. Statistically, scuba added far more names to the obituary column than any other type of accident in the history of North Sea diving.

Date	**Diver**	**Company**	**Cause of death**
Oct. 14, 1974	John Clarke	Comex	drowned
July 16, 1975	Peter Walsh•	Underwater Security Ltd.	sudden acute decompression of the head
July 16, 1975	Peter Carson•	Underwater Security Ltd.	drowned
May 12, 1976	Nicholas Hubert•	Not Known	pulmonary barotrauma
May 13, 1976	Christopher Dymott•	Not Known	drowned
July 14, 1976	Richard Dupuy	Comex	cerebral anoxia
Nov. 4, 1976	Howard Spensley	KD Marine	drowned
Nov. 4, 1976	Charles Meehan	KD Marine	drowned
Dec. 24, 1976	Michael Moore	Comex	drowned
Aug 20, 1977	D. Sansalone	Sub Sea Services	drowned
June 2, 1983	R. M. Wallace	Mobell (Marine)	drowned

• the accident occurred in territorial waters, but was investigated by the DOE.

As the table shows, eight of these men drowned which would have been unlikely had they been using full-faced masks. Two accidents involved the practice of "daisy chaining" divers, where the first diver was tethered to the dive station by a lifeline, while the second diver was merely connected to the first with a five-foot buddy line which he held in his hand.[30]

This buddy-line system was a Royal Navy diving technique,[31] and was employed ostensibly to comply with the regulations. In some settings it might be suitable, but to use it for commercial operations in the unpredictable waters of the North Sea was asking for trouble. If the line connecting the two divers parted, it would leave the second diver at the mercy of the elements. What's more, diving contractors apparently believed that the rope the second diver clutched in his hand satisfied the legal obligation "to provide an effective means of communication with every diver." It certainly did not satisfy the more specific requirement that the diver be able to communicate with his supervisor.[32]

In November 1976, K. D. Marine divers Howard Spensley and Charles Meehan died while using this technique. Like John Clarke in 1974, they were washed against an anchor bolster and drowned when their regulators came out of their mouths.[a]

Accordingly, the Inspectorate investigated the dual fatality and soon discovered there had been no standby diver on alert, that Meehan had not been securely connected to Spensley, and that he had entered the water with no communications. These were clear violations of the regulations which should have given the DOE sufficient grounds to initiate a prosecution. Instead, four months after the accident, Commander Warner dispensed another bromide in writing:

> Experience shows that if a diver is washed against the anchor bolsters serious damage to the diver and/or his equipment can occur.
>
> Extra care is required when deploying a diver through or near the surface interface to ensure that he is not put at risk by fouling or being washed against obstruction [sic]. [33]

A month after the deaths of Meehan and Spensley, the North Sea delivered yet another example of just how dangerously foolish this buddy-line system was. A little after 3 o'clock in the afternoon on Christmas Eve, Comex diver Michael Moore entered the water daisy chained to Trevor Murtough. They quickly completed the job without incident, but it was on their return trip to the recovery basket when the dive suddenly began to go wrong.

While trying to the enter the basket, which was only partially submerged, the line connecting the two divers slipped off Murtough's arm, and Moore began to drift away. Murtough reacted quickly and immediately paid out slack in his lifeline to hook his partner and pull him inside the basket. But after four unsuccessful attempts, an exhausted Moore finally gave up and fell back into the sea.

Now drifting freely in the water, Moore kept his wits about him, inflated his life vest, dropped his weight belt, and signaled a "thumbs up" to the men high above on the rig. He was okay, and an old fishing trawler on station for

---

a Apparently Spensley (who was tethered to the rig) was not wearing a full-faced mask during the dive. At the inquiry, the supervisor testified that, when Spensley was recovered, "His face mask was still on, but the mouthpiece was out of his mouth." *Source:* Spensley/Meehan FAI, p. 27.

such emergencies was nearby. The diving supervisor immediately radioed the vessel and a recovery operation got underway; Moore thought he was going to be picked up within minutes.

In the late winter afternoon darkness was falling fast and the crew of the trawler were experiencing difficulty getting a line on the bobbing figure. They tried throwing a life ring to him several times, but wind speeds had picked up to 25 knots, and the captain was having problems maneuvering the single-screw vessel due to it being low on fuel and fresh water.[34] The ship had a Fast Rescue Craft on board for such emergencies, but the crew decided not to launch it because the seas were "too rough."[35] In the waning twilight, they began to lose site of their objective. So Moore, uninjured and expecting to be rescued, drifted along with his arms raised, desperately trying to attract the attention of a vessel whose time had come to perform the job it was hired to do. The ship managed to circle two or three times, but within half an hour, its target had floated out of sight and into the darkness, never to be seen again. Sometime later, alone and frightened and with waves slapping at his face, Michael Robert Moore, age 29, married, and with a six month old baby boy at home, finally laid his head to his chest and gave in to the brutal cold.

As with the other accidents, the Inspectorate investigated this latest mishap, but there is no evidence to indicate that it took any corrective action. Of the nine recorded fatalities that year,[a] six occurred on scuba. Nine deaths within a twelve month period was considered unacceptable, but to have two thirds of them associated with a method of diving known for its increased risk should have set off alarm bells all the way from Aberdeen to London. What's more, none of these incidents involved a situation where the job would have been more hazardous using surface supplied diving gear. If anything, jumping these men on scuba in marginal sea conditions was the more dangerous course of action. While using surface gear cannot eliminate all risk, it is unlikely that these men would have drowned had they been using full-faced masks. And if Meehan and Moore had been equipped with communications, at the very least they would have been able to tell their supervisors what to do to assist them; certainly Moore would not have been lost in a drifting sea.

At this juncture, taking some form of action to stem the unprecedented tide of scuba deaths would have been a refreshing change in policy and consistent with the DOE's line that prevention of accidents was a serious priority. But nothing seemed to compel the agency to stop a practice which clearly had no business being used in the North Sea, so the deaths continued.

Eight months after Moore disappeared, an Italian diver by the name of Sansalone joined the growing gallery of scuba victims when he drowned while

---

a This includes the two territorial water fatalities that the Diving Inspectorate commented on.

diving from a pipelay barge. Again, he was using air tanks and a free mouthpiece. More than a decade later, Commander Warner opined about the equipment the young Italian was using that day:

> This, in my opinion, was quite the wrong equipment to be used by a man required to actually carry out work underwater.[36]

One more scuba death was recorded during the summer of 1983 off the coast of England, and then 11 years later, long after the oil rush was over, the AODC finally issued this statement:

> SCUBA (self-contained underwater breathing apparatus) has inherent limitations and it is strongly recommended that it should NOT be used for offshore diving operations in support of: oil/gas projects; marine construction; civil engineering or salvage.[37]

Thus, in 1994, the practice of using scuba was voluntarily stopped, but by then the damage had been done.

When it came to ranking priorities, the DOE emphatically stood by the assertion that its job was "to make the industry safer, to cut the death toll,"[38] and to prevent accidents from repeating themselves. This position was expressed numerous times by the Department as official policy and by Commander Warner himself.[39]

But the sad reality is, the Inspectorate was saying one thing, while doing another. In 1976, divers were dying in rapid sequence under the most preventable of circumstances. The Department investigated these accidents; it understood that scuba was more hazardous than surface supplied diving, and it issued Diving Safety Memoranda to that effect. It also knew, as a result of its investigations, that the AODC was not adhering to its earlier promise to use scuba only when absolutely necessary. If the number of fatalities was indeed "unacceptable" as claimed,[40] and if the government was serious about cutting the death toll, then what better way to achieve that goal than to eliminate one of the leading causes?

Industry advocates often assert that the North Sea's high fatality rate was the inevitable consequence of working close to the limits of human endurance, and that these fatalities formed an unavoidable "learning curve." That argument has some force to it; accidents which are unique and cannot be foreseen are part of a learning curve. And no one should expect a very small group of government safety inspectors to be able to police the entire North Sea or engage in crystal ball gazing in order to prevent an accident.

But the men who represent more than one third of a ten-year death list weren't dying because of some unforeseen danger or because the DOE and the AODC needed two decades to learn something about scuba that they didn't already know in 1974. By 1976, scuba had long since faded as a technological wonder; it was an invention created by Jacques Cousteau and Emile Gagnan in 1943.

On the contrary, men like Michael Moore were dying because the DOE had bargained away safety for the sake of reducing industry's operating expenses.

That bargain saved contractors time and money, but at what cost? Nearly a year after Moore vanished on Christmas Eve, his wife appeared at Aberdeen Sheriff Court to hear evidence on the disappearance of her late husband. During the inquiry she was caught attempting to tape-record the proceedings, an offense the sheriff would later characterize as "contempt of court."[41] After the tape was confiscated, she told the Court, "I have a baby only 15 months old and he is going to want to know what happened to his father."[42] In little more than two years time, these would be virtually the same words expressed by Jeanne Walker in order to find out what happened to Richard.

By the end of 1977, the Diving Inspectorate was conducting accident investigations in a manner that raises very serious doubts as to its independence and general level of expertise. Rather than ban unsafe practices, it permitted contractors to carry on in a "business as usual" manner, "subject to the divers being happy to dive." Bad law, once established, was neither amended nor repealed. It simply sat on the books, chiseled in stone. With full knowledge that employers were escaping their moral and legal responsibilities with back-door payment schemes to divers, the Inspectorate remained silent. And now it had failed to respond appropriately to a string of conspicuous repeat fatalities.

From the public's point of view, signs of an impending disaster at Thistle would not have been visible. The press reported these scuba fatalities, but the attention paid to them was understandably overshadowed by more catastrophic events involving capsized rigs, wellhead blowouts, and helicopter ditchings. In one sense, these were "quiet deaths," primarily drownings, which made it easy for officials to explain away their cause as the result of something other than bad safety policy. And, too, these accidents were scattered over a wide area: off the coast of Wales, northwest Scotland, the Orkney Islands, and far out in the North Sea. If a reporter had noticed a lethal pattern developing, it would have been because he was looking for it.

As time passed, Michael Moore and the others were forgotten. From the manner in which they died, no one could have foreseen what was on the horizon for Richard and Skip. But there was one further incident that should have given the DOE all the foresight it needed to prevent their deaths.

Chapter Twenty-nine

Canopus

Cut the hose! Cut the hose!

—— Supervisor Robert Kelly
to the divers in the bell[1]

On November 26, 1978, the Diving Support Vessel *Star Canopus* was dynamically positioned on the northeast side of the Beryl Alpha platform conducting diving operations. Three hundred thirty-four feet below, Michael Ward was working for Northern Divers about 60 feet off the seabed trying to connect a six-inch flowline to a riser flange protruding from the side of the concrete platform.

Oil production had declined that year,[2] and with winter conditions upon them, subsea work had fallen behind schedule.[3] There was "a lot of catching up to do," the contracts manager for Northern Divers would say later,[4] and Mobil Oil, like many operators, decided to allow the diving program to operate through the winter months to take advantage of available weather windows.

London Weather Center had predicted 40-knot gusts between 1800 and midnight.[5] That forecast proved to be correct. At 2030, and again at 2245, squalls reaching those speeds moved in and blew the *Canopus* off location.[6] Shortly thereafter, Captain Roy Forsyth terminated diving operations.[7]

But when the 12-hour shift change occurred at midnight, weather conditions had calmed. Coming on duty, Supervisor Robert Kelly was told by the superintendent that "the weather was reasonable to dive" and to continue with the dive program.[8] Kelly went to the bridge to review the weather forecast, which predicted there would be a change in wind direction from west to north, and that conditions "would worsen later in the day" with wind speeds gusting to 35 knots.[9]

The *Canopus* was only a year old, and although it represented the latest and best technology of that era, the limits of her dynamic positioning system were 18-20 knots with the wind hitting her broadside.[10]

On the bridge, Kelly noticed the wind speed indicator fluctuating between 15 to 20 knots. He walked outside to look at the weather firsthand. With the wind blowing from the west against his face, Kelly gazed across the dark sea. Standing alongside the Beryl platform, near where the *Canopus* would be commencing work, was the imposing silhouette of the *Haakon Magnus,* a Norwegian semi-submersible, with her massive anchor chains extended.

To Kelly, the weather looked suitable for diving. At 0240, he launched the bell with Tony Prangley and Michael Ward inside.[11] At 0312, Ward locked out and found the pipe flange protruding from the wall of the platform base. Eight minutes later Captain Forsyth called Kelly on the ship's intercom to alert him that a squall was approaching on the ship's radar.[12] Kelly ordered Ward back to the bell where he sat for the next 40 minutes. At 0400 Ward was back on the job. Then, at 0545, the wind suddenly shifted directions and began blowing from the north at a speed of 40 knots, hitting the *Canopus* on her beam, and overpowering her DP system. It was a sustained blow as one of the men would later testify.

"When this wind arrived, it arrived really without any warning and it was not like a sudden gust of wind or a passing squall which then died down, it started blowing suddenly at 40 knots and remained at that level for a considerable period of time thereafter. It was a very strange occurrence, to say the least."[13]

In the bell van, Supervisor Kelly was at the control panel when an amber warning light came on followed by the klaxon of the 'Dive Alert' alarm. Then, Captain Forsyth's voice came over the intercom: "Get the diver back in the bell please."[14] Kelly relayed the message to Ward who stopped what he was doing and swam back to the bell. In the trunking he took off his helmet as Prangley stacked his umbilical.

On the surface, the force of the wind had blown the *Canopus* sideways against the platform overhang, snapping off the ship's mast which came crashing down onto the deck with a loud bang. At the same time, the wind was driving the ship backwards towards the *Haakon Magnus.* Captain Forsyth quickly took the ship off DP and tried to swing the bow outboard to keep the vessel from grounding on the submerged pontoons of the *Magnus.* Then he pressed the 'Dive Abort' alarm and called dive control.

In dive control, Kelly was waiting for Prangley and Ward to seal the bell when the red 'Dive Abort' indicator suddenly lit up. Captain Forsyth came on the intercom again, this time with noticeable urgency in his voice. "Get the bell up as soon as possible," he said to Kelly.[15] But when Kelly called down to the bell, he discovered that Prangley was having difficulty pulling in Ward's umbilical; it was snagged on something outside the bell. He didn't know on what, but now there was no way to shut the inside hatch until the umbilical was cleared. Ward wanted to lock out to fix the problem, but Kelly flatly forbade him to do so. Instead, he told Ward to throw the dive hat out of the bell, then pull in the umbilical; perhaps that would free it. But when this was tried, the umbilical remained snagged.[16]

Meanwhile, Kelly received another urgent message from Captain Forsyth asking if he had started recovering the bell yet.[17] Kelly explained the delay and said he would notify the bridge when he started the recovery. Then he got on the comms and told the divers to stop what they were doing, and to cut the hose, and be quick about it. Several minutes later, as the bell smashed against the base of the platform, Prangley and Ward got a seal and notified topside.

In dive control, Kelly was watching his two divers on the video monitor while the bell was being raised. Prangley was calling out the depth every 10 meters. Eighty, seventy, sixty—when the bell was thirty meters from the surface, Prangley suddenly stood up, waved his fist shoulder high, and shouted. "All stop."[18] On the monitor, he appeared to be listening to something.

Kelly did as he was told, and from the bell van, he had a clear view of the moonpool and could see that something was wrong. The bell umbilical and clump weight guide wires were not hanging straight down through the moonpool; they were leaning towards the bow of the ship. What's more, the trolley that the wires were hanging from was violently shaking and vibrating.[19] Suddenly the video monitor blacked out. The *Star Canopus*, with her human payload dangling thirty meters below, had been blown across the path of a *Haakon Magnus* anchor chain, completely severing all connections to the bell, including the life-support umbilical, the main lift wire, and the clump weight/guide wire recovery system. In that instant, the bell, with Prangley and Ward inside, began plummeting to the seabed below.

Miraculously both divers survived the 294-foot fall, but now they faced a harrowing predicament. With their hot water supply completely severed, the internal temperature began to drop rapidly. If rescuers didn't arrive soon, their only recourse would be to release the drop weight. But Prangley and Ward couldn't release the weight because it was secured to the bell frame with a set of "secondary locking pins."[20] And because the bell was not equipped with a stage to keep the bottom hatch out of the mud, Prangley and Ward had no way to get out to release them. Nor was the bell equipped with a strobe flasher to help rescue divers find it in the dark; and just as the *Wildrake* crew would, the crew of the *Canopus* had removed the bell transponder some time before the dive.[21] What these two British divers were facing would be exactly what Richard and Skip would confront in nine months time. Helplessly trapped, with nothing to keep them warm, and no way to release the drop weight, all they could do was huddle together and pray that rescuers would find them in time.

But time was not on their side. With a storm raging on the surface, it took rescuers more than thirteen hours to recover the stricken bell to the deck of the *Canopus*. By that time, Ward had died of hypothermia and Prangley had drowned.[a]

Like all accidents, *Canopus* extended the DOE's "learning curve." But it also gave inspectors that rare and precious gift: it gave them 20/20 hindsight. They knew from their 1976 scuba investigations, that accidents can and do repeat themselves with uncanny similarities. And they were on notice that, even with the latest and best technology, diving bells could become separated from their motherships, that recovery could be delayed if the bell lacked a strobe light and transponder, and that divers would die from hypothermia if they were not found in time. At this point, the DOE should have stopped the practice of double securing drop weights, not because it could legitimately argue that Northern Divers had breached the drop-weight regulation, but because the practice represented a known threat to safety and contributed to the demise of the two divers.

Addressing these problems would have been consistent with the DOE's purported view that the prevention of similar accidents ranked high on its list of priorities. But that was not how London construed the matter. Other than issuing a mild warning about operating a DP vessel in close proximity to a fixed structure,[22] the Inspectorate took no action. There was no clarion call for contractors to begin equipping diving bells with dedicated transponders, strobe flashers, and bell stages. It did not prod industry to come up with some form of emergency heating for stranded divers. Nor did it send out an urgent message demanding contractors to stop securing drop weights to the outside of the bell. It took no action, in spite of *Canopus*, because the DOE believed that double securing drop weights was legal, and therefore acceptable.

At the *Canopus* inquiry a year later, when the attorney representing Northern Divers asked the investigating inspector whether he thought the drop weight modification "was in accordance with the regulations," he said, "That is correct."[23]

What makes this reply even more appalling is that it was made three months *after Richard and Skip died*. Thus, after investigating its second lost-bell accident where the drop weights had been externally secured, the sponsoring Department was still deeply invested in a practice it knew was manifestly unsafe.

Why would the Inspectorate continue to take this position in the wake of the *Wildrake* accident? Because, when it came to interpreting the law, the DOE invariably saw things through the eyes of the AODC. And because the DOE was unable to prevent its oil production concerns from influencing its oversight duties.

---

a Although Prangley was suffering from hypothermia, too, his death was ruled as a drowning. When the bell was initially raised off the seabed, he fell out of the trunking and floated to the surface.

That failure, plus the intense campaign to achieve self-sufficiency at breakneck speeds, ultimately led to the dislocation of its priorities with more emphasis placed on sponsorship than on enforcement. That a pronounced bias favoring industry would emerge from this dislocation is not surprising when one takes into account that promoting the needs of industry facilitated the government's goal, while aggressive enforcement would have led to confrontation. And the DOE was not tasked to be a confrontational agency. It was expected to rule within the policy framework of "benevolent enforcement" where solving conflict and promoting change through traditional enforcement strategies such as "duress" and "negative regulation" had been replaced by a doctrine which encouraged compliance through advice and persuasion. Consequently, inspectors came to view themselves not so much as offshore policeman, ferreting out acts of non-compliance and bringing lawbreakers to justice, but more as advisors assessing the adequacy of safety.

Furthermore, because there were gaps and omissions in the regulations, the confusion generated by these defects played no small part in allowing contractors to escape the consequences of the law, and whether or not the Inspectorate would proceed legally against those deemed to be at fault, if such judgments were even made at all. While inspectors had the power to shut down unsafe operations, and later, to impose administrative sanctions on suspected violators without reference to the courts, they never used these powers because the DOE's special relationship with industry discouraged them from doing so. In the end, inspectors tended to "play it safe," and to excuse certain diving practices even when it became apparent that they were patently unsafe. The Inspectorate's posture during the 1976 rash of scuba deaths is one example, and how it behaved following the *Canopus* accident is another. Instead of aggressively enforcing the basic "safe conduct" provision[a] of the regulations by ordering companies to stop the practice of double securing drop weights, the Department is seen extending quasi-official legitimacy to it.

This tendency to co-opt safety in favor of industry's needs was further reflected in the way in which the Inspectorate evaluated diving systems during routine inspections. From the agency's point of view, if an aspect of the diving system did not trigger a specific regulation, then it did not want to get involved sorting out vexing conundrums, even if one safety system had the potential to undermine another. Evidence for this attitude appeared at the *Canopus* inquiry when the inspector who investigated the accident was being questioned by Reg Christie, the attorney representing the relatives of Prangley and Ward. Seven months earlier the inspector had examined the *Canopus* diving system and found that, with some minor modifications, it complied with all the regulations.[24] But Christie wanted to know what the Department's position would be if it came across an issue relating to safety that fell outside the regulations; would the

a SI 1976 No. 923 Regulation 5 (3) (f).

Department be concerned about that? According to the inspector, it would not, as evidenced by the following exchange.

"So you would not have been concerned necessarily with any other aspects which may or may not affect the safety if they were not, in fact, covered by the Regulations or with which the Regulations were not covered?"

"That is correct."[25]

At this juncture, Christie was irresistibly drawn to the conclusion that if the DOE rigidly confined itself only to those aspects of safety covered by the regulations, and not with the broader picture of what made a diving system safe, then it could very well end up discriminating against its legal duty to stop acts which might result in, or contribute to, a casualty. To illustrate this point, Christie decided to take this issue to its logical conclusion by asking if there was a regulation which designated a safe minimum distance between the bottom of the bell and whatever the bell was resting on.

"There is no regulation," the inspector answered.[26a]

Since there was no regulation, it followed that the DOE saw itself free not to challenge whether lack of a bell stage would adversely impact upon the ability of the divers to release the external locking pins of the drop weight. When Christie suggested to his witness that it would be essential for the divers to have access to the pins to release the weight, the inspector avoided answering the question and began paraphrasing one of the regulations.

"As far as the Regulation states, they have to have a second means of bell recovery, and, if they have that second means of bell recovery, then it is acceptable as far as the Regulations are concerned."

Dissatisfied with that answer, Christie pursued.

"I was not really referring to the Regulations when I posed that question, all I am putting to you, I think, is the obvious, that if you are to have a second means of discharging the weight on the outside [of the bell], then it is essential that the diver has access to the outside at the time when he wishes to discharge the weight, is it not?"

Realizing he had been trapped by a rationale that elevated safety over other priorities, the inspector stubbornly refused to yield to it.

"That is not necessarily so."[27]

The unfortunate consequence of holding such a narrow perspective regarding safety is that the DOE saw no reason to impose upon industry a device which would permit the diver to save himself. Thus, when the bell was suspended during operations, and the divers could enter and exit without difficulty, the law was without defect. But when the bell came into direct contact with the seabed, with the divers unable to exit to release the locking pins, suddenly the law fell into the "gray zone" and became murky. Seventeen years after the *Wildrake* accident,

---

a The law merely states that divers must be able to enter and leave the bell "without difficulty."

Commander Warner was still finding it difficult to be unequivocal about whether contractors were violating the law by conducting operations without bell stages. "Difficult to say, the law was not entirely clear and this comes back to the old business of a learning curve."[28]

Learning curves and murky laws aside, the Inspectorate's inability to view safety through the eyes of the common diver paid double dividends for contractors. It kept them in good standing with authorities when things went terribly wrong, and it saved them the time, trouble and expense of equipping bells with underframes when things were going right. Once again, as was the case with scuba, there was an underlying economic benefit to industry (with a net loss to safety) if the Inspectorate saw things as industry did.

Thirteen months later, the DOE would shift its view. But in the wake of *Canopus*, it was firmly wrapped around a temperament that interpreted the law in favor of industry while dismissing critical aspects of safety that fell outside the language of the statutes. Unfortunately, this bias also interfered with its obligation to improve standards of health and safety.

When it became known that Prangley and Ward had been diving without a backup heat source, the subject of hypothermia, and what to do about stranded divers in an HE-O_2 environment, stirred up a national debate when the press brought the issue to the public forum. A panel of medical experts was assembled and everyone agreed that something should be done. Recommendations were put to the government, but that's as far as the matter progressed. According to Professor J. Nelson Norman, a noted expert on fatal diving accidents and head of Aberdeen University's Institute of Offshore Medicine, the issue fell victim to bureaucratic languor:

> ...we identified the problem of what to do with a diving bell that has lost its power supply and a lot more should have been done to solve it. But the industry said it was the government's responsibility and the government said it was the industry's responsibility.[29]

Thus, reform did not arrive in time for Richard and Skip because authorities in London were unable to recognize that changes were necessary, but because industry and government got into a quarrel over whose responsibility it was to effect those changes.

Seven years earlier, a government report concluded that there was too much reliance upon the state to intervene in what were essentially industry-generated problems.[a] It advocated a more self-regulated system where advances in safety standards would be implemented through better cooperation between management and labor, rather than through a policy of rigorous enforcement and negative regulation.[30] Self-regulation was a system that Commander Warner applauded:

a The 1972 Robens Report.

> Looking back, one of the best things to emerge from the 1960s and 1970s was the A.O.D.C. (Association of Offshore Diving Contractors), in essence a trade association which must take considerable credit now for the improved standards of professionalism in diving. The A.O.D.C. accepted that an industry should manage and regulate itself, and should oversee the application of safe working practices.[31]

Under this system, government claimed no direct legal responsibility for offshore safety. That burden rested firmly on the shoulders of industry. Industry created the risk, therefore industry was required by law to control it. As time passed, and as fatal accidents highlighted the growing need for such things as strobe flashers, transponders and bell stages, government expected industry to proactively incorporate these devices into future operations without much official prodding.

But in actual practice, this expectation vested far too much faith in the belief that a profit-motivated industry would willingly spend money on safety devices not required by law. To the heads of the diving companies, cost was not some theoretical abstraction; it was a very real part of the profit-and-loss equation that determined survival in a highly competitive industry. It inspired unscrupulous contractors to pay divers through bogus offshore companies, and it provided the underlying basis for the use of scuba in the North Sea. As one former diving contractor candidly admitted: "The AODC tried to police themselves but the problem with them was cost. Safety entails considerable expense."

Still, if industry could not be cajoled into voluntarily upgrading safety, especially after serious accidents such as *Canopus*, then regulators were free to impose more rule making and stiffer sanctions to compel contractors to raise safety standards. But time and again the DOE demonstrated that sanctions of any kind would not be imposed. And, if industry was reluctant to spend money on safety, and government reluctant to impose changes, then that left the reform process bogged down in a stalemate which is why nothing was done during the nine month period after *Canopus* and before *Wildrake*. It also meant that the DOE's devotion to law enforcement was nothing more than a façade to give the illusion that safety was the number one priority, when in fact oil production had been occupying that position all along. And nowhere did this façade reveal itself more clearly than when wrongdoers were brought before the Courts. Here, the legislature had made a mockery of the law because the real subtext of benevolent enforcement was that violators were not meant to be punished at all.

Chapter Thirty

Privileged Offenses[1]

The public see these fines as the value the courts put on human life. Widows and families are shattered by the size of them. What do sums of a few hundred pounds mean to these companies? They spend more in the bar afterwards than they pay in fines.

—— Roger Lyons
Burgoyne Committee, *Glasgow Herald*, June 24, 1981

In Scotland, it is a matter of public policy for prosecutors to use their office to stamp out crime.[2] They see the arrest and conviction of lawbreakers as part of their duty to safeguard the public interest, and punishment "a valuable component of the criminal justice process, largely because of its capacity to deter future crime."[3] But not all violations of the criminal code are viewed equally. Murder, rape, robbery, for example, occupy a higher echelon than say public drunkenness or driving without a license. And when it comes to crimes against the industrial labor force, Carson says Britain has had a long history of placing corporate crime into a special category where it was "rarely thought of as 'real' crime."[4]

Moreover, as this view penetrated the modern day offshore arena, tolerance towards infringement of the regulations became so pervasive and "institutionalized at a fairly high level,"[5] that when oil companies were actually caught violating the law, those violations were "only rarely subjected to criminal prosecution."[6]

Indeed, the Diving Inspectorate had the power to prosecute wrongdoers, but preferred instead to rely upon exhortation to convince contractors to mend their ways.[7] Added to this dynamic were other troubling factors. Carson discovered that the legal branch of the Department of Energy and the Scottish legal system did not always see eye to eye "over whose responsibility it was to lay the groundwork for possible prosecution."[8] And London was not always anxious to cooperate with the investigations of Scottish prosecutors,[9] refusing in one case to comply with a fiscal's request for a copy of a lease which would have established the identity of a concession owner,[10] and managing to persuade Scottish authorities not to prosecute an offender, in another.[11]

This tension between the two legal powers spilled over into another fundamental law enforcement area. DOE officials believed it was the job of the police to investigate criminal activity and the coroner's inquest to decide whether there were grounds for prosecution. But north of the border, Carson interviewed a fiscal who said it was not the job of the police officer to go offshore "looking for offences." That was the job of the DOE inspector.[12]

Still, even if inspectors were inspired to look for lawbreakers, Parliament had removed any chance of catching them in the act. Unlike their Norwegian counterparts, who had the power to make sudden inspections without notice, UK inspectors were required to give the oil operator advance notice when they intended to fly out to have a look.[13]

The Department's reluctance to prosecute is not surprising when one takes into account that bringing an oil company to court was inherently at odds with its role as sponsor. When Energy Secretary Tony Benn held up a bottle of crude oil in front of press reporters in 1975 and declared, "I hold the future of Britain in my hand,"[14] he wasn't just rallying the country with an inspirational slogan; he was pledging official support. To men like Commander Warner who worked under Secretary Benn, that declaration must have resonated deep within their spines because they had to know that shutting down operations and hauling rule breakers off to court was not going to help get the country out of its economic mess. As Commander Warner best describes the situation, "there were no 'Brownie points' for convictions."[15]

Lastly, one must never discount the power of the Association of Offshore Diving Contractors. Had Chief Inspector Warner taken a harsh stand against offenders, the AODC would not have taken the matter lying down. Complaints from the Association would have rained down upon the oil companies, who in turn would have brought pressure to bear upon Warner's superiors. When this scenario was put to one of the leading diving contractors of that era, he was unable to suppress a smile: "We would have worked it that way, so none of us are totally innocent."

While the DOE was reluctant to prosecute, Carson discovered that the agency was not alone in that feeling.[16] Against the usual rhetoric about "safeguarding the public interest," and prosecution being "a deterrent," it turns out that Scottish authorities also ranked offshore crime far down on its list of illegalities. Among government records, from 1977 to 1980, Carson found 29 reported fatalities, 168 serious accidents, and 368 dangerous occurrences involving all aspects of offshore work, diving and non-diving alike.[17] And yet, Carson was able to find only 13 cases where prosecutors had initiated criminal proceedings.[18] These 13 cases (involving 23 companies or individuals) concluded with ten convictions, four not guilty, four deserted, and five dismissed as incompetent.[19]

Moreover, even when the accused were found guilty, the level of fines was so deplorably low that it made the conviction virtually meaningless. In the 1977 K. D. Marine case, for example, when divers Howard Spensley and Charles Meehan drowned, the company and the diving supervisor were fined £200 and £25 respectively after pleading guilty to one of the charges. In 1978, after Philip Baxter of Dundee fell 80 feet to his death through a badly fixed grating, Shell Oil was convicted of negligence and paid a fine of £200.[20] These fines were near the level of what might be imposed for a serious motoring offense in Scotland, such as driving while drunk or without insurance.

Besides illustrating the degree to which negligent acts were being punished, the magnitude of these fines reflects the prosecutor's view of the seriousness of the crime and his judgment of the accused. In Scotland, he plays no part in recommending to the sheriff what sentence should be imposed upon a convicted offender. But he does affect the sentencing when he chooses what prosecutorial procedure to employ. With the exception of the Infabco trial, all the defendants in Carson's study were prosecuted by the less time consuming, less expensive procedure of Summary Complaint rather than by the more serious process of Indictment. This decision was not insignificant because Parliament had restricted the maximum penalty upon conviction by Summary Complaint to £400, whereas a company convicted on Indictment could have received a larger fine at the discretion of the sheriff. Thus, when the fiscal chooses summary procedure, he not only limits the sentencing options of the judge, he is also "in effect helping to define what constitutes serious crime," as legal experts Moody and Tombs point out.[21] And it's clear from the record that during the 1970s, authorities did not view the types of infringement going on offshore as serious crime.

As for the accused, the likelihood of serving jail time was more a legal fiction than a reality. In 1979, when Richard and Skip died, the Crown was not indicting company directors for their managerial actions. According to Health and Safety expert Gareth Watkins, more than a decade would have to pass before Britons would hear of a managing director in the UK being found guilty of manslaughter.[22] And it would not be until 1997 that a person working in the North Sea oil industry,[a] would receive a prison term for committing a crime offshore.[23] Furthermore, had Infabco been convicted in 1980, it is unlikely that the court would have imposed an exemplary fine on Masterson's company. More likely Infabco would have received a small fine as was the case when Sub Sea Offshore lost two divers in 1984.

---

a A diving supervisor in the Bradley Westell case.

On February 21, David Bowmar and Thomas Mackey died while performing a welding test in a hyperbaric chamber at Sub Sea's base of operations on Greenwell Road in Aberdeen. The supervisor neglected to analyze one of the gas mixtures to the diving system, and in the course of the simulated dive, Bowmar and Mackey were fed, not air as was intended, but pure nitrogen. Sub Sea and the company that supplied the gas were subsequently prosecuted under Indictment, and after pleading guilty to the charges, paid £2,500 and £1,000 respectively for the dual fatality.[24]

To officials at the Crown Office in Edinburgh, it was not the seriousness of the accident that determined the scale of the fine, it was the "degree of negligence." In their view, oil companies pay a premium for all the bad publicity and lawsuits generated by the loss of life.[25] That attitude coincided with the Department of Energy's. The year Richard and Skip died, it made this comment to the Burgoyne Committee regarding the current level of penalties:

> Prosecutions are rare but where they have been undertaken, particularly in Scotland, it has been the adverse publicity which has hurt rather than the available penalties. It must be remembered that the fines which can be imposed can never themselves have much financial impact on a multi-million pound international oil company.[26]

Indeed they cannot if the actions of the legislature and prosecutors prevent the courts from imposing adequate penalties. As for using adverse publicity to backstop abuse and neglect: given the fact that the oil industry operates as a closed society far out at sea, it seems unreasonable to rely upon press coverage to act as a substitute penalty system.

What was needed to stop the flow of preventable deaths was an independent oversight agency, with the will to enforce the law. Had the Diving Inspectorate not been encumbered by conflicting interests, then it would have stopped the rash of scuba deaths in 1976, and it would have instituted reforms after *Canopus* and before Richard and Skip went on their fatal dive. But those actions did not occur because the DOE was unable to keep the priority of safety separate from the political goal of pushing forward the national oil policy.

Although improvements in the offshore enforcement system would eventually arrive on the scene, they did not appear as the result of 20/20 hindsight, nor at a time when more lives could have been spared. Instead, reform came about through external pressures. Between 1974 and 1977, the death toll for divers was substantially higher than the two years preceding the *Wildrake* accident. These were the years when fatalities were regarded by authorities as unacceptable and when pressure for reform would have been at its highest peak. But the doctrine of benevolent enforcement was in force and so the DOE sat on the sidelines and watched as fatal accidents continued to occur, sometimes repeating themselves. Initial reaction to these accidents by the press was mild, but as time passed, the headlines got bigger and the stories began moving closer to the front page.

When Tony Prangley and Michael Ward died in late 1978, public attention was naturally drawn to the high-profile drama because the bell had fallen nearly 300 feet to the seafloor and there was a race to save the two men. The press asked questions after the failed rescue, but the safety issues surrounding the double securing of the drop weight and lack of a bell stage went unnoticed. That allowed the DOE to continue its support for these "gray zone" industry positions.

But then Richard and Skip died under nearly identical circumstances, and their deaths began to generate the kind of publicity that the AODC and the DOE were not used to, and didn't want. Investigative reporting became more focused. Reporters tracking the offshore diving industry drew immediate attention to the back-to-back tragedies and the similarities between them. Professor Norman and other experts were quoted about the lack of government response after the *Canopus* accident, and within a few months, a reporter for the *Sunday Times* would hunt down Professor Carson in Edinburgh to be told that the diving industry's accident rate over the past five years was "35 times worse than that of the British construction industry."[27]

Meanwhile, Jeanne's friends from California, Attorney Bob Parker and Fos Campbell had traveled to the UK to conduct their own inquiries and urged government officials and Members of Parliament to conduct a thorough investigation. At the same time, rumors concerning a possible bribery link between BNOC and Infabco began to surface resulting in an investigation by the police and BNOC's internal security division. Several months later, the *Canopus* inquiry opened with Attorney Reg Christie pressing the investigating inspector about the wisdom of externally securing the drop weight to a bell with no stage. The only conclusion that could be drawn from the inspector's testimony was that the government condoned a practice it believed was "in accordance with the regulations," yet barred the divers from saving themselves.

Then, in June 1980, a milestone decision arrived which promised to shine an even brighter light on the internal policies and attitudes of the Diving Inspectorate. Aberdeen Fiscal Malcolm MacNeill decided to prosecute Infabco Diving Services in front of a jury. If previous DOE behavior was any guide, the upcoming trial was likely to see the Inspectorate defending the position that lashing drop weights on the *Wildrake* bell was no less "in accordance with the regulations" than the way the drop weight was fixed on the *Canopus* bell.

But the Diving Inspectorate realized it could no longer support such a position. The controversy surrounding the *Wildrake* accident had picked up too much momentum, and MacNeill did not view the relevant safety issues through the eyes of the AODC. In fact, the leading charge in his Indictment specifically contradicted London's position that the combination of externally securing drop weights and no bell stage was consistent with the employer's duty to provide a diving plant in good working order. What's more, MacNeill believed that the regulation requiring the diver be able to enter and leave the bell without difficulty included emergency situations as well, one of those gray-zone conundrums that the Diving Inspectorate had remained silent about for years.

Also weighing onto the scene was the fact that Richard and Skip's relatives had gone out and hired a smart San Francisco attorney who was keeping close tabs on how UK authorities were handling the case. Infabco had caused the accident, then botched the rescue. Rumors about abuse and corruption were reaching the halls of Parliament. The tide was turning, and amid all the unwanted press coverage, the DOE clearly saw that the *Wildrake* accident had blown up into a big deal. The agency's current policies and views were due to be examined once more in a public forum, a place where no oversight agency wanted to be seen standing on the wrong side of the safety line. Nor did it wish to have its connotations about drop weight modifications subjected to another round of embarrassing questions that Reg Christie had applied during the *Canopus* Inquiry.

Sometime during the summer of 1980, before MacNeill brought Infabco into court, the Diving Inspectorate reversed its position. At the trial and *Wildrake* inquiry, the same inspector who testified at the *Canopus* inquiry now took the position that securing drop weights on the outside of the bell did not comply with the regulations. He said the practice "would not have been approved by the Department of Energy,"[28] and that whatever means the diving company adopted to prevent accidental release, it should not interfere with the operation of the drop weight system.[29]

There were laws, there were inspectors, there were investigations, and even the occasional trial to project the image that safety was being taken seriously. But standing behind that image was the constant push for oil. The government wanted diving inspectors to enforce the regulations, but only in a manner that kept the oil companies happy and the oil flowing, no matter what the added risks. By establishing a set of ambiguous laws with gaps and overlaps, by limiting the quality and number of inspectors, by restricting free access to inspect facilities without notice, by failing to shut down unsafe operations and issue Improvement and Prohibition Notices, and by failing to provide credible sanctions to hold offenders accountable, the government had, in effect, granted the oil industry the right to violate the law.

Wanting to be seen on the side of safety, but unwilling to exercise control over the risks that industry was taking, the Diving Inspectorate essentially became an adjunct of the AODC, a defender of corner cutters, adopting its views and interpretations of the statutes, while at the same time giving the appearance that it was in control of safety when in fact the real power lay in the hands of the AODC.

To some observers, the DOE's behavior might be explained away as the unintended lumbering of a government bureaucracy. But Carson reminds us that, "Actors do not always perform in complete ignorance of the constraints which shape their action."[30] The people on the front lines knew that diving contractors were cutting corners on safety while shielding their legal exposure when establishing payment schemes in places like Jersey and Panama; and they knew that the penalties for violating the law were absurd.

The DOE may have advised, and they may have threatened, but ultimately the AODC did as it pleased.[a] Tired of exhorting the companies without results, Commander Warner eventually resorted to threats in an attempt to scare contractors into thinking he would legislate to force compliance.[31] But contractors were not scared at all, and they were not deterred by the language of his Diving Safety Memoranda, nor by token fines. They knew full well that the threats emanating from London were empty and that Warner's Watchdogs were not really watching at all, but keeping guard over a rapid exploitation policy. The lack of devotion to law enforcement and complacency towards safety explains why contractors continued to go unchallenged in the years preceding the *Wildrake* accident, and it explains why a man like Brian Masterson was able to operate out in the middle of the North Sea for 14 months without being detected.

---

a The DOE and AODC today, see Appendix.

Chapter Thirty-one

Slaying the Dragon

I would work the rest of my whole life, give everything I ever owned, to have my brother back.

—— Sheila Borsari

All I can say is I miss him every day.

—— Roger Guiel

There is no 'we' anymore.

—— Jeanne Walker

In 1982, as Professor Carson's book was being released in the UK, the civil case against Infabco and the other defendants was just getting started. Up until this point, the attempt to hold Masterson's company accountable had met with little success. The criminal case had failed in Aberdeen. On the other hand, the US civil case in Los Angeles had succeeded, but only as a moral victory. The Guiels and the Walkers were in possession of a $550,000 federal default judgment, but had no way to collect because Infabco and Møllerodden had no assets in the United States, nor could a US judge order a seizure in Scotland and Norway.

In San Francisco, Attorney Gerald Sterns remained undeterred by this outcome. He knew it would not be easy, but he decided to try to have the US claim enforced by a UK court. Unlike in the United States where one state will enforce a judgment from another state, Sterns says there is no such compact between nations. Each country makes its own rules.

"Now the United States will enforce a lot of foreign judgments as a matter of comity, but the reverse is not true, and the British system is notorious. The rule has been in Britain for a long time, unless you can prove you have jurisdiction over the defendants by British standards you cannot enforce the foreign judgment."[1]

In the fall of 1981, Sterns hired a London firm to get a second opinion regarding the feasibility of a UK court enforcing the Los Angeles default judgment. When the opinion came back it confirmed Sterns's suspicions. The experts in London believed a UK court was unlikely to enforce an undefended judgment obtained in the United States. They advised Sterns to do two things: 1) attempt to have the California judgment enforced anyway to turn up the "heat" on the defendants to try to facilitate an early settlement, and 2) file a new action in Britain and ask for a "summary judgment" on the grounds that there was no defense to the action. This might prompt the defendants to settle in order to avoid the risk of an adverse ruling. If the summary judgment motion failed, then Walker and Guiel could always proceed to trial on the merits of the case, which they expressed were good.[2]

By early 1982, Sterns was informed that his request to have the US judgment enforced had been denied. This left him with the only option to bring a new action in Britain. Through a contact he had with Elborne Mitchell and Company in London, Sterns was advised to seek the aid of Simpson and Marwick in Edinburgh,[3] where the case eventually landed on the desk of Peter Anderson.

Anderson, a tall blonde-haired solicitor originally from Grangemouth, looked over the stack of papers relating to the death of the two American divers. Among them were Colin Sanderson's summary notes of the Infabco trial, the 2,000-page transcript of the Fatal Accident Inquiry, and Sheriff Risk's critical Determinations. Although Anderson was barred from introducing Risk's findings as evidence, they did provide him with a solid foundation from which he could mount an attack against the defendants. After studying the material and looking at the date of the accident, he realized that, if he was going to file an action against anyone, he'd better get moving. In Scotland, plaintiffs have three years after the alleged negligent act in which to begin proceedings to claim damages. August 8, 1982 was rapidly approaching and Anderson had to work with great speed to make that deadline.

On July 23, he filed two lawsuits with the Court of Sessions in Edinburgh. One action was on behalf of Jeanne and Marisa Walker, the other for Sheila, Roger and Shirley Guiel. Three days later, a Messenger-At-Arms by the name of Robert Dick walked into the offices of Infabco Diving Services on Greenwell Road in Aberdeen and left a package.[4] Inside were copies of the Summons, ordering the company to appear in court within two weeks to give cause as to why a decree should not be pronounced against them. Over the next several days, British National Oil Corporation, Møllerodden A/S (the manufacturers of the *Wildrake* diving system), Wilhelmsen Marine Services A/S (the owners of the *Wildrake*) and Continental Oil Corporation (part owner of the block production license where the accident took place), were also served.

Although twelve Defendants were named in the lawsuit,[a] in reality there were only five, as the other seven were subsidiaries or companies either owned, directed and/or in some way controlled by the same entity through the use of a proxy. By including BNOC and Conoco in the lawsuit, Anderson was deliberately casting a wide net to ensure that, at the end of the day, if there was no Infabco insurance fund he could lay his hands on, then he would test the provision in the regulations which imposed the duties and responsibilities of "the employer of divers" onto the "owner of the pipe-line."[5] Anderson simply wanted no "empty chairs" on the defense side of the table.

Within the 12-page Summons, Anderson summarized the accident and charged the defendants with negligence by breaching their duty to provide and maintain a safe diving plant and equipment. He repelled the Infabco Diving Services defence that it was not the employer of Richard and Skip. Instead, he insisted that IDS (or alternatively Infabco Group or Offshore Co-ordinators), was the employer,[6] and that it was their duty to take reasonable care for the safety of Richard and Skip whether they were employees or not.[7] The Summons then listed at length all of the material defects in the diving system that contributed to the accident and deaths of Richard and Skip. It also singled out Brian Masterson personally for failing to take reasonable care for the safety of Richard and Skip during the time when the bell became jammed against the side of the SALM base.[8]

Anderson also accused Møllerodden for providing an unsafe diving system,[9] and shipowners Wilhelmsen Marine Services for failing to ensure that the equipment and machinery used in the diving operation "was suitable and fit for its intended purpose."[10] He charged the captain of the *Wildrake* for breach of statutory duty because it was his responsibility "to secure that no operations or activities which might be a danger to any person engaged on those diving operations," were carried out from or on the vessel.[11] And lastly, BNOC and Conoco were faulted for failing to ensure that the other defendants were complying with the regulations as required by law.[12] On behalf of his clients, Anderson was asking the Court to award £250,000 to Jeanne Walker, £75,000 to her daughter Marisa, and £50,000 each for Roger, Sheila and Shirley Guiel.

In October the defendants presented their defenses in writing. Not surprisingly, they attempted to lay the responsibility for the accident elsewhere. Infabco, reiterated its earlier refrain that Richard and Skip were not employees of the company, but self-employed consultants of Offshore Co-ordinators Ltd.[13] It blamed Møllerodden and Wilhelmsen for supplying an "awkward" and

---

a The Defendants: 1) Infabco Diving Services Ltd; 2) Infabco Group Ltd.; 3) Offshore Co-ordinators Ltd.; 4) Møllerodden A/S; 5) Wilhelmsen Marine Services A/S; 6) BNOC (Development) Ltd.; 7) British National Oil Corporation; 8) Continental Oil Company Ltd.; 9) Conoco Ltd.; 10) Conoco (UK) Ltd.; 11) Conoco Leasing Ltd.; 12) Conoco North Sea Incorporated.

"dangerous" diving system,[14] and were it not for their negligence and failure to fulfill "the duties incumbent on them as they ought to have done, the accident would not have occurred."[15]

Infabco also pointed the finger directly at Richard and Skip by charging them with causing their own deaths.[16] It was the deceased, Infabco asserted, as well as "other Divers" who "insisted" on having the drop weights lashed to the bell frame.[17] (Later, this answer would be amended to say that Richard and Skip would not have dived if the weights "had not been lashed.")[18] Furthermore, it was Richard and Skip's duty to take reasonable care for their own safety and not to "acquiesce" to a situation that would expose themselves unnecessarily to the risk of injury or death.[19] Instead, they should have refused to dive in a bell with lashed drop weights, and had they done so, they would not have died as they did.[20]

Finally, Infabco contended that the plaintiffs' assertions were "unfounded," that the monetary claims were "grossly excessive" and "should be repelled," and that it should be absolved of any blame in the matter.[21] As for the other defendants, they all denied that the action against them was necessary, and urged the Court to dismiss the case.[22]

In Edinburgh, Anderson carefully studied the answers, none of which dispelled his unshakable belief that this was a case he couldn't lose. The degree of negligence was just too great. And in late January 1983, Lord Kincraig at the Court of Sessions saw the situation as did Anderson. Kincraig found that Infabco Diving Services Ltd. and Infabco Group Ltd. were insured against the claim and that:

> Mrs. Walker would be certain to prove a breach of duty and establish liability. She would almost certainly succeed in getting an award of damages against one [or more of] the defenders without any substantial finding of contributory negligence by her husband.[23]

And then Lord Kincraig ordered the two companies to pay Jeanne interim damages of £30,000.

While the interim payment emphasized the merits of Jeanne and Marisa's case, it underscored that the Guiels' case was essentially washed up as Sterns had warned.

Two weeks later, Masterson's attorneys attempted to appeal the Kincraig decision,[24] but in April, the appeal was rejected by Lord Cowie who modified his colleague's earlier decision and ruled that only Infabco Diving Services Ltd. should make the interim payment, not Infabco Group Ltd. Once again, Lord Cowie confirmed what Lord Kincraig had concluded: that Jeanne "would almost certainly succeed against one or more of the defenders."[25] After the decision, Anderson wrote to Jeanne to inform her of the good news:

> I have now written to those acting for Infabco requiring from them early settlement of that amount and I will send you on a cheque for your interim payment as soon as I can....

> In the meantime, this is a very positive step and I hope that the money will ease some of the difficulties in which you were finding yourself.[26]

For Jeanne, the interim payment couldn't have come at a better time. When Richard died in August 1979, the bills began to pile up fast. By late September she still hadn't received Richard's July and August paychecks, which she thought was odd because she remembered Masterson telling her at the funeral services "not to worry," that Infabco would continue to pay Richard's salary until the insurance company made a settlement. "But unfortunately, I find out pretty quickly, not only was he not going to do that, but he didn't even send Richard's paycheck that Richard had worked for."[27]

Jeanne ultimately turned to Tom Belcher to look into the matter. Offshore on the *Gulnare*, Belcher discussed Jeanne's problem with his crew then called the office in Aberdeen.

"This is what I'm doing," he said to the operation's manager, "I'm pulling these guys out [of sat] because they're having some psychological problems; and I think in the spirit of things...you really ought to pay her, and if you don't pay her, I really can't be responsible. This crew will know that you haven't paid her and they will respond appropriately....We have all lost someone out here that we were close to, so you guys have to do something."[28]

And that's how Belcher left it, as a threat, but not a hard threat. And Infabco got the message. They quickly issued Richard's backpay to Jeanne. Not long after, Tom resigned from Infabco. He didn't want anything more to do with them.

On December 15, 1979, Jeanne wrote to Brian Masterson referencing his earlier promise to help her financially until a settlement could be reached. Masterson turned the matter over to Bill Hemmings who wrote in late February 1980 that the company was willing to provide her with a loan which would be deducted from the eventual settlement decided by the court. "I am concerned that the company should do as much as possible for you and your daughter,"[29] he said in his letter.

But months would go by; Jeanne would write more letters to Hemmings, apologies would be extended, and still there was no progress on the matter of the loan. Eventually the amount was set at £10,000 ($23,400).[30] The agreement was drawn up, Jeanne signed it on October 8, 1980, but just before Christmas, Infabco canceled the offer.

Throughout that year and for the next several, Jeanne's financial woes continued to grow. She borrowed money when she had to and qualified for Richard's Social Security death benefits. But the money wasn't enough to cover the bills. At the same time, decisions about her financial future were somewhat influenced by the advice she was receiving from the people around her. Much of what she heard was, don't worry, you're going to be taken care of. And for a while she believed what everyone was telling her. But slowly it began to dawn on her that she had no money, that it wasn't going to fall out of the sky, and that, if she and Marisa were going to have any kind of life, she'd better get busy.

Jeanne heard about a part-time seamstress job at an upholstery shop. She applied and got the job, but within a year the business failed. Out of work again, she was hired at another upholstery shop. But the money was not enough to make ends meet and the hours she worked were odd.

In May 1983, Marisa turned five and was due to start kindergarten in the fall, which meant she would have to go to after-school daycare until her mother got off work. And Jeanne was opposed to that nightmare scenario, so she decided to go into business for herself. She bought a used sewing machine, approached interior designers for work, and began making cushions, pillows, bedspreads and furniture slip covers at night while continuing to work for the upholstery shop during the day. Then, on June 2, Anderson wired £27,000 of the interim judgment to Sterns. Three weeks later, Sterns sent $20,000 of it to Jeanne followed by an additional $3,901.97 on July 6, after deducting some of his costs. By October Jeanne was making enough money from her business to quit her daytime upholstery job to take care of Marisa.

Back in Edinburgh, the judge's assessment that Jeanne would succeed with her claim might have delivered a fatal blow to Infabco's defence team. Instead, it only seemed to harden their attitude. A month after Jeanne quit her job, it amended its defenses and blamed Richard and Skip for dying of hypothermia, alleging that hot water was flowing to the bell through a "repaired" hose, and that they had "shut off the valve" that controlled the supply from the surface.[31] In addition, Infabco claimed Richard and Skip failed to avail themselves of the thermal blankets and "extra warm clothing" contained within two emergency diving packs in the bell.[a] Had they done so, Infabco said, Richard and Skip would not have died.

Throughout 1983, the record was opened, amended, and closed more than half a dozen times by the various parties. This pattern would continue for many years to come, and it would be during this period of prolonged litigation that the emotional struggle to overcome the tragedy would turn the relatives' lives upside down. And if there was one emotion which ruled them all, it was anger. Not just resentment and indignation, but a deep seething anger against a system that had failed to achieve justice for the lives of Richard and Skip. To whatever extent this anger was justified, it was now interfering with their work, with their personal relationships with other people, and with their futures.

In Massachusetts, Roger's depression over his brother's death had mushroomed since the Fatal Accident Inquiry. Discovering the magnitude of negligence that had killed Skip intensified the hatred he felt for the people who

---

a Richard and Skip had used the space blankets provided. However, evidence of an emergency pack with "extra warm clothing" was never presented at the Fatal Accident Inquiry.

brought this grief upon him. Without Skip he saw no reason to exist except to claim retribution. At night, he dreamed "in color" about getting a gun past British customs to take care of business. But as satisfying as these dreams were, he knew he hadn't drifted far enough from sanity to carry out the threat, and so he vented his anger in other ways.

One day, while driving south out of West Springfield, he was riding Skip's red BMW motorcycle. It was a fast bike and he was thinking about the accident and his brother. Suddenly, the rage took control and he opened the throttle. The more he accelerated the bike, the more it wanted to go faster. And for a long time, as Roger screamed down the highway with the wind beating against his chest, he dared life itself to rid him of the inconsolable grief he could no longer bear. At 130 mph he nearly ended up crashing and killing himself. And that's how Roger lived during these years, dangerously suicidal, with an "I-don't-care" attitude because he was still "so pissed" about losing Skip.

Sheila was angry, too, and tormented. Tormented by the images of Skip trapped in the bell, and by the feelings of guilt and remorse for not saying good-bye to him before he was buried. In a conversation with her husband almost twenty years later, she talked about these feelings that were still hot, and still just beneath the surface.

"I tell you, my head was screwed up for like three or four years. I mean it really did a number on me...I was still waiting for him to come out of the diving bell...You know, because I never saw my brother dead. No one allowed me to see my brother dead...That's why I had a hard time accepting the fact that my brother died!"

"There was no reason for you to go through that," Rick says.

"I needed to see Victor dead. It took me a long time to accept the fact that he died."

"I realize that, but—I'm going to tell you something Sheila—I'd rather have you with a picture of your brother as he was—just the way I think of Victor now, sitting at my table eating all my food—right?"

"I don't know. You know—but I needed to see him. I needed to see him."

And then Rick tried to console Sheila by talking about his grandfather on his deathbed with a mass of tubes sticking out of him.

"But my brother was already dead," Sheila replies.

"Yeah, but that's a different thing. Listen, your last thoughts should be seeing your brother there..."

Sheila, now weeping softly, interrupts her husband with a voice cracking with emotion: "You know what my last thoughts of my brother were? Him suffering in that frickin' diving bell."

"Listen, no, no, no. Your last thoughts should be of your brother over with us in Maine having a good time."

"Yeah, well, it doesn't happen that way. My last thoughts of my brother are him looking out that bell, seeing somebody there and giving a thumbs up and thinking he's okay now, and he wasn't." Sheila is crying deeply now.

"It hurts me too," Rick answers, his face loaded with pain. "You know that."

"I know, I know. You hurt just as much as me."

"That's right I do."

"I know you do."

And then Rick begins to cry. He gets up from the table, wipes away the tears, and moves around the kitchen towards his wife. Then, in a raspy voice barely audible, he whispers, "It's not easy, I tell you."[32]

On the other side of the country, Jeanne was struggling with her own demons and with the responsibility of raising her daughter alone. Richard was not there to share in the decisions about Marisa's welfare and she worried about the long-term effect this would have on her daughter's life:

> You can't give your child what you want to give your child because you're only half there, and their other parent isn't there at all. So it's very hard to try to feel like they're going to have what you want them to have, in the way of an upbringing and a life, when physically it's impossible to do that. And you don't have the moral support and companionship of this person that you decided to spend your life with....You kind of have to re-evaluate what you're planning to do because, of course, all of the plans that you've made involved someone else, and they're not there anymore.[33]

From choosing Marisa's schools to matters of discipline, Jeanne was faced with all the decisions. At PTA meetings, she felt awkward and conspicuous as a single parent surrounded by the mothers and fathers of Marisa's classmates. It was a scene filled with a quantum of sadness and surrealism because, in the back of her mind, this was not how life was supposed to be; Richard was missing from the picture. And it was during these moments, when the depression took over completely, when she felt she just couldn't deal with the overwhelming responsibilities anymore, that she would hear herself saying to her daughter, "I can't go to the PTA meeting" or, "We're not having Christmas this year. Forget Christmas."

But it wasn't Jeanne doing the talking, it was the depression.

Friends and family did their best to help her overcome the emotional trauma of losing Richard. But mending deep wounds required time and Jeanne was discovering that, if there was any healing taking place at all, it was being drained away by the demands of the lawsuit. Although she knew it wasn't healthy for her to carry around the tragedy in her heart day after day, she sensed that her lawyers did not want her to let go of her "victim status:"

> They really didn't want me to forget because they wanted to be sure I was there when the day in court came. And I was there, as I had been for however many years before. And so that was kind of a big problem because emotionally you can't put it behind you because it's always there. Every day it's there.

This constant drain on her emotions sometimes interfered with her personal relationships with people. In certain situations she would find herself responding in a "negative" or "dangerous way." These difficulties transformed Jeanne, and over time, she lost her connection to the person she used to be. The laughter and the smiles disappeared. Simple joys, like cooking and eating a meal, vanished. She would spend time in the ocean as before, but "it wasn't the same." A dark cloud had moved over her life and it was there to stay. Richard was dead, and life wasn't fun anymore. But more than anything else, the accident and the aftermath had eroded her confidence in people. She didn't know whom to trust. Jeanne was slowly dying inside, and the only thing that kept her from fading away was the care she owed to her daughter.

Nineteen eighty-four and five slipped by, and on the eve of every New Year, Jeanne would write to Sterns and ask how the case was coming along; why hasn't it been settled? And every year Sterns would write back: "These things take time."

Sterns says the reason for this is because the insurers want to hang on to the money as long as they can, and because of "the power of money in compound interest."[34] As the case drags on, the attorneys continue to collect their fees, while the money, earmarked for settlement, is earning compound interest which Sterns says "will pretty much pay for the lawyers"[35]

In the meantime, until a judgment is handed down, a lot of things can happen. The plaintiff might grow weary of the lawsuit and reduce the claim. She might quit the case altogether, or even die—there is always that chance—and with no plaintiff, there is no case.

When July 1986 arrived, seven years had passed since the accident and, of the twelve original defendants, now there were only eight.[a] British National Oil Corporation had been dismantled by the Thatcher government in the spring of 1985, its oil assets and liabilities taken over by Britoil. Four of the Conoco companies were let out, but Conoco (UK) Ltd. was not.

In Edinburgh, Anderson was lodging documents with the Court relating to Jeanne's economic loss, and making efforts to locate witnesses to the *Wildrake* accident. On that score he was having limited success because most of the UK divers had moved. On July 21, he wrote to Sterns asking for help locating divers with US addresses. "Ideally, I would also like statements from them and their confirmation that they will come to Scotland for the Hearing commencing on the 14th of October." At the end of his letter, Anderson made this final observation:

> It does seem clear that the defenders are going to take this case the very least to the door of the Court and may even fight the action all the way. Accordingly, we are gearing up our preparations on that expectation.

---

a Defendants in 1986: 1) Infabco Diving Services Ltd; 2) Infabco Group Ltd.; 3) Offshore Co-ordinators Ltd.; 4) A/S Møllerodden; 5) Wilhelmsen Marine Services A/S; 6) Britoil (Development) Ltd.; 7) Britoil plc; 8) Conoco (UK) Ltd.

He was right about the defendants' intent on playing the game of brinkmanship. In late July, instead of throwing in the towel, Infabco amended its answers, once again, blaming Richard and Skip for failing to use emergency lights to attract the attention of the rescue divers.[36]

In San Francisco, Sterns hired Investigative Consultants International to begin the process of finding the *Wildrake* divers living in the United States.[37] Out of a list of 18 divers, the detective agency made contact with about a dozen. Rod Pelzel remembers their call.

"Somebody from San Francisco called and wanted to know if I'd testify, if I'd be willing to fly to Scotland to give testimony, or San Francisco for a deposition. I said, 'Certainly whatever it takes,' and I never heard from them again."[38]

Meanwhile, Anderson's search for UK witnesses had extended itself across the globe. Near the seaside resort of Vung Tau, Vietnam, former *Wildrake* Supervisor Don Cox was working for an international aid organization called ISOT. It was September and Cox was clearing some culverts for a nearby commune. He'd been on the job for about two days and was standing with the local representative for the People's Committee when the rep pointed across the field and said, "Oh look, the postman."

When Cox looked up he saw a skinny little man dressed in a singlet and pair of shorts running barefoot down a dirt path holding a raised bamboo stick in his hand. At the end of the stick, a piece of paper flapped in the breeze. When the runner approached the group, Don says he began "jabbering away" in Vietnamese and asking for the Englishman. The Vietnamese liaison then turned to Cox and said, "Mr. Don there's a Telex for you."

The Telex was from Simpson and Marwick and it read:

> Walker and Guiel v. Infabco Diving Services Ltd. We act as solicitors for relatives of Richard Walker and Victor Guiel. Divers killed in accident in North Sea August 1979. Action raised on behalf of relatives for compensation with hearings beginning October 14, 1986.[39]

After reading the message, Don thought, "*Christ, I've got to get back.*" Shortly thereafter, he sent his reply:

> Tell Jeanne Walker, no problem. Will advise upon arrival in UK soonest. Best regards, D. Cox.

Two days later, on Monday September 22, Infabco filed a "Minute of Tender" with the Court in which it offered to share liability for the accident with Møllerodden and Wilhelmsen. This was an offer amongst the defendants, it was not directed at Simpson and Marwick. In the document, Infabco proposed that, in the event the Court awarded damages to the plaintiffs, it would pay 80 percent, and suggested that Møllerodden pay 12.5 percent, and Wilhelmsen 7.5 percent.[40] But Møllerodden was unable to pay any damages at that time. It was in liquidation and its estate had been "sequestered for the benefit of creditors."[41]

Eleven days later, on the third of October, the remaining defendants offered to pay £10,000 to the Guiels, as well as court costs.[42] This was followed four days later with an offer to pay Jeanne and Marisa £262,500, from which the defendants would deduct the 1983 interim payment.[43] This offer was £62,500 less than the original amount Anderson had been suing for, and he promptly rejected it.

For the next seven days, efforts to get Infabco and the others to pay the full £325,000 continued under intense negotiations, but without agreement. Thus, as the October 14 hearing date approached, it looked as though another court was about to listen to the tragic story of how Richard Walker and Skip Guiel had died on the bottom of the North Sea.

When the 14th arrived, Anderson dispatched another telegram to Don Cox:

> Your evidence as witness very important. Contacted ISOT who would be agreeable to you attending if we meet your expenses. Please contact by telephone.[44]

Unknown to Anderson, Cox was already in route to the UK and arrived late on the afternoon of the hearing at Heathrow Airport. After disembarking from the plane, he telephoned Anderson's office and was told that the case had been canceled due to the parties agreeing to an eleventh-hour settlement.

It was a straightforward business deal and it had taken seven years to effect. Anderson believed he obtained a good offer, and better than the one he thought he could get if he had to run the case before the Court and justify the damages. The two sides ended up splitting the difference. Jeanne and Marisa would receive £293,000. Converted to dollars it would eventually amount to $422,066.50. Subtracting out Sterns costs and fees ($123,908.91), that left $298,157.59 of which $74,539.40 would be set aside in a trust for Marisa.

The Guiels also received a settlement, but the amount, when it arrived on the doorstep of Shirley Guiel's residence, would remain a foggy memory in Roger's mind: two guys in business suits ringing the door bell, handing his mother a check and then leaving. Today, Roger remembers that the sum was "insulting," so low, that it represented nothing more than a slap in the face. Sheila says it was approximately $8,000, and that, had it not been for Jeanne's insistence that Shirley be included in the settlement, her mother wouldn't have received anything.[45]

In the spring of 1987, payments began arriving in Santa Barbara "in bits and pieces." A year later Anderson managed to extract another £10,000 from the defendants to help offset Sterns's fees, but Jeanne's final disbursement would not come until October 6, 1988.

From the day of the accident, it had taken nine years of torment and tears for the case to reach its final conclusion. The relatives had won the war—Infabco Diving Services was no longer in business and its insurers were forced to settle—but in one respect it had been a hollow victory. The settlement had produced no statements of fault, no acknowledgment that anyone did anything wrong, no sense that justice had been achieved. Far from admitting responsibility for causing the accident, Infabco fought to the bitter end, repeatedly and deliberately pointing the finger of blame at Richard and Skip, thereby inflicting more pain and anguish on the families. And this is why forgiveness is not an issue with them today, and why talk about achieving true closure in cases like this is fatuous nonsense.

Unable to inflict the same amount of pain and anger on the other side, the fight had been reduced to a battle over money, which was not the issue for the Walkers and the Guiels. In most wrongful death cases, it never is. Along the way, they realized they had to take control of the anger that had fed on their souls for so many years. There was no precise moment when they began to emerge from the darkness. It wasn't like turning on a light; it was like snow melting. When asked how she's been able to overcome the tragedy, Jeanne gives a cynical laugh and says, "Well, we're still working on that." And then adds that, although she's been able to repair a lot of the damage done to her life, she still has this place in her heart that will never be resolved:

> For a long time I thought what we needed was the answer—what happened? Well, we're never going to know, I mean, not really. We won't ever have the questions answered about what were they thinking. I mean, roughly, I could tell you...but I just feel like a lot of people kind of destroyed any kind of evidence that was left. And we don't really know just whose fault stuff was. We don't really know. And in the beginning we wanted to find out; we wanted to have somebody to blame, but I don't think we will ever have an answer to that. And I had to get to a point where I had to realize that there is no answer to that. A lot of people did the wrong thing. A lot of people tried to help to have it come out a different way, but it didn't. It just happened. After it goes enough times around in your head I think that you either have to come to terms with the fact that that's something you're really not going to know, or you could go crazy, because it's not ever going to stop going around in your head.

Sick of marinating in her misery, she ultimately decided being a victim wasn't working for her. Breaking free of that self-image, however, was no easy task because for a long time the anger and the search for truth was what drove her to fight so hard for Richard's memory. It kept her connected to him, and confirmed her love for him, but it also prevented her from moving on with her life and becoming the woman she wanted to be and the mother Marisa needed. Ultimately, after years of struggle and some therapy, she began to question what she was doing to herself.

"How long do you think you should suffer for this? Is this helping me? Because it isn't hurting them. It's hurting me. What am I doing? It's making it worse. They're succeeding more fully."

Mixed with this internal dialogue were words of reproach: "*You're going to have to sit up and start paying attention here because this is my life and I am an adult, and I care; you're going to have to get over it!*"

But the discipline required "to get over it" would be life altering. "You kind of, like, have to become a different person," Jeanne says looking back:

> You have to let that person that you were just kind of die, too, because you just can't go on saying, 'We have this dream,' because we doesn't exit anymore. It's like, 'Well, we want to live on the ranch,' Well, that's nice, but it doesn't work. Under some circumstances, yes, maybe it would, but it's not we anymore...now the whole picture's changed. Can you do this by yourself? What can you do by yourself? Because you're all by yourself, right now!

When she got her first job after the accident that was helpful because it made her feel as though she had some control over her life. She started attending PTA meetings again and engaging with people. Just as Richard reinvented himself when he moved to Santa Barbara after his army days, Jeanne slowly crawled back into society and made a new life for herself.

Across the country, Roger and Sheila had come to the same conclusion. Roger realized he had to stop feeding the hatred and living his life as though he didn't care. He got married, and, in December 1989, had a son. And it was his son that gave him hope for the future. Eventually, the anger subsided, and the sadness began to fade. Normalcy returned in small ways. On the floor of his garage were the remnants of an old V-Twin Harley Davidson motorcycle that he'd purchased and given to Skip. But after the accident it sat in pieces because Roger refused to touch it. Then, in 1991, he decided it was time to put it together.

Like Roger, Sheila's salvation came about by focusing on how Skip would have wanted her to behave, and by channeling her love to her three children. He would have wanted her to pick herself up and never forget him. And Sheila has done that. Roger says his sister invokes Skip's name in many of the conversations she has with her children.

"Uncle Victor wouldn't have done that!"

Sheila says she does it so her kids will grow up knowing who their uncle was and what he stood for. "I think it's a really big important part, that when you lose somebody, you don't lose more than just their physical being," she says. "You keep them alive in your heart, and I think that's what keeps you going."

The long term consequences of this tragedy, however, still reverberates through their lives. Rick says his wife won't watch underwater films. "It's still in the back of her mind, she avoids it....It just brings back bad memories, which, probably until the day she dies, will never be gone."

"Even now I have difficulty talking about it," Sheila says. "I can go to the cemetery now and not leave all the time in tears." She says her brother's death was meaningless, and that it didn't have to happen—then adds a final reflective thought. "I think that as I get older, it was murder."

Today, Jeanne finds it impossible to speak about her relationship with Richard because looking back on those days is too frightening, the trail of sadness so long, that if she stares too hard at it, she'll see herself going mad the day she was told he was dead. In the end, she accepted the loss of Richard as an indelible mark on her life that shaped her future:

> It always affects your life. That's just the way it is. I don't think anybody's telling you the truth if they say it doesn't. A lot of other people—they would get remarried—they would have another family. Well, I didn't, for whatever reason, I didn't....So I don't know how they dealt with it. But I can't really think that they just shut out the whole past.

That past is housed in a cardboard box she has difficulty digging through. It's a box full of screams, and the papers inside, illustrations of her nightmares. They trigger the anger and the rage, and give glimpses of the life lost. Still, it's a box she is unable to throw away. Sheila and Roger have a cardboard box full of screams, too, and keep it as proof of what happened to Skip. The news clippings, the letters from Jeanne, the log of the dive, the TV Eye documentary—it's all there. There are other cardboard boxes. Some are collecting dust in the Scottish Records Office in Edinburgh, others are in the Sheriff Courthouse in Aberdeen. It's appropriate, and necessary, for the relatives to keep their records because, according to British authorities in London, the records of the *Wildrake* accident have gone missing.[46]

Four years after Anderson sent Jeanne her last payment, Ken Duell, Brian Masterson's former business partner, was at his home in Scotland watching a BBC-1 "Panorama" documentary in his den. It was June 29, 1992, and the program was about a pension fund scandal currently brewing in Britain. Ken's wife, Yvonne, was in the kitchen. Seven minutes into the report, Yvonne heard her husband shouting, "Watch the television! Watch the television!" When she entered the room, there, on the screen, was the image of Brian Masterson staring back at her.

The story of how Masterson had come to the attention of the BBC began eight months earlier when the body of a sixty-eight-year-old man was found floating in the Atlantic Ocean about 20 miles off the coast of Grand Canary Island. The son of Jewish parents murdered in the Holocaust, he had been born Jan Ludvik Hoch in Czechoslovakia. But Hoch was only his birth name. His real identity, the one Britons knew him by, was that of Robert Maxwell, the publishing tycoon and former Member of Parliament.

Several hours earlier, Maxwell had been reported missing from his 180-foot yacht, the Lady Ghislaine. After his death became public, Maxwell's financial empire came under quick scrutiny. Investigations by various news organizations revealed that the publishing magnate was deeply in debt, estimated to be in the region of $3.9 to $4.6 billion.[47] To solve his cash flow problems he had been selling off his assets, Pergamon Press and 49 percent of Mirror Group,[48] and "borrowing," then reselling $700 million worth of securities from the pension funds of the companies he controlled.[49]

This revelation prompted authorities in Britain's Serious Fraud Office to begin an investigation into the disappearance of the pension fund money. It also inspired the BBC to look into the matter which resulted in the documentary that Ken and Yvonne were now watching. BBC discovered there were 19 million people currently invested in a variety of company pension fund schemes, with nearly £500 billion at stake, and yet there was "next to no regulation" to prevent the Maxwells of this world from raiding the funds and using the proceeds as if they were his own.[50]

As Ken and Yvonne listened, BBC began presenting evidence of pension funds that went missing leaving pensioners with next to nothing and no recourse to recoup their loss. One pension fund lawyer BBC interviewed said there was nothing in the current system to prevent an employer from setting up a pension fund scheme exactly as he saw fit. He could draft the documentation according to his wishes, determine the appointment of trustees, the contribution rate, the circumstances under which the scheme could be amended or terminated.[51] With no laws to say otherwise, the lawyer said, "the whole edifice rests entirely on goodwill."[52]

Nor did the law prevent the company from using the pension fund to invest in itself.[53] This practice of "self-investment" was dramatically demonstrated in the case of a struggling shoe manufacturer that had sold all of its factories to its pension fund for £6.2 million. Nine months later it generated more capital by selling its shoemaking machinery to the fund for over a million pounds. Then it siphoned off an additional £8.5 million surplus, and borrowed £1.5 million under a loan agreement.

All of these transactions were claimed to have been made to "prevent the company's collapse," and yet BBC reported that "the firm failed nonetheless, putting fifteen hundred people out of work."[54] The purpose of the pension fund was to make money for its pensioners, but as one former employee queried, "how on earth can you have such a return from manufacturing machinery?"

Furthermore, the law allowed pension funds to be used as collateral to guarantee loans made by banks to the company. But this put the fund at risk if the company went under because it would be used to pay off the loans to the bank.[55]

It was while the Duells were watching the second case in the documentary that Masterson's photo suddenly appeared on the television screen. Sometime after leaving the North Sea, Masterson created a company called Energy and Marine from former subsidiaries of British Shipbuilding. According to the BBC, in April 1987, the company's sixty-five employees were moved from the British Shipbuilders Pension Scheme to a new scheme run by the Crown Life Insurance Company.[56]

In 1989, Crown Life closed down the policy because of a "huge shortfall in payments." Three years later, a firm of independent trustees was brought in to investigate and discovered that, where £200,000 should have been paid into the pension fund over the first two years of its life, only about £100,000 could be accounted for.[57]

Pension funds are normally managed by a board of trustees who should be independent of the company and act as the "whistle blower" to protect the pensioner in the event that the company tries to commit fraud. But Parliament put no restrictions on who the trustees could be. At Masterson's company there was only one trustee for its pension fund which was the company itself.[58] In such a circumstance, as one BBC expert pointed out, why would the company blow the whistle on its own actions?

When one of Masterson's pensioners failed to receive his annual statement, he suspected something was wrong. He sent telexes to company officials reminding them that they were responsible for issuing certified statements, but he got "no reaction at all from them, no telephone calls, no correspondence, nothing at all."[59]

When he wrote to the Occupational Pensions Board, he was informed that they were not a watchdog organization and that it "had no powers to intervene in the way in which a pension scheme is run."[60] Just as in the case of diving, there was little or no regulation to prevent exploitation. The Energy and Marine pensioners could sue for the shortfall, but as one expert said, "such an action would be unsuccessful because the pension fund will simply be an unsecured creditor along with all the other unsecured creditors." When BBC contacted Masterson to get his side of the story, "he refused to comment."[61]

When the program was over, Ken and Yvonne just looked at each other and said, "He's at it again."[62]

In late September 1997, on wind-swept promontory on the north coast of Scotland, a tall, deeply suntanned American women, now 43, began suiting up to compete in the World Kayak Surfing competition. It was a small tournament, but it had an international cast. People had traveled there from all over the world to slide down the massive storm-generated waves of Thurso. But that year, a high pressure system had blanketed the entire region and conditions in Thurso were eerily quiet, the sea flat calm. Temperatures were in the mid-70s and the September sky was as blue as it was in the woman's home town of Santa Barbara, California. With no waves in Thurso, tournament officials moved the contest to Betty Hill, and there atop a cliff overlooking the mile-wide cove and sprawling sandy beach, contestants were joyfully stripping off their clothes to take in the uncommon heat. It was as if the weather of Santa Barbara had followed the woman to the contest site.

It had been 16 years since Jeanne had last set foot on Scottish soil. Sixteen years since the day she was questioned by Bryan Gould about the events of August 8, 1979 and read her husband's diary on camera. It seemed like a lifetime ago, and it was an age when there were dragons to slay.

Still unmarried, Jeanne had no answers for why her life had turned out the way it did. It just happened. She could have been destroyed by it, she could have refused to set foot in the country that had taken her husband. Instead, she was thrilled to be there. On the cliff top she laughed and smiled and spoke freely with her friends until she heard her name called. It was her turn to compete.

Jeanne had always thought that if Richard were ever to be reborn he would come back as either a dolphin or a seal—especially a seal, she says—because they have those mustaches and large eyes that appear to be intelligent and able to communicate with her. Carefully, she set her kayak on the water and began paddling out to sea. The sea below her was crystal clear, and as the wind brushed her hair while she stroked the water, there was a moment when her heart skipped a beat, when she knew that the spirit of the poet diver was out there, somewhere, swimming among the dolphins and the seals.

Been a gas
But it has to pass...

Conductors, nodes,
Anodes and cells

Doodads on magnets
cradling the bell

Back to back jacks...
Out on the diver's slack!

Though it's been fun
I just have to run,

Cash the old paycheck
and catch me some sun

RW

Courtesy of Jeanne Walker

Richard at Hollister Ranch.

Photo by Author

The *Wildrake,* renamed *Felinto Perry*, just outside Rio de Janeiro, 1996.

Photo by Author

In 1996, the Brazilian Navy was using the *Wildrake* for submarine rescue operations.

Acknowledgements

This story has been beating in my heart for more than twenty years. But its undertaking would not have been possible without the full support and cooperation of the Guiel family and Jeanne Walker. They allowed me to intrude into their lives, drag up the past, and sift through the unspeakable pain that still shadows their lives today. It is my sincere hope that this book will bring some measure of peace to them and to those who still have questions about the deaths of Richard and Skip.

Many thanks to Commander Zanella of the Brazilian Navy for giving me access to his ship, the *Felinto Perry* (formerly the *Wildrake),* and to rummage through the diving system's records. Special thanks to the *Wildrake* divers who, without reserve, took me back to that terrible day, opened their log books and diaries, and offered personal photographs to document this story. It was their unquestionable honesty and the information they provided, that filled the gaps of the official record.

Thanks also to Alex Rosie in Scotland who located and arranged my meeting with Captain Allan Mellor, a man of impeccable integrity, who still feels the heartache of losing Richard and Skip as though they were his own sons. To Malcolm MacNeill, Peter Anderson, Reg Christie, and Sheriff James Risk, I give my heartfelt thanks for patiently explaining Scots Law to me. To the Scottish Inquiry System, that serves as a catalyst for change while searching for the truth. To Dr. Stephen J. Watt of the Department of Environmental and Occupational Medicine at the University of Aberdeen who sat down with me and detailed the stages of hypothermia. To Marcus Taylor in London who miraculously saved the *Wildrake* accident photos and gave me permission to use them. To Ken Duell and Ed Goltz who came forward and revealed a long hidden secret about the company called Infabco Diving Services. To Chuck Alexander, project engineer of the Thistle SALM, who provided me with the history and schematics of the loading buoy, along with his point of view of the accident from the *Wildrake* bridge.

I give thanks and appreciation to my close friend, Dr. Victor Udwin of the University of Tulsa, who read the first draft of the manuscript and didn't pinch his nose, then had to suffer through years of stubborn argument while I rejected some, but not all, of his advice. Never once has he wavered from being my friend. To Skip Gordon who has been my sounding board throughout the process of research and development of this book. He spent untold hours discussing the safety issues with me and offered me a place to stay in Aberdeen while I attacked the City Library for information.

I also want to express my sincerest appreciation to the rescue divers Phil Kasey-Smith and Eddy Frank who invited me into their homes to tell me of their struggles while trying to save the lives of Richard and Skip. Their courage and bravery in the face of overwhelming odds was remarkable, extraordinary, and left me with a sense of awe that will never fade.

Lastly, I reserve special thanks for Christopher Swann, author of *The History of Oilfield Diving: An Industrial Adventure*—someone who did not know me—yet took it upon himself to rescue me from the despair of years of dejection when I thought I'd failed. He rekindled the flame and urged me to push on to add this story to the historical record of those who have died on the bottom of the sea. Without his expert advice and guidance, this book would still be sitting on my shelf collecting dust. I owe him a debt of gratitude that will last for the remainder of my life. For those of you who wish to give thanks for this story coming to light, thank Chris. He's the person who certainly deserves it.

Appendix

The log entry at 0245

On the morning of August 8, 1979, BNOC diving consultant Jim Burgess made the following entry in his log:

> 0245 - Bell cable either parted or came loose from bell. Bell has a seal, and comms. Both divers inside. Depth 510'—on way up bell umbilical jumped off tire, and wedged between tire and side plate.

This entry, along with testimony at the inquiry, led Sheriff Risk to conclude that Supervisor Brian Reid tried to raise the *Wildrake* bell *without first consulting with Superintendent Peter Holmes*, and in doing so, jammed the umbilical in the power block, causing damage to it.[1] This supposition was further supported by Holmes, who not only repeated the allegation,[2] but speculated that it was probably Reid's inexperience in operating the power block that caused the mishap.[3] But, on the basis of other evidence, there are several compelling reasons why Sheriff Risk's conclusion cannot be correct.

First, when confronted by this charge at the inquiry, Reid denied it emphatically and insisted that he did nothing during the first few minutes of the accident except to get Richard safely back inside the bell and under seal. Thereafter, he says he immediately began alerting other members of the crew, including Holmes.[4] The decision to raise the bell using the umbilical wheel, Reid insisted, was made by Holmes after he came onto the scene.

Second, Reid's version is corroborated by crew members who were not adduced as witnesses at the inquiry, and who were either in the bell van at the time of the accident or were called onto the scene immediately after.

Third, there was virtually no window of opportunity for Reid to attempt the recovery unobserved before Holmes arrived in the bell van.

And fourth, the 0245 entry in the dive log, which the attorneys at the inquiry relied upon to make the allegation against Reid, was misinterpreted.

On the night of the fatal dive, John Brook was standing in the bell van with Brian Reid *at the exact moment* the main lift wire separated from the bell. Of all the people on board the *Wildrake*, Brook had to know if Reid intended on raising the bell, because, as deck foreman, he would have been involved in that operation, and because it would have been impossible for Reid to recover the bell by himself. Brook acknowledges that the umbilical became entangled in the umbilical wheel, but he says it didn't happen within minutes after the accident; it happened after "everybody had come on deck."[5] Unfortunately, Brook was never called to give evidence at the inquiry, but he did testify at the trial, where he never alluded to the scenario that Holmes put forward. In fact, his testimony supported Reid's account in every detail.

Another person first on the scene was diver Tim Newson. After being alerted by Brook, Newson arrived in the bell van, where Reid informed him he'd tried calling Holmes in his cabin but got no answer.[6] Reid then asked Newson to go and physically locate Holmes.

Had Reid already attempted to hoist the bell at that point? Surprisingly, no one put this question to Newson. But years later, he said during interview that "nothing had been done" to the bell when he left to get Holmes, that it was still hanging by the umbilical, and that Reid had no intention of doing anything until Holmes arrived at the scene.[7]

But beyond Newson and Brook's evidence, there is this. Reid testified that he'd called a saturation technician to begin setting up the through-water comms system. This was Gary Legere. Like Brook, Legere was not subpoenaed to testify at the inquiry, but confirms that, very soon after the wire separation, Reid called him to the bell van.

"We didn't really know what had happened yet," Legere says, looking back. "I think [Reid] was asking me about the comms and why there was so much gas flow. The thing I can remember clearly, I said, 'We should get the comms going again.' So I took that radio [transducer wire] over the side."[8]

Notifying the other shift and setting up the underwater radio are not the actions of a man in a hurry to raise the bell.

At the inquiry, Supervisor Don Cox testified that upon arriving in the bell van, he found Holmes discussing rescue options with Reid. He was also asked whether anyone had operated the umbilical winch at the time of his arrival and said no.[9] And while he was not in the dive van when Reid supposedly tried to raise the bell, Cox never heard any talk or rumors to that effect.

There is also the testimony of BNOC engineer Ron Wilson who was the first off-duty person to arrive in the bell van *before* Peter Holmes and *before* Don Cox.[a] During the inquiry, it was intimated that Wilson had been an eyewitness to Reid's attempt to recover the bell without first consulting with Holmes.[10] But examination of the inquiry transcript does not support this claim. When Wilson was asked what Reid was doing when he arrived, he said Reid was "requesting someone to go and waken up several other people, people whom he realized should be out and about because of the accident."[11] Again, this evidence supports Reid's version of the events. Most likely, the "someone" Wilson was referring to was either Newson or Legere.

Perhaps Reid had enough time to raise the bell before he spoke with Newson and Legere? Again, deck foreman John Brook says no such attempt was made until after "everybody had come on deck."

Still, there was the testimony of Peter Holmes who said that, when he arrived in the bell van, Reid was at the controls of the umbilical wheel with the power block "still turning."[12] This suggests that Reid was still in the process of raising the bell when Holmes walked in on him. In fact, Holmes says he had to actually

a According to the inquiry transcript, the sequence of people entering the bell van after the accident was Ron Wilson, Peter Holmes, Don Cox, Brian Masterson.

order Reid to stop what he was doing. "I said, 'You have got to stop it, you are just destroying the umbilical,'" he told the Court.[13]

Yet, no witness could be found to corroborate that account. Ron Wilson, who arrived there before Holmes, said Reid was busy raising the alarm. And when Newson was asked whether the power block was turning when he first looked at the bell umbilical, he said, "No, the wheel of the power block was not turning." Was it turning *after* he came back from informing Holmes of the emergency? Again, Newson answered, "It was not turning."[14]

Even at face value, there is something inherently implausible about a story which commits Reid to calling out his superintendent, and then, for some unknown reason, puts him at the hydraulic controls, panic-stricken and intent on raising the bell by the umbilical, knowing that his superintendent is on the way. The two halves of that story don't fit. One half says, "I need help, I'm calling my superior," while the other half says, "I can't wait on him, I'll just go ahead and raise the bell anyway."

It is also inconceivable that any supervisor with ten years experience in the industry would suddenly lose his head and try to get his divers back on board without first notifying his superintendent that he had a serious emergency. With his bell down, and the potential of losing two guys looming over his head, Reid had everything to gain by seeking help, and everything to lose if he acted alone.

Lacking a shred of corroborating evidence, Holmes's account probably would have been dismissed as a flagrant attempt to deflect blame but for the 0245 log entry:

> 0245 - Bell cable either parted or came loose from bell. Bell has a seal, and comms. Both divers inside. Depth 510'—on way up bell umbilical jumped off tire, and wedged between tire and side plate.

On the face of it, this appears to be damning evidence against Reid. But there is a fundamental problem with this entry. It lists three critical events—1) the separation of the main lift wire, 2) Richard and Skip sealing themselves inside the bell, and 3) the failed bell recovery—under a single time designation, suggesting that they either happened simultaneously or within the same time frame.

But the accident did not occur at 0245, and the umbilical did not become jammed in the power block until much later—possibly 20 to 25 minutes later—when everyone was out on deck, including Holmes.

Understandably, the first hour of the accident was fraught with confusion. A routine dive had suddenly become a shipwide emergency. People were appearing in the bell van asking a lot of questions and wanting a lot of answers. Ron Wilson arrived early, having been alerted to the emergency by Jim Burgess. Initially, Wilson said he and Burgess stood back out of the way and watched as the drama unfolded. But as time passed, he noticed that no one was keeping a record of the transpiring events. He suggested to Burgess that he resume his log, but by that time, an entire hour had passed and the rescue was well underway.

Burgess was an experienced diver and certainly familiar with the practice of log taking. He knew there were certain rules to follow when recording multiple events under a single notation. For example, if he looked out the window and saw the *Stena Welder* 200 yards off the port bow while the deck crew was pulling in the bell umbilical, it would be perfectly acceptable to record these two events together because they happened simultaneously and are unrelated to each other. On the other hand, it is not acceptable to combine consecutive events which are clearly related *and occur at different times* as in the case of the wire separation and Richard returning to the bell to get a seal.

Burgess understood these rules because he'd been following standard procedure for the previous three and one half hours, averaging one notation every seven minutes and carefully recording each event with a single time stamp—the last one being when Richard moved the *Wildrake* 15 feet forward at 0220.

Then suddenly, there is an unprecedented twenty-five minute gap in the log (0220-0245). If Richard was working during this time, then what was he doing? And why aren't there any notations to that effect? The answer is, Richard wasn't working: he had just come face to face with every diver's nightmare.

Then the log comes alive with the 0245 entry, where Burgess does something he's never done before: he makes this very long, eccentric notation, listing three consecutive events together. And then incredibly the record goes silent again—this time for 40 minutes (0245-0325).

At 0325, Burgess continues the log. More significantly, he's back to following standard procedure—one event, one time stamp—and averaging one notation about every six minutes over the next six hour period.

So why the detour from standard procedure?

Because Burgess had become distracted by the emergency. Supervisor Reid should have been maintaining his own record, but Wilson noticed that he had his hands full with other duties.[15] So when Burgess took up the log again at 0325, he knew he had to mark the inception of the accident in some way. Reflecting back upon "the missing hour," he estimated the time of the accident, lumped together whatever salient events he'd just observed, and logged them under a single notation at 0245. He had no idea that 21 months later it would cause so many fingers to point at Brian Reid.

It is also no surprise that this entry conflicts with the time Holmes says he was notified of the emergency at 0240, which is five minutes *before* the purported time of the accident. Either Holmes is wrong or the dive log is wrong.

Assuming for the moment that Holmes is wrong, then all of the succeeding events—from Richard returning to the bell and pressuring up, to the unobserved failed bell recovery, to Legere setting up the comms link, to Holmes being notified, to the discussion in the dive van about rescue options, to Don Cox secretly going to the *Wildrake* bridge to speak with the *Stena Welder* and then returning to find Masterson, Reid and Holmes still discussing what they were going to do—would have to be magically squeezed into a tiny 12 minute window (from 0245 to 0257) in order for Holmes to be on the bridge declaring the *Wildrake* emergency to Thistle Alpha at 0257.

But those events cannot crush themselves into that time frame because they would collide like jalopies in a demolition derby. With no way to judge exactly when the umbilical had become fouled, Sheriff Risk naturally assumed that what the log said was correct. But the 0245 entry was just *a guess*, and its content the scant remembrances of an hour's worth of confusion.

At the inquiry, Reid never once denied that he had jammed the umbilical in the power block: it just didn't happen the way Holmes said it did. After Holmes and Cox appeared in the bell van, there was a discussion to use the power block to raise the bell to the surface. But before that plan was executed, Holmes went to the bridge to call the platform to declare the emergency. When he returned, he gave Reid permission to initiate the lift with the result that parts of the umbilical became entangled in the machinery. This recovery attempt, which occurred sometime after 0300 but before 0325, was what Burgess observed during that missing hour and included in the 0245 entry.

Amid the accusations, Reid tried to get the Court to believe his account, but to no avail. The evidence to support the theory that he'd "jumped the gun" appeared to be too compelling. And, too, Reid was not his own best witness. From the start of his testimony, he seemed calm and collected and delivered forthright answers. But once it was suggested that he'd somehow panicked at the helm, Reid became increasingly flustered and faltered occasionally in his testimony, due partly to his tendency to exaggerate on minor points, and partly because he failed to remember the exact sequence of key events.

It is certainly true that Reid was not the most popular supervisor in the North Sea. To some of his colleagues, this shaggy-haired Englishman had a "bombastic" and "obnoxious" nature. But when the accident occurred in the early morning hours of August 8, 1979, Brian Reid did not panic and forget his duty. He did what any supervisor should do under the circumstances: he got his diver safely back inside the bell, and called for help.

The length of bell tethers

On April 8, 1980, the DOE recommended that "as a general rule" bell umbilicals should not exceed 100 feet.[a] In practice, diving contractors ignored this recommendation, particularly on construction jobs, where bell tethers were typically 150-165 feet in length.

---

a Diving Safety Memorandum 5/1980.

A year later, the DOE amended this position by saying that bell umbilicals could be more than 100 feet, but that they should not be "deployed" out of the bell further than 100 feet.[a] Then, on April 26, 1982, the DOE issued another memorandum:

> Due to the variety in size of equipment and techniques employed, this advice [given in DSM 5/1980] has not necessarily been of value.

The DOE now maintained that the maximum safe distance should be determined by the "circumstances prevailing at the time plus the equipment in use."

> The establishment of this distance is the responsibility of the diving contractor and the on site supervisor remembering that the diver should always be placed as close to his task as is reasonably practical.[b]

Here again, diving contractors were determining the standard of safety in the North Sea, not the Diving Inspectorate. That effectively ended the debate on setting a fixed length for bell umbilicals; contractors could now make up tethers of any length. And over the years they kept getting longer and longer reaching lengths as much as 72 meters (236 feet).[c]

The DOE and AODC today

In 1991, as a result of a recommendation from Lord Cullen's report on the 1988 Piper Alpha catastrophe, the Health and Safety Executive took charge of regulating and enforcing occupational health and safety standards in the North Sea.

Four years later, in 1995, the AODC merged with the Dynamically Positioned Vessel Owners Association (DPVOA) to form the International Marine Contractors Association (IMCA).

---

a Amendment to Diving Safety Memorandum 5/1980, June 30, 1981.

b Diving Safety Memorandum 11/1982.

c Case No. T960640, *Regina v Kenneth Roberts*: On July 30, 1995, Bradley Westell was killed in a thruster accident while diving from the DP vessel *Orelia*. The distance from the bell to the nearest thruster was 98 feet. Westell was out of the bell on the end of a 236-foot umbilical.

Endnotes

Chapter 1: Notes

1 From a note written to Marisa Walker on her eighteen birthday by Mary Ellen Fletcher.
2 Letter from Richard Walker to Mary Ellen Fletcher, October 1, 1973.
3 Sir Leslie Munro, *Hugo Grotius on Freedom of the Seas,* United Nations Review, vol. 4, February 1958, p. 8.
4 Time Magazine, *International Law: The Three-Mile Limit*, May 5, 1958: p. 27.
5 Yearbook of the United Nations 1958, *The Law of the Sea*, p. 379.
6 Yearbook of the United Nations 1958, *The Law of the Sea*, p. 379.
7 W. G. Carson, *The Other Price of Britain's Oil*, p. 116, 140.
8 Petroleum (Production) Act 1934.
9 D. I. MacKay and G. A. Mackay, *The Political Economy of the North Sea*, p. 20-21.
10 Richard Walker's curriculum vitae.
11 SI 688, Factories: The Diving Operations Special Regulations, 1960, sections 5, 6.
12 Health and Safety Executive records, OSD Reference No. 10560, 10559, 10563-KJAADC.
13 Htun Minn Fatal Accident Inquiry (FAI), March 28, 1972; *The News Line*, April 15, 1977.
14 Letter from Richard Walker to Mary Ellen Fletcher, September 16, 1973.
15 Letter from Richard Walker to Mary Ellen Fletcher, September 26, 1973.
16 *US Navy Diving Manual*, 1973, Section 12 p. 26. In terms of volume, the bell was only 80 cubic feet or 2.26 cubic meters.
17 Letter from Richard Walker to Mary Ellen Fletcher, September 26, 1973.
18 Submex, *The Professional Diver's Handbook*, p. 10; Telephone Interview, Michael Mangan, September 11, 1997.
19 Letter from Richard Walker to Mary Ellen Fletcher, September 26, 1973.
20 Letter from Richard Walker to Mary Ellen Fletcher, September 26, 1973.
21 Letter from Richard Walker to Mary Ellen Fletcher, October 1, 1973.
22 Letter from Richard Walker to Mary Ellen Fletcher, October 23, 1973.
23 Letter from Richard Walker to Mary Ellen Fletcher, October 23, 1973.

Chapter 2: Notes

1 *New York Times*, January 11, 1973.
2 *New York Times*, February, 23, 1973.
3 *New York Times*, January 19, 1973.
4 *New York Times*, November 14, 1973.
5 *New York Times*, October 18, 1973.
6 *New York Times*, October 18, 1973.
7 *New York Times*, October 22, 1973.
8 *New York Times,* October 21, 1973.
9 Daniel Yergin, *The Prize*, p. 623.
10 *New York Times,* October 8, 1973.

11 Daniel Yergin, *The Prize*, p. 623-624.
12 *New York Times*, December 14, 1973.
13 *New York Times*, December 14, 1973.
14 *New York Times*, November 14, 1973.
15 *New York Times*, November 14, 1973.
16 *New York Times*, December 11, 1973.
17 *The Europa World Year Book*, 1996, p. 215.
18 *Press and Journal*, January 22, 1974.
19 *The Scotsman*, January 17, 1974.
20 Adrian Hamilton, *Oil: The Price of Power*, p. 147-149.
21 Department of Trade and Industry, *North Sea Oil and Gas: A Report to Parliament*, January 25, 1973, p.14.
22 D. I. MacKay and G.A. Mackay, *The Political Economy of the North Sea*, p. 84.
23 D. I. MacKay and G.A. Mackay, *The Political Economy of the North Sea*, p. 80.
24 Scottish Development Department, *North Sea Oil Discussion Paper*, April 1973.
25 D. I. MacKay and G.A. Mackay, *The Political Economy of the North Sea*, p. 83.
26 Guy Arnold, *Britain's Oil*, p. 38, 48.
27 D. I. MacKay and G.A. Mackay, *The Political Economy of the North Sea*, p. 74.
28 Charles Woolfson, John Foster, and Matthias Beck, *Paying for the Piper*, p. 30.
29 Guy Arnold, *Britain's Oil*, p. 35.
30 Charles Woolfson, John Foster, and Matthias Beck, *Paying for the Piper*, p. 20.
31 Department of Energy, *North Sea Costs Escalation Study*, Energy Paper No. 7, page 55.
32 Department of Energy, *North Sea Costs Escalation Study*, Energy Paper No. 7, page 10.
33 Bella Bathurst, *The Lighthouse Stevensons*, p. 8.
34 Jackie Warner and Fred Park, *Requiem for a Diver*, p. 45.

Chapter 3: Notes

1 Burt Davis, Personal Interview, August 8, 1996.
2 Rod Pelzel, Personal Interview, November 12, 1995.
3 Burt Davis, Telephone Interview, February 23, 1998.
4 Undated letter from Richard Walker to Mary Ellen Fletcher. Burt Davis says Richard's first sat came in the spring of 1974.
5 SI 1229, The Offshore Installations (Diving Operations) Regulations 1974, Regulation 16(5) Schedule 2.
6 The Burgoyne Report, *Offshore Safety*, Command Paper 7841, London HMSO, p. 10-11.
7 W. G. Carson, *The Other Price of Britain's Oil*, p. 23-24.
8 *Press and Journal*, January 18, 1974. The police later said "a technical fault" with the bell led to the tragedy.
9 Health and Safety Executive records, OSD Reference No. 10535.
10 Jackie Warner and Fred Park, *Requiem for a Diver*, p. 38-39.
11 Dimmer FAI, p. 56.
12 Dimmer FAI, p. 32.
13 George F. Bond, *Papa Topside: The Sealab Chronicles of Capt. George F. Bond, USN*, p. 81.

14 Dimmer FAI, p. 102.
15 Dimmer FAI, p. 15.
16 Dimmer FAI, p. 132.
17 Dimmer FAI, p. 132-133.
18 Dimmer FAI, p. 133-134.
19 Dimmer FAI, p. 106, 135.
20 Dimmer FAI, p. 15.
21 Dimmer FAI, p. 144.
22 Dimmer FAI, p. 89.
23 Dimmer FAI, p. 135.
24 Dimmer FAI, p. 87.
25 Dimmer FAI, p. 20.
26 Dimmer FAI, p. 37, 145.
27 Dimmer FAI, p. 94. Asked at the Fatal Accident Inquiry if he had ever attended any course on decompression or diving accidents, the doctor answered, "No. Four times I have been engaged when the courses were on."
28 Dimmer FAI, p. 150.
29 Tony Benn, *Conflicts of Interest: Diaries 1977-80*, p. 3.
30 Department of Energy, *United Kingdom Offshore Oil and Gas Policy*, Command Paper 5696, p. 4.
31 UK Offshore Operators Association Limited, *Exploration and Development of UK Continental Shelf Oil*, Energy Commission Paper Number 17, October 1978, p. 13.
32 UK Offshore Operators Association Limited, *Exploration and Development of UK Continental Shelf Oil*, Energy Commission Paper Number 17, October 1978, p. 13.

Chapter 4: Notes

1 *Press and Journal*, November 3, 1975.
2 Department of Trade and Industry, *North Sea Oil and Gas: A Report to Parliament*, 1973, p. 2; *The Scotsman*, January 17, 1974. Hamish B. Grant, Regional Secretary, Scottish Office, Confederation of British Industry projected that within 10 years the country would be self-sufficient.
3 Carson, W. G., *The Other Price of Britain's Oil*, p. 106.
4 Jackie Warner and Fred Park, *Requiem for a Diver*, p. 45.
5 Craig Roberts, Personal Interview, June 7, 1996.
6 Scottish Development Department, *North Sea Oil Discussion Paper*, p. 27; By 1977 that figure was upgraded to 500,000 barrels a day, Guy Arnold, *Britain's Oil*
7 *Press and Journal*, *The Forties Story*, November 3. 1975.
8 Laing Offshore in Teeside built Alpha, Highland Fabricators in Nigg Bay built Charlie. Forties Alpha went in during June 1974, Charlie was installed during August. Bravo and Delta were scheduled for the following year. The platforms were built in record time.
9 *Press and Journal*, November 3. 1975.
10 *Press and Journal*, November 3. 1975.
11 Charles Woolfson, John Foster, and Matthias Beck, *Paying for the Piper*, p. 549.

12 *Press and Journal*, *The Forties Story*, November 3. 1975.
13 Tom Belcher, Personal Interview, September 5, 1995.
14 *Press and Journal*, *The Forties Story*, November 3. 1975.
15 Department of Energy, *Development of the oil and gas resources of the United Kingdom,* (*Brown Book*), 1980, p. 42.
16 Eugene Gordon, Personal Interview, May 24, 1996.
17 Undated letter to Mary Ellen Fletcher.
18 John Martin killed February 6, 1975; George Turner killed June 14, 1975, Unknown diver killed in the Dutch Sector. *Source:* Health and Safety Executive.

Chapter 5: Notes

1 Roger Guiel, Personal Interview, July 21, 1996.
2 Affidavit preparation of Roger Guiel.
3 September 22, 1971, USS Howard W. Gilmore AS, Report of Enlisted Performance Evaluation: NAVPERS 792.

Chapter 6: Notes

1 Ellis FAI, p. 198.
2 Ellis FAI, p. 198.
3 Ellis FAI, p. 261.
4 Ellis FAI, p. 228.
5 Ellis FAI, p. 47, 204.
6 Ellis FAI, p. 264.
7 Ellis FAI, p. 355. There was no mention in court records about compensating the bell umbilical as well.
8 Ellis FAI, p. 233.
9 Ellis FAI, p. 80-81.
10 Ellis FAI, p. 302.
11 Ellis FAI, p. 302-303.
12 Ellis FAI, p. 412.
13 Ellis FAI, p. 321.
14 Ellis FAI, p. 306.
15 Ellis FAI, p. 112.
16 Ellis FAI, p. 113, 121, 124, 165, 169, 181.
17 Carson, W. G., *The Other Price of Britain's Oil*, p. 69.
18 *Press and Journal,* January 19, 1976.
19 Ellis FAI, p. 242, 379.
20 Ellis FAI, p. 297.
21 Department of Trade and Industry, *North Sea Oil and Gas: A Report to Parliament*, 1973, p. 2.
22 Petroleum and Submarine Pipe-lines Act 1975, Part I section 13; *Roustabout* April 1978 volume 82; *Oil and Gas International Yearbook* 1977/78, p. 149.
23 *Source*: British Petroleum.
24 Letter to Skip's parents, October 3, 1976.
25 Letter to Skip's parents, October 12, 1976.
26 Letter to Roger Guiel, October 9, 1976.

27 Letter to Rick and Sheila Borsari, November 1976.
28 *Brown Book*, 1977, page 2.
29 Kerry Kanemitsu, Telephone Interview, March 4, 1997.
30 February 1977 letter to Roger Guiel.
31 Letter to Debbie Bulewich, May 15, 1977.
32 In 1977 Eugene Gordon was hired by Sub Sea as a diver/tender although he had already achieved diver rank with Ocean Systems International. Sub Sea knew this, but offered him the lower position instead with the promise to move him up to diver if he did his apprenticeship in the Placid Field for a couple weeks. Gordon agreed only to discover at the end of the season, that the office had continued to pay him as a diver/tender through the entire summer. In the office he pointed out the error to the personnel manager who acknowledged the mistake, but then said, "Oh well, we'll make you up to diver now, and don't worry about it, from now on you'll be paid as a diver."
33 E-mail from Eugene Gordon to the author, February 5, 1998. Gordon was the superintendent on the job.
34 The incident described occurred on board the *Lunde Senior* in Stavanger harbor between the months of August and November 1980. The author was one of the principal divers and witnessed the events.
35 Letter to Debbie Bulewich, May 15, 1977.

Chapter 7: Notes

1 Ken Duell, Telephone Interview, October 29, 1996.
2 Guy Arnold, *Britain's Oil*, p. 165.
3 John Paterson, *Behind the Mask*, p. 122.
4 Chuck Alexander, *Thistle A, Single Anchor Leg Mooring Project Assessment, March 1977*, p. 22.
5 *Brown Book*, 1978, p. 21.
6 Guy Arnold, *Britain's Oil*, p. 156.
7 Ric Wharton, Personal Interview, May 29, 1996.
8 Ken Duell, Telephone Interview, October 29, 1996.
9 Ken Duell, Telephone Interview, October 29, 1996.
10 Craig Roberts, Personal Interview, June 7, 1996.
11 Ed Goltz, Personal Interview, August 7, 1996.
12 Ed Goltz, Personal Interview, August 7, 1996.
13 Chris Roper, Personal Interview, November 9, 1995.
14 Chris Roper, Personal Interview, March 3, 1996.
15 Letter to Rick and Sheila Borsari, July 5.
16 Undated letter from Skip Guiel to Debbie Bulewich.
17 Ken Duell, Telephone Interview, October 29, 1996.
18 Ed Goltz, Personal Interview, August 7, 1996.
19 Ed Goltz, Personal Interview, August 7, 1996.
20 Ed Goltz, Personal Interview, August 7, 1996.
21 Ken Duell, Telephone Interview, October 29, 1996.
22 Ken Duell, Telephone Interview, October 29, 1996.
23 Ken Duell, Telephone Interview, October 29, 1996.

24 Ken Duell, Telephone Interview, October 29, 1996.
25 Burt Davis, Personal Interview, August 8, 1996.
26 Undated letter to Mary Ellen Fletcher.
27 Doug Campbell, Personal Interview, August 6, 1996.
28 Burt Davis, Personal Interview, August 8, 1996.
29 Doug Campbell, Personal Interview, August 6, 1996.
30 Tom Belcher, Telephone Interview, March 12, 1998.
31 Jeanne Walker, Personal Interview, August 7, 1996.
32 Tom Belcher, Personal Interview, September 5, 1995; Walker/Guiel FAI, p. 48-49.
33 Rod Pelzel, Telephone Interview, October 23, 1995.
34 *Press and Journal*, November 27, 1978; *The Scotsman*, November 27, 1978.

Chapter 8: Notes

1 Crude Oil prices, www.ioga.com/Special/crudeoil_Hist.htm.
2 *Brown Book*, 1978, p. 48. *Brown Book*, 1980, p. 43.
3 *Brown Book*, 1980, p. 11, 43.
4 The Brent, Argyll, Auk, Beryl, and Montrose Fields, are examples.
5 Chuck Alexander, *Thistle A, Single Anchor Leg Mooring Project Assessment, March 1977*, p. 128.
6 Chuck Alexander, *Thistle A, Single Anchor Leg Mooring Project Assessment, March 1977*, p. 129.
7 Chuck Alexander, *Thistle A, Single Anchor Leg Mooring Project Assessment, March 1977*, p. 63.
8 Chuck Alexander, *Thistle A, Single Anchor Leg Mooring Project Assessment, March 1977*, p. 20.
9 Chuck Alexander, *Thistle A, Single Anchor Leg Mooring Repair and Reinstallation, November 1979*, p. 14.
10 Chuck Alexander, *Thistle A, Single Anchor Leg Mooring Repair and Reinstallation, November 1979*, section 1.1.1.
11 Chuck Alexander, *Thistle A, Single Anchor Leg Mooring Repair and Reinstallation, November 1979*, section 1.1.
12 Chuck Alexander, Personal Interview, July 26, 1998.
13 Chuck Alexander, Telephone Interview, February 21, 1996.
14 Chuck Alexander, *Thistle A, Single Anchor Leg Mooring Repair and Reinstallation, November 1979*, Section 2.1.
15 Chuck Alexander, *Thistle A, Single Anchor Leg Mooring Repair and Reinstallation, November 1979*, Section p. 9.
16 Chuck Alexander, *Reinstallation Procedure for the Thistle Field SALM, 1979*, section 6.4.1.
17 Chuck Alexander, Telephone Interview, February 24, 1996.
18 Chuck Alexander, *Thistle A, Single Anchor Leg Mooring Repair and Reinstallation, November 1979,* p. 9.
19 Chuck Alexander, *Thistle A, Single Anchor Leg Mooring Repair and Reinstallation, November 1979,* p. 16.
20 Chuck Alexander, *Thistle A, Single Anchor Leg Mooring Repair and Reinstallation, November 1979,* p. 16.

21 Richard Walker's 1979 diary; Rod Pelzel's 1979 diary.
22 Rod Pelzel's 1979 diary.
23 Rod Pelzel's 1979 diary.
24 Affidavit of Roger Guiel, March 30, 1981, United States District Court in and for the Central District of California.
25 Colin J. Sanderson's Trial Notes, p. 30; Thames Television, TV Eye documentary, *The Last Dive*, December 17, 1981.
26 Craig Roberts, Personal Interview, June 7, 1996.
27 Richard Walker's 1979 diary.
28 Colin J. Sanderson's Trial Notes, p. 3; Court of Session, Closed Record November 9, 1983, p. 65.

Chapter 9: Notes

1 Gary Legere, Personal Interview, September 23, 1995.
2 Walker/Guiel FAI, p. 1446; Chuck Alexander. *Reinstallation Procedure for the Thistle SALM, 1979*, Figure 6.2.; BNOC Telex, June 1, 1979.
3 Walker/Guiel FAI, p. 1763.
4 Craig Roberts, Personal Interview, June 7, 1996; Tim Newson, Personal Interview, May 28, 1996.
5 Walker/Guiel FAI, p. 1774.
6 Walker/Guiel FAI, p. 1472-1473, 1578.
7 Walker/Guiel FAI, p. 1774.
8 Ted Roche, Personal Interview, August 6, 1996.
9 Colin J. Sanderson's Trial Notes, p. 24; Court of Session, Closed Record, November 9, 1983, p. 17.
10 Estimated time frame. *Source*: Chuck Alexander. *Reinstallation Procedure for the Thistle SALM, 1979*, Figure 6.2.
11 Walker/Guiel FAI, p. 1446, 1736; John Joly, Telephone Interview, January 10, 1997.
12 Walker/Guiel FAI, p. 1446-1447, 1617, 1643. Superintendent Peter Holmes testified that the *Wildrake* diving system had never been used to carry out any practice dives or experimental dives prior to July 4, 1979. Tim Newson testified that no diving was done from the *Wildrake* until it went out to the Thistle Field.
13 Dave Lane's 1979 Diary; Walker/Guiel FAI, p. 1136.
14 Dave Lane's 1979 Diary; Dave Lane, Personal Interview, June 11, 1996.
15 Chuck Alexander, *Reinstallation Procedure for the Thistle SALM, 1979*, section 6.4.2.
16 Dave Lane's 1979 Diary.
17 Ted Roche, Personal Interview, August 6, 1996.
18 Dive log, John Brook and Dave Lane.
19 Dave Lane's Dive log. Decompression began at 1307 on July 19, 1979.
20 Walker/Guiel FAI, p. 1312.
21 Walker/Guiel FAI, p. 602, 1148-1149, 1154, 1234; Swivel diagram 2-0437A, October 31, 1978; Ted Roche, Personal Interview, August 6, 1996; John Joly, Telephone Interview, October 11, 1996.
22 Dave Lane, Personal Interview, June 11, 1996; Colin J. Sanderson's Trial Notes, p. 9; Walker/Guiel FAI, p. 1737, 1763-1767, 1885. Production 1, Photograph

D. The open socket on the end of the lift wire could not be directly attached to the top of the bell. The bell pad-eye was designed to operate as a bell wire release mechanism which was activated by a 13-inch stainless steel turn wheel from inside the bell. In order for the first shackle (a safety bolt anchor shackle) to be attached, the turn wheel had to be activated to open the jaws of the bell pad-eye. The safety shackle was inserted into the jaws, and then it was closed. The second shackle (a screw pin chain shackle) was then attached to the safety shackle, bolt to bolt. Craig Roberts would later testify that when he left the *Wildrake* prior to July 3, 1979, the swivel was "on the [bell] pad-eye," that he never saw the shackle arrangement, and would not have approved of it.

23 Dave Lane, Personal Interview, June 11, 1996.

24 Walker/Guiel FAI, p. 1765.

25 John Brook, Personal Interview, May 3, 1996; Colin J. Sanderson's Trial Notes, p. 17-18.

26 John Joly, Telephone Interview, October 11, 1996.

27 Walker/Guiel FAI, p. 1418.

28 Court of Session, Closed Record, November 9, 1983 p. 53. Smedvik's contract with Møllerodden was dated February 3, 1978.

29 John Joly, Telephone Interview, October 11, 1996.

30 Roger Palmer, Personal Interview, September 23, 1995; Lacey Lahren, Telephone Interview, August 27, 1995.

31 Ted Roche, Personal Interview, August 6, 1996; Lacey Lahren, Telephone Interview, August 27, 1995.

32 Frank Montafia, Telephone Interview, September 27, 1996.

33 Steve Tegt, Telephone Interview, January 8, 1996; John Brook, Personal Interview, May 3, 1996; Dave Lane, Personal Interview, June 11, 1996; Steve Lobban, Telephone Interview, October 3, 1996.

34 Tim Newson, Personal Interview, May 28, 1996.

35 Kurt Baukau, Telephone Interview, May 22, 1996. Baukau was Wilhelmsen's Offshore Manager. Møllerodden drawing 1-0420, August 23, 1978, shows the original design with a length of chain and turnbuckle to connect each drop weight to its release mechanism. The purpose of the turnbuckle was to keep the weight cinched up tight against the bell frame (to prevent it from sagging). The object of the modification was to take the strain off the release mechanisms. The turnbuckle/chain assembly was removed and replaced by chain only. To prevent the weight from sagging backwards out of its cradle (and applying tension against the release mechanisms), pad-eyes were welded to the tops of the drop weights and to the frame of the bell. Nylon rope was then run between the two pad-eyes to take the strain off the release mechanisms. John Nadeau declined to be interviewed, June 20, 1996.

36 Court of Sessions, Scotland, Answers by the Fourth Defenders to Minute of Amendment for the Fifth Defenders, 1986, p. 1.

37 Court of Session, Closed Record, November 9, 1983, p 29-30; Walker/Guiel FAI, p. 850-852.

38 Commander "Jackie" Warner, Personal Interview, May 4, 1996.

39 The Submarine Pipe-lines (Diving Operations) Regulations 1976, regulation 17 (1).

40 Colin J. Sanderson's Trial Notes, p. 23.
41 Walker/Guiel FAI, p. 1526.
42 Colin J. Sanderson's Trial Notes, p. 23-24.
43 Colin J. Sanderson's Trial Notes, p. 22.
44 Dave Lane, Personal Interview, June 11, 1996.
45 John Joly, Telephone Interview, October 11, 1996.
46 Department of Energy, Diving Safety Memorandum, 1/1974.
47 *Press and Journal*, *The Forties Story*, November 3, 1975.
48 The death of Comex diver Clay Ellis is a good example. By modifying the bell from which Clay Ellis was diving, Comex was forcing Ellis to accept a higher level of risk for the benefit of the client. See Chapter 6 for details.
49 Walker/Guiel FAI, p. 185.
50 Ric Wharton, Personal Interview, May 29, 1996.
51 Telex from Joe Singletary, June 1, 1979.
52 Ric Wharton, Personal Interview, May 29, 1996.

Chapter 10: Notes

1 John Joly, Telephone Interview, October 11, 1996.
2 Hale and Hamlin letter to Skip Guiel, May 15, 1979.
3 Rick and Sheila Borsari, Personal Interview, July 20, 1996.
4 Harry Tranghese, Personal Interview, July 21, 1996.
5 Walker/Guiel FAI, p. 918.
6 Chuck Alexander, *Thistle A, Single Anchor Leg Mooring Repair and Reinstallation, November 1979*, Section 6.1.4.
7 Lacey Lahren's diary.
8 Infabco Services Organization Chart, Thistle SALM Installation. July 24, 1979.
9 Walker/Guiel FAI, p. 611. Brian Masterson would testify at the inquiry that the *Wildrake* "was the most advanced diving system at that time"
10 Computed from entries in Dave Lane's dive log. July 20-21, 1979.
11 Chuck Alexander, *Thistle A, Single Anchor Leg Mooring Repair and Reinstallation, November 1979,* p. 52.
12 Chuck Alexander, *Thistle A, Single Anchor Leg Mooring Repair and Reinstallation, November 1979,* Section 6.4.8.
13 Walker/Guiel FAI, p. 540.
14 Walker/Guiel FAI, p. 388.
15 Chuck Alexander, *Thistle A, Single Anchor Leg Mooring Repair and Reinstallation, November 1979,* Section 6.4.4.
16 Dave Lane's dive log; John Brook's dive log. July 24, 1979.
17 Lacey Lahren, Telephone Interview, August 27, 1995.
18 Walker/Guiel FAI, p. 263.
19 John Joly, Telephone Interview, October 11, 1996.
20 Walker/Guiel FAI, p. 263-267, 274-275, 282.
21 Walker/Guiel FAI, p. 290, 286.
22 Lacey Lahren, Telephone Interview, March 15, 1996.
23 Walker/Guiel FAI, p. 275.
24 Walker/Guiel FAI, p. 275-276.

25 Walker/Guiel FAI, p. 305.
26 Walker/Guiel FAI, p. 294.
27 Walker/Guiel FAI, p. 295.
28 Walker/Guiel FAI, p. 301-302.
29 John Joly, Telephone Interview, October 11, 1996.
30 Walker/Guiel FAI, p. 285.
31 Harry Street, *The Law of Torts*, p. 164.
32 Harry Street, *The Law of Torts*, p. 167.
33 Harry Street, *The Law of Torts*, p. 167.
34 Chuck Alexander, *Thistle A, Single Anchor Leg Mooring Repair and Reinstallation, November 1979,* Section 6.4.6.
35 Chuck Alexander, *Thistle A, Single Anchor Leg Mooring Repair and Reinstallation, November 1979,* Section 6.4.6.
36 Chuck Alexander, *Thistle A, Single Anchor Leg Mooring Repair and Reinstallation, November 1979,* Section 6.4.7.
37 Chuck Alexander, *Thistle A, Single Anchor Leg Mooring Repair and Reinstallation, November 1979,* Section 6.4.8.
38 Chuck Alexander, *Thistle A, Single Anchor Leg Mooring Repair and Reinstallation, November 1979*, Section 6.4.8.
39 Lacey Lahren, Telephone Interview, August 27, 1995.
40 Chuck Alexander, *Thistle A, Single Anchor Leg Mooring Repair and Reinstallation, November 1979*, Section 6.4.8.
41 Mark Hansen, Personal Interview, November 4, 1995.
42 Based upon the Infabco Services Organization Chart, Thistle SALM Installation. July 24, 1979.
43 Chuck Alexander, *Thistle A, Single Anchor Leg Mooring Repair and Reinstallation, November 1979*, Section 6.4.1.
44 Richard Walker's diary; Lacey Lahren's dive log; Roger Palmer's dive log.
45 Dave Lane, Personal Interview, June 11, 1996.
46 Walker/Guiel FAI, p. 1464.
47 Walker/Guiel FAI, p. 1460, 1462.
48 Mark Hansen, Personal Interview, November 4, 1995.
49 Roger Palmer's dive log, Lacey Lahren's dive log; Chuck Alexander, *Thistle A, Single Anchor Leg Mooring Repair and Reinstallation, November 1979*, section 6.4.10.
50 Dave Lane's diary. August 4, 1979.
51 Lacey Lahren, Telephone Interview, August 27, 1995.
52 Dave Lane's diary. August 4, 1979.
53 Lacey Lahren, Telephone Interview, August 27, 1995.
54 Steve Lobban, Telephone Interview, January 24, 1996.

Chapter 11: Notes

1 Roger Palmer, Personal Interview, September 23, 1995.
2 Chuck Alexander, *Thistle A, Single Anchor Leg Mooring Repair and Reinstallation, November 1979*, Section 6.4.11.
3 Walker/Guiel FAI, p. 527.
4 SI 923, The Submarine Pipe-lines (Diving Operations) Regulations 1976, Regulation 7(1(a)) and 7(2).

5 Chuck Alexander, *Thistle A, Single Anchor Leg Mooring Repair and Reinstallation, November 1979*, Section 6.4.11.
6 Dave Lane, Personal Interview, June 11, 1996; John Brook, Personal Interview, May 3, 1996.
7 Roger Palmer, Personal Interview, September 23, 1995.
8 Chuck Alexander, *Thistle A, Single Anchor Leg Mooring Repair and Reinstallation, November 1979*, Section 6.4.11.
9 Chuck Alexander, *Thistle A, Single Anchor Leg Mooring Repair and Reinstallation, November 1979*, Section 6.4.10.
10 Roger Palmer, Personal Interview, September 23, 1995.
11 Walker/Guiel FAI, p. 1414-1415.
12 Roger Palmer's dive log, August 6, 1979.
13 Walker/Guiel FAI, p. 1669; Dave Lane, Personal Interview, June 11, 1996.
14 Dave Lane's 1979 diary; Tim Newson, Personal Interview, May 28, 1996; Walker/Guiel FAI, p. 1919-1920.
15 Don Higham's 1979 diary.
16 Don Higham's 1979 diary.
17 *The Press and Journal*, July 27-28, 31, 1979, *Glasgow Herald*, July 31, 1979; Rod Pelzel's 1979 diary.
18 *The Press and Journal*, July 31, 1979.
19 Rod Pelzel's 1979 diary.
20 Rod Pelzel's 1979 diary.
21 Rod Pelzel, Telephone Interview, September 27, 1996.

Chapter 12: Notes

1 Ted Roche, Personal Interview, August 6, 1996.
2 Don Cox, Personal Interview, May 7, 1996; Jim Hurney, Telephone Interview, July 14, 1996; Michael Mangan, Telephone Interview, October 7, 1995.
3 Eddy Frank, Personal Interview, May 10, 1996; Alan Cameron, Telephone Interview, June 3, 1996.
4 Walker/Guiel FAI, p. 969, 1122.
5 *Stena Welder* log.
6 Letter to Doug Campbell, August 4, 1979.
7 Burgess/Wilson dive log. *Wildrake* internal bell check list.
8 Don Cox, Personal Interview, May 7, 1996.
9 Walker/Guiel FAI, p. 793; Colin J. Sanderson's Trial Notes, p. 9.
10 *Wildrake* external bell check list.
11 Walker/Guiel FAI, p. 1578.
12 Walker/Guiel FAI, p. 1391.
13 Burgess/Wilson dive log.
14 Walker/Guiel FAI, p. 114.
15 Walker/Guiel FAI, p. 1392, 1516.
16 Walker/Guiel FAI, p. 1516- 1517.
17 Walker/Guiel FAI, p. 1221, 1248-1249.
18 John Brook, Personal Interview, May 3, 1996; Colin J. Sanderson's Trial Notes, p. 14; Walker/Guiel FAI, p. 1517, Superintendent Peter Holmes acknowledged

at the inquiry that the transponder was taken off the bell before the dive. He testified, however, that the transponder was not for the bell but for the SALM base. But that testimony is contradicted by the external bell check list which specifically lists a transponder for the bell. Moreover, Supervisor Don Cox said the *Wildrake* bell was diving with a dedicated transponder (Interview: May 7, 1996). Thames Television, TV Eye documentary, *The Last Dive*, December 17, 1981, describes the bell as having a transponder on it when it first went into the water.

19 John Brook, Personal Interview, May 3, 1996.
20 Lacey Lahren, Telephone Interview, August 27, 1995.
21 Burgess/Wilson dive log.
22 John Brook, Personal Interview, May 3, 1996.
23 Kurt Baukau, Telephone Interview, May 22, 1996.
24 Burgess/Wilson dive log.
25 Eddy Frank, Personal Interview, May 9, 1996.
26 Jim Hurney, Telephone Interview, July 14, 1996.
27 Walker/Guiel FAI, p. 657.
28 Burgess/Wilson dive log.
29 Walker/Guiel FAI, p. 1175.
30 Kurt Baukau, Telephone Interview, May 22, 1996.
31 Kurt Baukau, Telephone Interview, May 22, 1996.
32 Walker/Guiel FAI, p. 1392.
33 Tim Newson, Personal Interview, May 28, 1996; Walker/Guiel FAI, p. 1624.
34 Colin J. Sanderson's Trial Notes, p. 14; John Brook, Personal Interview, May 3, 1996.
35 John Brook, Personal Interview, May 3, 1996.
36 Walker/Guiel FAI, p. 1176, 1250; Colin J. Sanderson's Trial Notes, p. 5.
37 Walker/Guiel FAI, p. 1175-1176.
38 Walker/Guiel FAI, p. 1176.
39 Walker/Guiel FAI, p. 1251.
40 Walker/Guiel FAI, p. 1261
41 Colin J. Sanderson's Trial Notes, p. 14.
42 Colin J. Sanderson's Trial Notes, p. 14; John Brook, Personal Interview, May 3, 1996.
43 Colin J. Sanderson's Trial Notes, p. 19.
44 Walker/Guiel FAI, p. 1177.
45 Walker/Guiel FAI, p. 1178.
46 Walker/Guiel FAI, p. 1624; Tim Newson, Personal Interview, May 28, 1996.
47 Walker/Guiel FAI, p. 1625.
48 Tim Newson, Personal Interview, May 28, 1996; Walker/Guiel FAI, p. 1625.
49 Colin J. Sanderson's Trial Notes, p. 19.
50 Tim Newson, Personal Interview, May 28, 1996; Walker/Guiel FAI, p. 1626-1627.
51 Tim Newson, Personal Interview, May 28, 1996.
52 Tim Newson, Personal Interview, May 28, 1996; Walker/Guiel FAI, p. 1394. Holmes testified at the inquiry that had to put his clothes on.
53 Walker/Guiel FAI, p. 1180; Gary Legere, Personal Interview, September 23, 1995, April 4, 1998.

54 Walker/Guiel FAI, p. 658, 674, 698.
55 Walker/Guiel FAI, p. 658.
56 Walker/Guiel FAI, p. 667.
57 Walker/Guiel FAI, p. 733; Don Cox, Personal Interview, May 7, 1996; Dave Olson, Telephone Interview, January 21, 1996.
58 Walker/Guiel FAI, p. 1193.
59 Dave Olson, Telephone Interview, January 21, 1996; Dave Lane, Personal Interview, June 11, 1996; Don Cox, Personal Interview, May 7, 1996.
60 John Brook, Personal Interview, May 3, 1996; Walker/Guiel FAI Determinations p. 15.
61 Walker/Guiel FAI, p. 734-739, 828-831; Don Cox, Telephone Interview, October 27, 1997.
62 Walker/Guiel FAI, p. 736; Don Cox, Telephone Interview, October 27, 1997.
63 Walker/Guiel FAI, p. 736-737.
64 Walker/Guiel FAI, p. 831.
65 Walker/Guiel FAI, p. 1193-1194, 1196; Don Cox, Telephone Interview, October 27, 1997.
66 Walker/Guiel FAI, p. 497.
67 Don Cox, Telephone Interview, October 27, 1997.
68 Walker/Guiel FAI, p. 1396-1397, 1421-1422, 1521, 1566, 1598.
69 Walker/Guiel FAI, p. 1157, 1396, 1566, 1598.
70 Colin Sanderson's Trial Notes, p. 6.
71 Walker/Guiel FAI, p. 1275, 1278, 1396-1397, 1473; Colin Sanderson's Trial Notes, p. 24.
72 Walker/Guiel FAI, p. 1157-1158.
73 Walker/Guiel FAI, p. 1159-1160; Karm High Pressure Hydraulic Equipment, *Umbilical Wheel Control Console*, January 18, 1979, p. 2.
74 The Submarine Pipe-lines (Diving Operations) Regulations 1976, regulation 6 (1).
75 Walker/Guiel FAI, p. 1191.
76 Don Cox, Personal Interview, May 8, 1996.
77 Walker/Guiel FAI, p. 1194, 1403.
78 Walker/Guiel FAI, p. 498.
79 Walker/Guiel FAI, p. 500.
80 Walker/Guiel FAI, p. 500.
81 Walker/Guiel FAI, p. 417-418.
82 Walker/Guiel FAI, p. 740, 1194,1197. Reid testified that just before Holmes exited the bell van, the superintendent said, "Just hold fire for a minute, we'll try and get communications and see just how they are." And then Holmes "went to the bridge to get a standby boat to come, to see what was in the vicinity in the field before deciding to make a lift."
83 Walker/Guiel FAI, p. 1489.
84 Walker/Guiel FAI, p. 1489-1491.
85 Walker/Guiel FAI, p. 511, 1408, 1490.
86 Walker/Guiel FAI, p. 1490.

Chapter 13: Notes

1 John Joly, Telephone Interview, October 4, 1996.
2 Don Cox, Telephone Interview, October 27, 1997; Michael Mangan, Telephone Interview, October 7, 1995.
3 *Stena Welder* dive log.
4 Eddy Frank, Personal Interview, May 9, 1996.
5 Jim Hurney, Telephone Interview, July 14, 1996.
6 Tom Belcher, Personal Interview, September 5, 1995.
7 Walker/Guiel FAI, p. 1723; Craig Roberts, Personal Interview, June 7, 1996.
8 Walker/Guiel FAI, p. 452-453, 1485; Gunter/Almon Police Statement.
9 Gunter/Almon Police Statement; Walker/Guiel FAI, p. 1036, 1094; Rod Pelzel, Personal Interview, November 12, 1995.
10 Roy Giles, Personal Interview, May 5, 1996.
11 Evan L. Lloyd, *Hypothermia and Cold Stress*, p. 320.
12 Norwegian Underwater Institute, *Deep-Ex '80, Project II, Thermal Studies in Diving*, p. 2-3.
13 Evan L. Lloyd, *Hypothermia and Cold Stress*, p. 320.
14 Alfred A. Bove and Jefferson C. Davis, *Diving Medicine*, p. 102.
15 Dave Olson, Telephone Interview, January 21, 1996.
16 Walker/Guiel FAI, p. 1487-1488, 1492-1495, 1500. The tape was entered into evidence at the FAI as Production 14.
17 Walker/Guiel FAI, p. 1197.
18 Walker/Guiel FAI, p. 1285.
19 Walker/Guiel FAI, p. 677-678, 1185-1186; Colin J. Sanderson's Trial Notes, p. 7.
20 Walker/Guiel FAI, p. 676; Colin J. Sanderson's Trial Notes, p. 16.
21 John Brook, Personal Interview, May 3, 1996.
22 Court of Session, Answers, IDS, July 29, 1986, p. 5.
23 Walker/Guiel FAI, p. 1601-1602.
24 Walker/Guiel FAI, p. 678; Colin J. Sanderson's Trial Notes, p. 16; John Brook, Personal Interview, May 3, 1996.
25 Walker/Guiel FAI, p. 1632.
26 Walker/Guiel FAI, p. 421-422; Tim Newson, Personal Interview, May 28, 1996.
27 Walker/Guiel FAI, p. 1630-1634.
28 Colin J. Sanderson's Trial Notes, p. 16.
29 John Brook, Personal Interview, May 3, 1996.
30 Walker/Guiel FAI, p. 1648-1650, 1675-1676; John Brook, Personal Interview, May 3, 1996; Colin J. Sanderson's Trial Notes, p. 16.
31 Don Cox, Personal Interview, May 7, 1996.
32 Walker/Guiel FAI, p. 1650.
33 Walker/Guiel FAI, p. 1215.
34 John Brook, Personal Interview, May 3, 1996.
35 *Press and Journal*, December 17, 1980; *Glasgow Herald*, December 17, 1980; Tim Newson, Personal Interview, May 28, 1996.
36 Tim Newson, Personal Interview, May 28, 1996.
37 Walker/Guiel FAI, p. 1273.
38 Walker/Guiel FAI, p. 1399-1401.

39 Burgess/Wilson dive log.
40 Walker/Guiel FAI, p. 659-661, 666-667.
41 *Wildrake* Daily Vessel Report, August 8, 1979.
42 Burgess/Wilson dive log.
43 Walker/Guiel FAI, p. 1122.
44 Gary Legere, Personal Interview, September 23, 1995.
45 John Brook, Personal Interview, May 3, 1996.
46 Walker/Guiel FAI, p. 1200, 1306; Dave Lane, Personal Interview, June 11, 1996.
47 Kurt Baukau, Telephone Interview, May 22, 1996.
48 *Wildrake* log, BNOC compilation.
49 Walker/Guiel FAI, p. 680, 1399, 1063-1064; Kurt Baukau, Telephone Interview, May 22, 1996. Baukau described the rope as being in a continuous loop tied to the umbilical in a girth hitch.
50 Walker/Guiel FAI, p. 1545.
51 Steve Lobban, Telephone Interview, December 13, 1995.
52 Dave Lane, Personal Interview, June 11, 1996; Walker/Guiel FAI, p. 1064, 1199. Reid testified that he thought the crew were going to take a pick off the 10-ton strength member of the umbilical and not around the whole umbilical.
53 Steve Lobban, Telephone Interview, January 24, 1996.
54 Walker/Guiel FAI, p. 1181, 1192, 1268, 1324, 1400-1401.
55 Dr. Stephen J. Watt, Personal Interview, May 30, 1996.
56 Surgeon Commander Frank Golden, *Hypothermia: a Problem for North Sea Industries*, Journal of Occupational Medicine, 1976, p. 87.
57 P. Marcus, D. Robertson, and R. Langford, *Metallised Plastic Sheeting for Use in Survival*, 1977.
58 Rod Pelzel, Personal Interview, November 12, 1995.
59 Walker/Guiel FAI, p. 1268, 1364, 1494.
60 Walker/Guiel FAI, p. 1723; Rod Pelzel, Personal Interview, November 12, 1995; Chris Roper, Personal Interview, November 9, 1995.
61 Walker/Guiel FAI, p. 1006-1007, 1038; Gunter/Almon Police Statement. Within hours after the accident, Randy Gunter and Tom Almon would co-author a statement to the police saying it was Peter Homes who ordered them out of the bell van. This is confirmed by Chris Roper's account in an interview on November 9, 1995. However, during the FAI, almost two years after the accident, Gunter would testify, perhaps erroneously, that Reid told him "to get the fuck out of the bell van."
62 Chris Roper, Personal Interview, November 9, 1995.
63 Rod Pelzel, Personal Interview, November 12, 1995.
64 Chris Roper, Personal Interview, November 9, 1995.
65 Walker/Guiel FAI, p. 1121; Burgess/Wilson dive log.
66 Walker/Guiel FAI, p. 1123; Thistle Radio log: "0345 - Wildrake requests Stena Welder launch bell for recovery operation."
67 Walker/Guiel FAI, p. 443, 741.
68 Don Cox, Personal Interview, May 7, 1996; Gunter/Almon Police Statement.
69 Walker/Guiel FAI, p. 1325, 1650-1651.

70 Gunter/Almon Police Statement. "The state of the umbilical at this point was severed—hot water and power. I, Mr. Gunter, could see that the water was coming out and I also saw the electric's arcing [in the water]. At this time the bell still had internal and external pneumo and main gas supply."

Chapter 14: Notes

1 Dave Olson, Telephone Interview, October 7, 1995.
2 Walker/Guiel FAI, p. 1006, 1084; Chris Roper, Personal Interview, November 9, 1995.
3 Walker/Guiel FAI, p. 1007.
4 Dave Olson, Telephone Interview, October 7, 1995.
5 Walker/Guiel FAI, p. 1011.
6 Chris Roper, Personal Interview, November 9, 1995.
7 Dave Olson, Telephone Interview, January 21, 1996.
8 Greg Hansen, Telephone Interview, January 23, 1996.
9 *The Scotsman*, August 29, 1973; *Press and Journal*, August 29, 1973.
10 Greg Hansen, Telephone Interview, January 23, 1996.
11 Rod Pelzel, Personal Interview, November 12, 1995; Gunter/Almon Police Statement.
12 Rod Pelzel, Personal Interview, November 12, 1995.
13 *Wildrake* log, partial section recovered by Chris Roper, August 9, 1979.
14 Olson, Dave. *Report from Sat Technicians of Their Activities on August 8, 1979.*
15 Telephone Interview. Alan Mellor, July 5, 1996. Dunlin was seven miles directly south, Stratjord eleven miles southeast, and Thistle Alpha a mile away.
16 Walker/Guiel FAI, p. 1123.
17 *Stena Welder* dive log.
18 Walker/Guiel FAI, p. 446, 448.
19 Submex. *The Professional Diver's Handbook*, 1982, p. 253.
20 Walker/Guiel FAI, p. 1408; Jim Hurney, Telephone Interview, July 14, 1996.
21 Walker/Guiel FAI, p. 445-446, 779, 1409-1410.
22 Walker/Guiel FAI, p. 1410.
23 *Wildrake* dive log.
24 Burgess/Wilson dive log.
25 Thistle Alpha radio log. "0515 - Jon Viking at Gulnare. Re: *Stena Welder* anchor buoys."
26 Burgess/Wilson dive log.
27 Walker/Guiel FAI, p. 1124-1125; Telephone Interview. Alan Mellor, July 5, 1996.
28 Walker/Guiel FAI, p. 1125.
29 Burgess/Wilson dive log; Walker/Guiel FAI, p. 481.
30 Walker/Guiel FAI, p. 561; *Wildrake* dive log.
31 Rod Pelzel, Personal Interview, November 12, 1995.
32 Craig Roberts, Personal Interview, June 7, 1996.
33 Chuck Alexander, Telephone Interview, September 9, 1996.
34 Chris Roper, Personal Interview, November 9, 1995.
35 Walker/Guiel FAI, p. 1602.

36 Chris Roper, Personal Interview, November 9, 1995.
37 Walker/Guiel FAI, p. 1013.
38 Walker/Guiel FAI, p. 1013, 1100.
39 Walker/Guiel FAI, p. 1041-1042.
40 Walker/Guiel FAI, p. 1013.
41 Walker/Guiel FAI, p. 1041, 1654.
42 Burgess/Wilson dive log.
43 Gunter/Almon Police Statement.
44 Burgess/Wilson dive log; Walker/Guiel FAI, p. 1068.
45 *Wildrake* log, partial section recovered by Chris Roper, August 9, 1979.
46 Craig Roberts, Personal Interview, June 7, 1996.
47 Walker/Guiel FAI, p. 1047.
48 Walker/Guiel FAI, p. 1048, 1537.
49 Dave Lane, Personal Interview, June 11, 1996.
50 Gunter/Almon Police Statement.
51 Walker/Guiel FAI, p. 1316.
52 Kurt Baukau, Telephone Interview, May 22, 1996.
53 Walker/Guiel FAI, p. 1439, 1563.
54 Walker/Guiel FAI, p. 1044; Chris Roper, Personal Interview, November 9, 1995.
55 Walker/Guiel FAI, p. 1727.
56 Walker/Guiel FAI, p. 1485, 1602.
57 Walker/Guiel FAI, p. 1734.
58 Walker/Guiel FAI, p. 1721,1722,1734,1744-1745.
59 Walker/Guiel FAI, p. 1728.
60 Craig Roberts, Personal Interview, June 7, 1996.
61 Walker/Guiel FAI, p. 1728.
62 Walker/Guiel FAI, p. 685-688, 720-721.
63 Walker/Guiel FAI, p. 1045; Kurt Baukau, Telephone Interview, May 22, 1996.
64 Steve Lobban, Telephone Interview, January 24, 1996.
65 Walker/Guiel FAI, p. 393.
66 Walker/Guiel FAI, p. 426-428, 587-588.
67 Chuck Alexander, Telephone Interview, January 9, 1996; Walker/Guiel FAI, p. 573.
68 Walker/Guiel FAI, p. 427.

Chapter 15: Notes

1 Walker/Guiel FAI, p. 1425.
2 *Stena Welder* dive log; Burgess/Wilson dive log.
3 Burgess/Wilson dive log.
4 Phil Kasey-Smith, Personal Interview, April 29, 1996.
5 *Stena Welder* dive log; Burgess/Wilson dive log.
6 *Stena Welder* dive log.
7 *Stena Welder* dive log.
8 *Stena Welder* dive log; Walker/Guiel FAI, p. 784.
9 *Stena Welder* dive log.
10 Walker/Guiel FAI, p. 1127; *Stena Welder* dive log.

11 Phil Kasey-Smith, Personal Interview, April 29, 1996; Walker/Guiel FAI, p. 835, 1126. Don Cox would testify that the problems Kasey encountered began to manifest themselves within one half hour to one hour after the rescue dive began.
12 Burgess/Wilson dive log; Walker/Guiel FAI, p. 705.
13 *Stena Welder* dive log.
14 Shirley Guiel, Personal Interview, July 18, 1996.
15 Walker/Guiel FAI, p. 987-988.
16 Roy Giles, Personal Interview, May 5, 1996.
17 Marcus Taylor photograph # 65196 shows the hinge for the bottom door at 9 o'clock. The upper pad-eyes modifications were welded to the bell frame at approximately 8 o'clock and 1 o'clock. The lower pad-eye modifications were welded to the drop weights in the 0630 position and the 0030 position. The distance to the ropes from the edge of the trunking was the same. Moreover, photograph # 65446 shows the bell recovered onto the deck with the bottom door pulled back and in no way obstructing access to the ropes in the 8 o'clock position.
18 Walker/Guiel FAI, p. 1831.
19 *Wildrake* log, partial section recovered by Chris Roper, August 9, 1979.
20 Dave Olson, Telephone Interview, January 21, 1996.
21 Gary Legere, Personal Interview, September 23, 1995.
22 Dave Olson, Telephone Interview, January 21, 1996.
23 Roger Palmer, Personal Interview, September 23, 1995.
24 Walker/Guiel FAI, p. 1299.
25 Walker/Guiel FAI, p. 1046.
26 Walker/Guiel FAI, p. 1300.
27 Dave Lane, Personal Interview, June 11, 1996.
28 Chris Roper, Personal Interview, November 9, 1995.
29 Walker/Guiel FAI, p. 1084.
30 Walker/Guiel FAI, p. 512, 515-516, 1049.
31 Walker/Guiel FAI, p. 1051.
32 Walker/Guiel FAI, p. 515.
33 Walker/Guiel FAI, p. 513-514.
34 Walker/Guiel FAI, p. 514.
35 Walker/Guiel FAI, p. 515.
36 Burgess/Wilson dive log.
37 Burgess/Wilson dive log.
38 *Stena Welder* dive log.
39 Eddy Frank, Personal Interview, May 9, 1996.
40 *Stena Welder* dive log.
41 Phil Kasey-Smith, Personal Interview, April 29, 1996.

Chapter 16: Notes

1 Don Cox, Personal Interview, May 7, 1996.
2 Phil Kasey-Smith, Personal Interview, April 29, 1996.
3 Walker/Guiel FAI, p. 1127.
4 Walker/Guiel FAI, p. 518.

5 Walker/Guiel FAI, p. 520.
6 Walker/Guiel FAI, p. 521.
7 Walker/Guiel FAI, p. 1299.
8 Walker/Guiel FAI, p. 1506.
9 Walker/Guiel FAI, p. 1506-1507.
10 Walker/Guiel FAI, p. 1507.
11 Walker/Guiel FAI, p. 1751.
12 Walker/Guiel FAI, p. 1752.
13 *Wildrake* log, partial section recovered by Chris Roper, August 9, 1979.
14 Walker/Guiel FAI, p. 1185.
15 Walker/Guiel FAI, p. 1127; *Stena Welder* dive log.
16 Burgess/Wilson dive log.
17 Walker/Guiel FAI, p. 684.
18 Walker/Guiel FAI, p. 778.
19 Walker/Guiel FAI, p. 451, 523-524, 778, 780-782, 788, 986.
20 Walker/Guiel FAI, p. 1557.
21 Walker/Guiel FAI, p. 1561.
22 Walker/Guiel FAI, p. 478, 524.
23 Walker/Guiel FAI, p. 524, 596. Masterson testified that he rejected the suggestion on advice from his experts.
24 Walker/Guiel FAI, p. 843.
25 Walker/Guiel FAI, p. 478-480.
26 Walker/Guiel FAI, p. 1411.
27 Walker/Guiel FAI, p. 1411, 1508, 1557.
28 Walker/Guiel FAI, p. 1753.
29 Walker/Guiel FAI, p. 1753-1754.
30 Walker/Guiel FAI, p. 971.
31 Walker/Guiel FAI, p. 1508, 1558.
32 Walker/Guiel FAI, p. 480.
33 Walker/Guiel FAI, p. 1753-1754.
34 Walker/Guiel FAI, p. 941.
35 *Stena Welder* dive log.
36 Chris Roper, Personal Interview, November 9, 1995.
37 Phil Kasey-Smith, Telephone Interview, June 9, 1997.
38 *Stena Welder* dive log.
39 *Stena Welder* dive log.
40 The *Stena Welder* log indicates there were three slings, but Tim Newson testified he put two slings on the hook, one was one inch in diameter and the other was one and one half inches in diameter. Walker/Guiel FAI, p. 1654.
41 Don Cox, Personal Interview, May 7, 1996; Jim Hurney, Telephone Interview, July 14, 1996.
42 Walker/Guiel FAI, p. 559.
43 Don Cox, Personal Interview, May 7, 1996.
44 Walker/Guiel FAI, p. 1127.
45 Allan Mellor, Telephone Interview, July 5, 1996.
46 Walker/Guiel FAI, p. 1128; *Stena Welder* dive log.

47 Walker/Guiel FAI, p. 1128-1129.
48 Walker/Guiel FAI, p. 1656; Tim Newson, Personal Interview, May 28, 1996.
49 Walker/Guiel FAI, p. 1657.
50 Walker/Guiel FAI, p. 1657.
51 Walker/Guiel FAI, p. 1657.
52 Chris Roper, Personal Interview, November 9, 1995.
53 Walker/Guiel FAI, p. 1658.
54 Walker/Guiel FAI, p. 1658.
55 Bob Gemmell, Telephone Interview, August 15, 1995.
56 Rod Pelzel, Personal Interview, November 12, 1995.
57 Rod Pelzel, Personal Interview, November 12, 1995.
58 Dave Lane, Personal Interview, June 11, 1996.
59 Chris Roper, Personal Interview, November 9, 1995.
60 Walker/Guiel FAI, p. 1081-1082.
61 Walker/Guiel FAI, p. 1082.
62 Walker/Guiel FAI, p. 1081-1084.
63 Chris Roper, Personal Interview, August 4, 1996.
64 *Stena Welder* dive log; Walker/Guiel FAI, p. 590.
65 Walker/Guiel FAI, p. 1426; Phil Kasey-Smith, Personal Interview, April 29, 1996.
66 Walker/Guiel FAI, p. 591; Burgess/Wilson dive log.
67 Walker/Guiel FAI, p. 591.
68 Walker/Guiel FAI, p. 591-592; *Stena Welder* dive log.
69 Phil Kasey-Smith, Personal Interview, April 29, 1996.
70 Dave Lane, Personal Interview, June 11, 1996.
71 Walker/Guiel FAI, p. 1834.
72 *Stena Welder* dive log; Walker/Guiel FAI, p. 592-593, 866-867, 924.
73 Walker/Guiel FAI, p. 1760.
74 Phil Kasey-Smith, Personal Interview, April 29, 1996; Walker/Guiel FAI, p. 866, 924, 1129.
75 Eddy Frank, Personal Interview, May 9, 1996; Walker/Guiel FAI, p. 529.
76 Walker/Guiel FAI, p. 533.
77 Walker/Guiel FAI, p. 1206; Burgess/Wilson dive log.
78 Walker/Guiel FAI, p. 1428, 1593-1594.
79 Chuck Alexander, Telephone Interview, January 9, 1996.
80 Walker/Guiel FAI, p. 556, 1017; Burgess/Wilson dive log.
81 Walker/Guiel FAI, p. 1017.
82 Rod Pelzel, Personal Interview, November 12, 1995.
83 Chuck Alexander, Telephone Interview, January 9, 1996.
84 Gary Legere, Personal Interview, September 23, 1995.
85 Chris Roper, Personal Interview, November 9, 1995.
86 Walker/Guiel FAI, p. 527, 529, 1129.
87 Chris Roper, Personal Interview, November 9, 1995.
88 John Brook, Personal Interview, May 3, 1996.
89 Walker/Guiel FAI, p. 1018, 1755.
90 Walker/Guiel FAI, p. 532-533, 553-554.
91 Walker/Guiel FAI, p. 557.

92 Walker/Guiel FAI, p. 554, 557.
93 Walker/Guiel FAI, p. 557-558.
94 Burgess/Wilson dive log. 1150 to 1212.
95 Walker/Guiel FAI, p. 693-694.
96 Rod Pelzel, Personal Interview, November 12, 1995.
97 Walker/Guiel FAI, p. 1429.
98 Walker/Guiel FAI, p. 597.
99 Walker/Guiel FAI, p. 1429.
100 Steve Lobban, Telephone Interview, December 13, 1995.
101 Thames Television, TV EYE documentary, *The Last Dive*, December 17, 1981; Walker/Guiel FAI, p. 1129-1130; Allan Mellor, Telephone Interview, July 5, 1996.
102 Walker/Guiel FAI, p. 1018, 1755-1756.
103 Walker/Guiel FAI, p. 1430.
104 Dave Lane, Personal Interview, June 11, 1996.
105 Rod Pelzel, Telephone Interview, September 27, 1996.
106 Chris Roper, Personal Interview, November 9, 1995.
107 Walker/Guiel FAI, p. 694, 1018.
108 Walker/Guiel FAI, p. 694.
109 Walker/Guiel FAI, p. 531-532, 574, 1538.
110 Burgess/Wilson dive log.
111 Rod Pelzel, Personal Interview, November 12, 1995.
112 Burgess/Wilson dive log.
113 Chris Roper, Personal Interview, November 9, 1995.
114 Dave Lane, Personal Interview, June 11, 1996.
115 Burgess/Wilson dive log.
116 Chuck Alexander, Telephone Interview, February 24, 1996.
117 Frank Surrell, Personal Interview, August 8, 1996.

Chapter 17: Notes

1 Walker/Guiel FAI, p. 1433.
2 Walker/Guiel FAI, p. 938-939, 1433-1434, 1749.
3 Written statement by Mike Allen; Walker/Guiel FAI, p. 1130
4 Burgess/Wilson dive log.
5 Walker/Guiel FAI, p. 534-535.
6 Colin J. Sanderson's Trial Notes, p. 18-19.
7 Press and Journal, December 17, 1980; Eddy Frank, Personal Interview, May 10, 1996.
8 Written statement by Mike Allen.
9 Walker/Guiel FAI, p. 95; *Thistle Alpha Helicopter Movement Report*, August 8, 1979.
10 Gary Legere, Personal Interview, September 23, 1995.
11 Written statement by Mike Allen; Walker/Guiel FAI, p. 944.
12 Written statement by Mike Allen.
13 Walker/Guiel FAI, p. 950.
14 Written statement by Mike Allen; Walker/Guiel FAI, p. 950-951, 1130.
15 Written statement by Mike Allen; Walker/Guiel FAI, p. 953-954, 1131; Phil Kasey-Smith, Personal Interview, April 29, 1996.

16 Written statement by Mike Allen; Walker/Guiel FAI, p. 955-956, 1131.
17 Walker/Guiel FAI, p. 1748-1749.
18 Craig Roberts, Personal Interview, June 7, 1996.
19 Walker/Guiel FAI, p. 956-957.
20 Burgess/Wilson dive log.
21 Walker/Guiel FAI, p. 18.
22 Sheila Borsari's diary notes.
23 Walker/Guiel FAI, p. 47.
24 Walker/Guiel FAI, p. 48; Ted Roche, Telephone Interview, March, 4, 1996.
25 Ted Roche, Personal Interview, August 6, 1996.
26 Written statement by Mike Allen.
27 Written statement by Mike Allen.
28 Olson, Dave. *Report from Sat Technicians of Their Activities on August 8, 1979.*
29 Ted Roche, Personal Interview, August 6, 1996; Tom Belcher, Telephone Interview, March 15, 1999.
30 Olson, Dave. *Report from Sat Technicians of Their Activities on August 8, 1979.*
31 Dave Olson, Telephone Interview, January 21, 1996.
32 Dave Olson, Telephone Interview, January 21, 1996.
33 Walker/Guiel FAI, p. 66-68.
34 Walker/Guiel FAI, p. 66.
35 Michael Mangan, Telephone Interview, September 11, 1995; Walker/Guiel FAI, p. 1132; Thistle Alpha OIM log.
36 Michael Mangan, Telephone Interview, October 7, 1995.
37 Burgess/Wilson dive log.

Chapter 18: Notes

1 Lacey Lahren, Telephone Interview, August 6, 1998.
2 Dave Lane, Personal Interview, June 11, 1996.
3 Burgess/Wilson dive log.
4 Walker/Guiel FAI, p. 1837; Marcus Taylor photograph # 65209.
5 Walker/Guiel FAI, p. 1838; Marcus Taylor photograph # 65183.
6 Walker/Guiel FAI, p. 1841.
7 Walker/Guiel FAI, p. 1020; Marcus Taylor photograph # 65454.
8 Walker/Guiel FAI, p. 1020, 1023, 1831.
9 Steve Lobban, Telephone Interview, January 24, 1996; Thames Television, TV EYE documentary, *The Last Dive*, December 17, 1981.
10 Burgess/Wilson dive log; *Wildrake* dive log; *Wildrake* Daily Vessel Report, August 8, 1979; Rod Pelzel, Personal Interview, November 12, 1995; Dave Lane, Personal Interview, June 11, 1996; Steve Lobban, Telephone Interview, January 24, 1996; John Brook, Personal Interview, May 3, 1996.
11 Rod Pelzel, Personal Interview, November 12, 1995.
12 John Brook, Personal Interview, May 3, 1996.
13 Walker/Guiel FAI, p. 90.
14 Walker/Guiel FAI, p. 100.
15 Walker/Guiel FAI, p. 90.
16 Sheila Borsari's diary notes from August 8, 1979.

17 Olson, Dave. *Report from Sat Technicians of Their Activities on August 8, 1979.*
18 Walker/Guiel FAI, p. 72.
19 Walker/Guiel FAI, p. 101.
20 Walker/Guiel FAI, p. 255.
21 Walker/Guiel FAI, p. 83-84, 255.
22 Walker/Guiel FAI, p. 259.
23 Walker/Guiel FAI, p. 85.
24 Walker/Guiel FAI, p. 256.
25 Walker/Guiel FAI, p. 78; E-mail from Lacey Lahren to the author, October 21, 1998.
26 Walker/Guiel FAI, p. 258.
27 Lacey Lahren, Telephone Interview, August 6, 1998.
28 Walker/Guiel FAI, p. 258.
29 Walker/Guiel FAI, p. 75-77.
30 Walker/Guiel FAI, p. 104.
31 Walker/Guiel FAI, p. 83.
32 Letter to the author from Dr. P. B. James, December 18, 1996.
33 Evan L. Lloyd. *Hypothermia and Cold Stress*, p. 51. Frankland, 1975; Freeman and Pugh, 1969; Golden, 1973a; Hillman, 1971; Leathart, 1971; Welton *et al.*, 1978; Zingg, 1966.
34 Evan L. Lloyd. *Hypothermia and Cold Stress*, p. 51.
35 Eddy Frank, Personal Interview, May 10, 1996.
36 Chris Roper, Telephone Interview, November 2, 1998.
37 Gary Legere, Personal Interview, September 23, 1995; Chris Roper, Telephone Interview, November 2, 1998.
38 Walker/Guiel FAI, p. 793; Colin J. Sanderson's Trial Notes, p. 9.
39 Colin J. Sanderson's Trial Notes, p. 17.
40 Kurt Baukau, Telephone Interview, May 22, 1996.
41 Don Cox, Personal Interview, May 8, 1996.
42 Rod Pelzel, Telephone Interview, October 23, 1995; Tom Belcher, Telephone Interview, April 5, 1996.
43 Rod Pelzel's 1979 diary.
44 Rod Pelzel, Personal Interview, November 12, 1995.

Chapter 19: Notes

1 *Press and Journal*, *Glasgow Herald*, *The Scotsman*, August 9, 1979.
2 Gary Legere, Personal Interview, September 23, 1995.
3 Dave Lane, Personal Interview, June 11, 1996.
4 Chuck Alexander, *Reinstallation Procedure for the Thistle SALM, 1979*, section 6.4.1.
5 Don Cox, Personal Interview, May 7, 1996; Telephone Interview, Roger Palmer, July 12, 1996. "I was involved in those conversations also," Palmer recalls. "One of his [Masterson's] concerns was how soon it was going to take to get that system back up to divable again, to running."
6 Gary Legere, Telephone Interview, July 13, 1996.
7 Lacey Lahren, Telephone Interview, July 12, 1996.
8 Dave Lane, Personal Interview, June 11, 1996.
9 Walker/Guiel FAI, p. 1865, 1913.

10 Walker/Guiel FAI, p. 1789.
11 Walker/Guiel FAI, p. 1790, 1792.
12 Walker/Guiel FAI, p. 1501.
13 Walker/Guiel FAI, p. 1802.
14 Walker/Guiel FAI, p. 1843.
15 Roy Giles, Personal Interview, May 5, 1996; Walker/Guiel FAI, p. 1803-1807, 1915-1918.
16 Roy Giles, Personal Interview, May 5, 1996.
17 Walker/Guiel FAI, p. 1797; Colin J. Sanderson's Trial Notes, p. 22.
18 Walker/Guiel FAI, p. 1895.
19 Walker/Guiel FAI, p. 1786-1787, 1910.
20 Commander "Jackie" Warner, Personal Interview, May 4, 1996; Roy Giles, Personal Interview, May 5, 1996.
21 Colin J. Sanderson's Trial Notes, p. 3. Testimony of Mrs. Sandra Percy, personal secretary to Commander Warner, Crown v Infabco Diving Services, 1980.
22 Roy Giles, Personal Interview, May 5, 1996.
23 Roy Giles, Personal Interview, May 5, 1996.
24 Chuck Alexander, *Reinstallation Procedure for the Thistle SALM, 1979*, section 6.4.1, 6.4.11.
25 Tim Newson, Personal Interview, May 28, 1996.
26 Transcription of a Tape Recording between BNOC legal advisor Bob McNaught/Frank Surrell and Chris Roper/Randy Gunter, at St. Machar House, Aberdeen, August 10, 1979.
27 Transcription of a Tape Recording between BNOC legal advisor Bob McNaught/Frank Surrell and Chris Roper/Randy Gunter, at St. Machar House, Aberdeen, August 10, 1979.
28 Transcription of a Tape Recording between BNOC legal advisor Bob McNaught/Frank Surrell and Chris Roper/Randy Gunter, at St. Machar House, Aberdeen, August 10, 1979.
29 Walker/Guiel FAI, p. 1023-1024, 1028-1030. Walker/Guiel FAI, p. 1024; Transcription of a Tape Recording between BNOC legal advisor Bob McNaught/Frank Surrell and Chris Roper/Randy Gunter, at St. Machar House, Aberdeen, August 10, 1979.
30 Chris Roper, Personal Interview, November 9, 1995; Transcription of a Tape Recording between BNOC legal advisor Bob McNaught/Frank Surrell and Chris Roper/Randy Gunter, at St. Machar House, Aberdeen, August 10, 1979.
31 Chris Roper, Personal Interview, November 9, 1995.
32 Walker/Guiel FAI, p. 792, 796, 808, 1147. John Brook, who testified at the criminal trial, but not the Fatal Accident Inquiry, also says there was a split pin in the clevis pin. Colin J. Sanderson's Trial Notes, p. 13-14, 17.
33 Walker/Guiel FAI, p. 1149, 1443. Production 1, Photograph D. The screw pin chain shackle was secured with seizing wire, the anchor safety shackle with a welding rod (Marcus Taylor Ltd. photograph #65179).
34 Report by Pellow Engineering Services, *Failure of the Wildrake Bell Connection*, October 28, 1998.

35 Walker/Guiel FAI, p. 1024; Transcription of a Tape Recording between BNOC legal advisor Bob McNaught/Frank Surrell and Chris Roper/Randy Gunter, at St. Machar House, Aberdeen, August 10, 1979.
36 David Hughes, *Underwater Systems Design*, April/May 1981, p. 14-16.
37 David Hughes, *Underwater Systems Design*, April/May 1981, p. 14-16.
38 Report by Pellow Engineering Services, *Failure of the Wildrake Bell Connection*, October 28, 1998.
39 Letter from Atherton Fuller, Law Offices of Hale and Hamlin, to Skip Guiel, August 13, 1979.

Chapter 20: Notes

1 Jeanne Walker, Personal Interview, August 7, 1996.
2 Rod Pelzel's 1979 diary.
3 Jeanne Walker, Personal Interview, August 7, 1996.
4 Undated notes from Jeanne Walker to Sheila Borsari.
5 Doug Campbell, Personal Interview, August 6, 1996.
6 Rick and Sheila Borsari, Personal Interview, July 20, 1996.
7 Joe Graveline, Personal Interview, July 19, 1996.
8 Bob Gemmell, Telephone Interview, August 15, 1995.
9 From the diary of Sheila Borsari.
10 Sheila Borsari, Telephone Interview, January 20, 1999.
11 Shirley Guiel and Sheila Borsari, Personal Interview, July 22, 1996.
12 Shirley Guiel and Sheila Borsari, Personal Interview, July 22, 1996.
13 Shirley Guiel and Sheila Borsari, Personal Interview, July 22, 1996.
14 Roger Guiel, Personal Interview, July 22, 1996.
15 Shirley Guiel and Sheila Borsari, Personal Interview, July 22, 1996.

Chapter 21: Notes

1 Doug Campbell, Telephone Interview, November 12, 1998.
2 Doug Campbell, Personal Interview, August 6, 1996.
3 Letter to Roger Guiel from Jeanne Walker, September 26, 1979.
4 Bob Parker's Trip Notes, September 10, 1979.
5 Bob Parker's Trip Notes, September 10, 1979.
6 Bob Parker's Trip Notes, September 9-13, 1979.
7 Bob Parker's Trip Notes, September 9-13, 1979.
8 *Roustabout*, volume 100, October 1979, p. 11, 13.
9 Sheila Borsari's diary notes, September 17, 1979.
10 Letter to Sheila Borsari from Jeanne Walker, November 6, 1979.
11 Letter from Gerald Sterns to Sheila Borsari, December 11, 1979.
12 Personal correspondence from Aberdeen Sheriff's Court, July 19, 1996, as quoted in an article by Sheriff Richard Scott.
13 A. V. Sheehan, *Criminal Procedure in Scotland and France*, p. 110.
14 The Sheriff Courts (Scotland) Act 1907.
15 Jurisdiction was decided by the Continental Shelf (Jurisdictional) Order 1968, which draws the border between Scotland and England, just north of Berwick, along the parallel 55° 50' North.

16 Malcolm T. MacNeill, Personal Interview, May 13, 1996.
17 Malcolm T. MacNeill, Personal Interview, May 13, 1996.
18 Susan R. Moody and Jacqueline Tombs, *Prosecution in the public interest*, p. 29.
19 The Submarine Pipe-lines (Diving Operations) Regulations 1976, section 23 (2 a).
20 The Submarine Pipe-lines (Diving Operations) Regulations 1976, section 23 (2 b).
21 Malcolm T. MacNeill, Telephone Interview, August 13, 2002.
22 Susan R. Moody and Jacqueline Tombs, *Prosecution in the public interest*, p. 24-26.
23 A. V. Sheehan, *Criminal Procedure in Scotland and France*, p. 138.
24 Susan R. Moody and Jacqueline Tombs, *Prosecution in the public interest*, p. 84.
25 Letter from Malcolm T. MacNeill to Sheila Borsari, June 11, 1980.
26 Letter from Malcolm T. MacNeill to Sheila Borsari, June 11, 1980.
27 Affidavit of Shirley Guiel, Civil Case No. 8003347 R, March 31, 1981, p. 4.
28 Gerald Sterns, Telephone Interview, September 16, 1999.
29 Moira Johnston, *The Last Nine Minutes: The Story of Flight 981*, p. 94.
30 Shirley Guiel *et al.*, vs. Infabco Diving Services *et al.*, Wrongful Death Case No. 8003347, United States District Court Central District of California, July 30, 1980, p. 8, 12.
31 Shirley Guiel *et al.*, vs. Infabco Diving Services *et al.*, Wrongful Death Case No. 8003347, United States District Court Central District of California, July 30, 1980, p. 13.
32 Shirley Guiel *et al.*, vs. Infabco Diving Services *et al.*, Wrongful Death Case No. 132961, Superior Court of the State of California in and for the county of Santa Barbara, July 30, 1980, p. 15.
33 Letter to Sheila Borsari from Jeanne Walker, October 14, 1980.
34 Jeanne Walker, Personal Interview, August 7, 1996.
35 MacNeill had listed the violation as Section 5 (3)(f) of The Submarine Pipe-lines (Diving Operations) Regulations 1976, but in fact the passage that deals with drop weights is Section 16 (4) schedule 2 Section 3 (b).
36 Criminal Procedure (Scotland) Act 1975, Section 67.
37 *Press and Journal*, November 29, 1980.

Chapter 22: Notes

1 Fenton Wyness, *City by the Grey North Sea: Aberdeen*, p. 197.
2 Francis H. Groome, *Ordnance Gazetteer of Scotland*, p. 18; Frances McDonnell, *The Adventures of Peter Williamson*, foreword. The events occurred between 1740-1744; Fenton Wyness, *City by the Grey North Sea: Aberdeen*, p. 214-215.
3 Alan Bain, Sheriff Clerk Depute, Aberdeen, Notes on "Open Doors Day," p. 1.
4 A. V. Sheehan, *Criminal Procedure in Scotland and France*, p. 106.
5 Robin M. White and Ian D. Willock, *The Scottish Legal System*, p. 233-234.
6 A. V. Sheehan, *Criminal Procedure in Scotland and France*, p. 106.
7 The Fatal Accident Inquiry of Clay Don Ellis FAI, June 29, 1977; and the prosecution of K.D. Marine, November 7, 1977.
8 Minutes of Proceedings, H. M. Advocate v. Infabco Diving Services, Limited.
9 *Press and Journal*, December 19, 1980; Minutes of Proceedings, H. M. Advocate v. Infabco Diving Services, Limited.
10 A. V. Sheehan, *Criminal Procedure in Scotland and France*, p. 160.

11 A. V. Sheehan, *Criminal Procedure in Scotland and France*, p. 167-168.
12 Norman Harper, Personal Interview, May 30, 1996.
13 Colin J. Sanderson's Trial Notes, p. 22.
14 Colin J. Sanderson's Trial Notes, p. 9, 11.
15 Colin J. Sanderson's Trial Notes, p. 2, 24.
16 *Glasgow Herald*, December 17, 1980; Colin J. Sanderson's Trial Notes, p. 20.
17 Colin J. Sanderson's Trial Notes, p. 17-18.
18 Malcolm T. MacNeill, Personal Interview, May 13, 1996; Jackie Warner and Fred Park, *Requiem for a Diver*, p. 55, 61.
19 Malcolm T. MacNeill, Personal Interview, May 13, 1996. MacNeill says he was aware of the bankers (Lazard Brothers in Jersey) before the trial, but not Offshore Co-ordinators.
20 Malcolm T. MacNeill, Personal Interview, May 13, 1996.
21 Colin J. Sanderson's Trial Notes, p. 7.
22 Colin J. Sanderson's Trial Notes, p. 7-8, 28; Walker/Guiel FAI, p. 138. The letter was dated June 14, 1979.
23 SI 923, The Submarine Pipe-lines (Diving Operations) Regulations 1976, regulation 5 (2).
24 Colin J. Sanderson's Trial Notes, p. 8.
25 Colin J. Sanderson's Trial Notes, p. 8.
26 Colin J. Sanderson's Trial Notes, p. 28. The letter was dated July 13, 1979.
27 W. G. Carson, *The Other Price of Britain's Oil*, p. 278-279.
28 Colin J. Sanderson's Trial Notes, p. 27.
29 *Press and Journal*, December 19, 1980.
30 Colin J. Sanderson's Trial Notes, p. 27.
31 Colin J. Sanderson's Trial Notes, p. 28.
32 Colin J. Sanderson's Trial Notes, p. 28.
33 Colin J. Sanderson's Trial Notes, p. 3; Walker/Guiel FAI, p. 165.
34 Telex from Joe Singletary, June 1, 1979.
35 SI 923, The Submarine Pipe-lines (Diving Operations) Regulations 1976, regulation 3 (2).
36 Stair Memorial Encyclopedia, *The Laws of Scotland*, Volume 9, p. 13.
37 SI 923, The Submarine Pipe-lines (Diving Operations) Regulations 1976, regulation 3 (2).
38 Colin J. Sanderson's Trial Notes, p. 3.
39 Colin J. Sanderson's Trial Notes, p. 28.
40 Colin J. Sanderson's Trial Notes, p. 28.
41 Colin J. Sanderson's Trial Notes, p. 23; Walker/Guiel FAI, p. 1862.
42 From July 10, 1978 to his last entry, June 10, 1979.
43 Colin J. Sanderson's Trial Notes, p. 31.
44 Colin J. Sanderson's Trial Notes, p. 29.
45 Colin J. Sanderson's Trial Notes, p. 28.
46 Colin J. Sanderson's Trial Notes, p. 28.
47 Colin J. Sanderson's Trial Notes, p. 31.
48 Malcolm T. MacNeill, Personal Interview, May 13, 1996.
49 Colin J. Sanderson's Trial Notes, p. 31.

50 Colin J. Sanderson's Trial Notes, p. 27.
51 Colin J. Sanderson's Trial Notes, p. 31.
52 Colin J. Sanderson's Trial Notes, p. 30.
53 Ready Mixed Concrete (South East) Ltd. v. Minister of Pensions and National Insurance.
54 Norman Harper, Personal Interview, May 30, 1996.
55 Graeme Smith, Personal Interview, May 31, 1996.
56 Jackie Warner and Fred Park, *Requiem for a Diver,* p. 61.
57 Jackie Warner and Fred Park, *Requiem for a Diver*, p. 61.
58 W. G. Carson, *The Other Price of Britain's Oil*, p. 277-278; Malcolm T. MacNeill, Personal Interview, May 13, 1996.
59 Baldwin/Holmes FAI, March 30, 1977, p. 10.
60 Howell FAI, June 6, 1977, p. 8.
61 Paycheck history of Sub Sea diver Eugene "Skip" Gordon, check no. 52 issued 7/21/77.
62 Ellis FAI, June 29, 1977, p. 7-8, 11-13.
63 Ellis FAI, June 29, 1977, p. 37-38; *Press and Journal* July 2, 1977.
64 Ellis FAI, June 29, 1977, p. 315.
65 Ellis FAI, June 29, 1977, p. 315.
66 *Glasgow Herald*, June 24, 1981.
67 Malcolm Williams, Personal Interview, September 23, 1998.
68 Walker/Guiel FAI, p. 26-28, 55, 169.
69 Malcolm Williams, Personal Interview, September 23, 1998.
70 Transcription of a Tape Recording between BNOC legal advisor Bob McNaught/Frank Surrell and Chris Roper/Randy Gunter, at St. Machar House, Aberdeen, August 10, 1979.
71 Chris Roper, Personal Interview, November 9, 1995.
72 William M'Millan, *Scottish Symbols*, p. 152.
73 Walker/Guiel FAI, p. 158, 190.

Chapter 23: Notes

1 Letter from Chris Roper to Sheila Borsari, December 21, 1980.
2 Jeanne Walker, Personal Interview, August 7, 1996.
3 Jeanne Walker, Personal Interview, August 7, 1996.
4 Walker/Guiel FAI, p. 30. The letter was dated December 20, 1978.
5 Letter from Infabco Diving Services to Skip Guiel, June 22, 1979.
6 Jeanne wrote to and received responses from Roger Palmer, John Nadeau, Ron Arscott, Tom Almon, and Dave Mitchell.
7 Letter from Sheila Borsari to Malcolm MacNeill, January 14, 1981.
8 Letter from Malcolm MacNeill to Sheila Borsari, January 20, 1981.
9 1981 diary of John Macadam, Barge Master of the *Uncle John*.
10 Letter from John Macadam to the author, October 23, 1996.
11 *Press and Journal,* January 22, 1981; 1981 diary of John Macadam.
12 Norwegian Underwater Institute Report No. 2-80: *Project Polar Bear I: Testing of diver thermal protection in a simulated "lost bell."*
13 Ric Wharton, Personal Interview, May 29, 1996.

14 Malcolm Williams, Telephone Interview, June 5, 1996.
15 Gerald Sterns, Telephone Interview, September 16, 1999.
16 Affidavit preparation of Roger Guiel.
17 Affidavit preparation of Shirley Guiel.
18 Letter from the Law Offices of Gerald Sterns to Sheila Borsari, September 29, 1981.
19 Letter from the Law Offices of Gerald Sterns to Sheila Borsari, September 29, 1981.
20 Letter from Malcolm MacNeill to Sheila Borsari, April 15, 1981.

Chapter 24: Notes

1 Reg Christie, Personal Interview, May 30, 1996.
2 Letter from Reg Christie to the author, September 5, 1995.
3 Letter from Reg Christie to Jeanette Shipman, June 23, 1981.
4 Ian Carmichael, *Sudden Deaths and Fatal Accident Inquiries*, p. 362.
5 Ian Carmichael, *Sudden Deaths and Fatal Accident Inquiries*, p. 349..
6 Jeanne Walker, Personal Interview, August 7, 1996.
7 Walker/Guiel FAI, p. 10-11.
8 Walker/Guiel FAI, p. 17.
9 Colin J. Sanderson's Trial Notes, p. 28-31
10 Walker/Guiel FAI, p. 137, 163.
11 Walker/Guiel FAI, p. 145, 150.
12 Walker/Guiel FAI, p. 157, 358-359.
13 Harry Street, *The Law of Torts*, p. 418.
14 Walker/Guiel FAI, p. 161-162.
15 Walker/Guiel FAI, p. 164.
16 Walker/Guiel FAI, p. 172.
17 Walker/Guiel FAI, p. 189.
18 Walker/Guiel FAI, p. 191.
19 Walker/Guiel FAI, p. 366-367.
20 Walker/Guiel FAI, p. 368.
21 Walker/Guiel FAI, p. 368.
22 Walker/Guiel FAI, p. 379.
23 Walker/Guiel FAI, p. 380-381.
24 Walker/Guiel FAI, p. 382-383.
25 Walker/Guiel FAI, p. 384.
26 Walker/Guiel FAI, p. 384-385.
27 A. V. Sheehan, *Criminal Procedure in Scotland and France*, p. 163; Ian Carmichael, *Sudden Deaths and Fatal Accident Inquiries*, p. 78-79.
28 Walker/Guiel FAI, p. 582.
29 Sheriff Douglas James Risk, Determinations of the Walker/Guiel FAI, p. 3.
30 Sheriff Douglas James Risk, Determinations of the Walker/Guiel FAI, p. 3-4.
31 Walker/Guiel FAI, p. 24.
32 Walker/Guiel FAI, p. 126-127.
33 Walker/Guiel FAI, p. 915-916.
34 Sheriff Douglas James Risk, Determinations of the Walker/Guiel FAI, p. 2.
35 Walker/Guiel FAI, p. 259; *Glasgow Herald* May 13, 1981.
36 Walker/Guiel FAI, p. 1601-1602.

37 Walker/Guiel FAI, p. 1593.
38 Walker/Guiel FAI, p. 1522.
39 Walker/Guiel FAI, p. 621-622.
40 Walker/Guiel FAI, p. 623-624.
41 Walker/Guiel FAI, p. 624.
42 Walker/Guiel FAI, p. 1021, 1088.
43 Walker/Guiel FAI, p. 1092.
44 Walker/Guiel FAI, p. 1170-1170, 1307.
45 Walker/Guiel FAI, p. 1308-1309.
46 Walker/Guiel FAI, p. 754.
47 Walker/Guiel FAI, p. 1891.
48 Roger Guiel, Personal Interview, July 21, 1996.
49 Walker/Guiel FAI, p. 533, 574.
50 Walker/Guiel FAI, p. 1750.
51 Walker/Guiel FAI, p. 600-601.
52 Walker/Guiel FAI, p. 575.
53 Walker/Guiel FAI, p. 1483-1484.
54 Walker/Guiel FAI, p. 1549.
55 Walker/Guiel FAI, p. 1547-1550.
56 Walker/Guiel FAI, p. 1542.
57 Walker/Guiel FAI, p. 1607-1609.
58 *Evening Express*, June 23, 1981.
59 *Glasgow Herald*, June 24, 1981.
60 *Evening Express*, June 23, 1981.
61 *Press and Journal*, June 24, 1981.
62 *Press and Journal*, June 24, 1981.
63 *Glasgow Herald*, June 24, 1981.
64 *Evening Express*, June 23, 1981.
65 David M. Walker, *The Scottish Legal System*, p. 287.
66 Ian Carmichael, *Sudden Deaths and Fatal Accident Inquiries*, p. 351-357.
67 *Glasgow Herald,* June 24, 1981; *Press and Journal*, June 24, 1981.
68 Ian Carmichael, *Sudden Deaths and Fatal Accident Inquiries*, p. 357.
69 Ian Carmichael, *Sudden Deaths and Fatal Accident Inquiries*, p. 127.
70 Sheriff Douglas James Risk, Determinations of the Walker/Guiel FAI, p. 2.
71 Sheriff Douglas James Risk, Determinations of the Walker/Guiel FAI, p. 9.
72 Sheriff Douglas James Risk, Determinations of the Walker/Guiel FAI, p. 9.
73 Sheriff Douglas James Risk, Determinations of the Walker/Guiel FAI, p. 13.
74 Sheriff Douglas James Risk, Determinations of the Walker/Guiel FAI, p. 20.
75 Sheriff Douglas James Risk, Determinations of the Walker/Guiel FAI, p. 18.
76 Sheriff Douglas James Risk, Determinations of the Walker/Guiel FAI, p. 5.
77 Sheriff Douglas James Risk, Determinations of the Walker/Guiel FAI, p. 21.
78 Sheriff Douglas James Risk, Determinations of the Walker/Guiel FAI, p. 19.

Chapter 25: Notes

1 Bryan Gould, Telephone Interview, March 19, 2000.
2 Bryan Gould, Telephone Interview, March 19, 2000.

3 Joe Singletary, Telephone Interview, October 7, 1998.
4 Thames Television, TV Eye documentary, *The Last Dive*, December 17, 1981.
5 Ed Goltz, Personal Interview, August 7, 1996.
6 Ken Duell, Telephone Interview, October 29, 1996.
7 Ken Duell, Telephone Interview, October 29, 1996.
8 Ken Duell, Telephone Interview, October 29, 1996.
9 Ken Duell, Telephone Interview, October 29, 1996.
10 Ken Duell, Telephone Interview, October 29, 1996.
11 Ken Duell, Telephone Interview, October 29, 1996.
12 Malcolm Williams, Personal Interview, September 23, 1998.
13 Telex from Joe Singletary, June 1, 1979.
14 Letter from Bryan Gould to Mike Townson, November 25, 1981.
15 Walker/Guiel FAI, p. 185.
16 Richard Walker's diary.
17 Malcolm Williams, Telephone Interview, June 5, 1996.
18 Ed Goltz, Personal Interview, August 7, 1996.
19 Ed Goltz, Personal Interview, August 7, 1996.
20 Letter from Bryan Gould to Mike Townson, November 25, 1981; Bryan Gould, Telephone Interview, March 19, 2000.
21 Roger Chapman, *No Time On Our Side*, p. 66.
22 Roger Chapman, *No Time On Our Side*, p. 90.
23 Roger Chapman, *No Time On Our Side*, p. 166.
24 Roger Chapman, *No Time On Our Side*, p. 72-73.
25 *Press and Journal*, August 31, 1973.
26 *Press and Journal*, August 31, 1973.
27 Andre van Heerden, Telephone Interview, March 18, 2003.
28 *Press and Journal*, March 10, 1978.
29 Andre van Heerden, Telephone Interview, March 18, 2003.
30 *Press and Journal*, March 11, 1978.
31 Walker/Guiel FAI, p. 501.
32 Walker/Guiel FAI, p. 503.
33 Walker/Guiel FAI, p. 586-587.
34 Walker/Guiel FAI, p. 504.
35 Walker/Guiel FAI, p. 585.
36 Walker/Guiel FAI, p. 561, 585.
37 Walker/Guiel FAI, p. 1409.
38 Walker/Guiel FAI, p. 559-562.
39 Walker/Guiel FAI, p. 512.
40 Walker/Guiel FAI, p. 500.
41 The Burgoyne Report, *Offshore Safety*, p. 262.
42 The Burgoyne Report, *Offshore Safety,* p. 141.
43 Walker/Guiel FAI, p. 515.
44 Letter from Bernard Chauvelin to the author, November 12, 1996.
45 Walker/Guiel FAI, p. 515.
46 Walker/Guiel FAI, p. 512.
47 Walker/Guiel FAI, p. 514.

48 Walker/Guiel FAI, p. 598, 1125. At the inquiry Mr. Masterson could not recall the suggestion being made to him by the *Stena Welder*.
49 Walker/Guiel FAI, p. 1049.
50 Walker/Guiel FAI, p. 559-560.
51 Walker/Guiel FAI, p. 506; Chuck Alexander, Telephone Interview, February 24, 1996; Frank Surrell, Personal Interview, August 8, 1996.
52 Walker/Guiel FAI, p. 478, 499, 505, 507, 513-514, 516, 524, 587, 643-644.
53 Walker/Guiel FAI, p. 499, 516.
54 Walker/Guiel FAI, p. 506, 514.
55 Walker/Guiel FAI, p. 478.
56 Walker/Guiel FAI, p. 524.
57 Walker/Guiel FAI, p. 507.
58 Walker/Guiel FAI, p. 504-506, 599.
59 Ric Wharton, Personal Interview, May 29, 1996.
60 The Burgoyne Report, *Offshore Safety*, p. 262.
61 Walker/Guiel FAI, p. 443.
62 Walker/Guiel FAI, p. 448, 563.
63 Allan Mellor, Telephone Interview, February 29, 2000.
64 Telex from Joe Singletary, June 1, 1979. The telex indicates three reimbursable rates: Working day rate, Standby day rate, Weather standby day rate.
65 Ric Wharton, Personal Interview, May 29, 1996.
66 Ric Wharton, Personal Interview, May 29, 1996.
67 Yvonne Duell, Telephone Interview, September 13, 1997.
68 Ken Duell, Telephone Interview, October 29, 1996.
69 Ric Wharton, Personal Interview, May 29, 1996.
70 *Brown Book*, 1980, p. 50-52.
71 Guy Arnold, *Britain's Oil*, p. 160; *Press and Journal*, July 27, 1979.
72 See Chapter 4.
73 Guy Arnold, *Britain's Oil*, p. 160.
74 Frank Surrell, Personal Interview, August 8, 1996.
75 Malcolm Williams, Personal Interview, September 23, 1998.
76 Edward (Jack) Geer, Telephone Interview, May 18, 1998.
77 Chuck Alexander, Telephone Interview, October 3, 1998; Frank Surrell, Personal Interview, August 8, 1996.
78 Joe Singletary, Telephone Interview, October 7, 1998.
79 E-mail from Allan Mellor to the author, May 4, 2000.
80 Chuck Alexander, Personal Interview, July 26, 1998.
81 Terry Slusher declined to be interviewed, April 22, 1997.
82 Joe Singletary, Telephone Interview, October 7, 1998.
83 Joe Singletary, Telephone Interview, October 7, 1998.
84 Bill Philliber, Telephone Interview, October 16, 1998.
85 Bill Philliber, Telephone Interview, November 12, 1998.
86 Don Shimmon, Telephone Interview, October 20, 1998.
87 Bill Philliber, Telephone Interview, February 26, 2000.
88 Joe Singletary, Telephone Interview, October 7, 1998.
89 Bill Philliber, Telephone Interview, October 16, 1998.

90 Joe Singletary, Telephone Interview, October 7, 1998.
91 Bill Philliber, Telephone Interview, October 16, 1998.
92 Don Shimmon, Telephone Interview, October 20, 1998.
93 Joe Singletary, Telephone Interview, October 7, 1998.
94 The Burgoyne Report, *Offshore Safety*, p. 195.
95 Harry Street, *The Law of Torts*, p. 124.
96 Bryan Gould, Telephone Interview, March 19, 2000.

Chapter 26: Notes

1 W. G. Carson, *The Other Price of Britain's Oil*, p. ix.
2 W. G. Carson, *The Other Price of Britain's Oil*, p. 91.
3 W. G. Carson, *The Other Price of Britain's Oil*, p. 9.
4 *Brown Book*, 1977, p. 2.
5 Peter Hennessy, *Whitehall*, p. 446.
6 W. G. Carson, *The Other Price of Britain's Oil*, p. 166-167.
7 W. G. Carson, *The Other Price of Britain's Oil*, p. 165.
8 W. G. Carson, *The Other Price of Britain's Oil*, p. 173.
9 Charles Woolfson, John Foster, and Matthias Beck, *Paying for the Piper*, p. 265.
10 Charles Woolfson, John Foster, and Matthias Beck, *Paying for the Piper*, p. 266.
11 W. G. Carson, *The Other Price of Britain's Oil*, p. 203.
12 W. G. Carson, *The Other Price of Britain's Oil*, p. 175.
13 The Burgoyne Report, *Offshore Safety*, p. 58.
14 W. G. Carson, *The Other Price of Britain's Oil*, p. 209.
15 W. G. Carson, *The Other Price of Britain's Oil*, p. 209.
16 W. G. Carson, *The Other Price of Britain's Oil*, p. 247-248.

Chapter 27: Notes

1 Jackie Warner, and Fred Park, *Requiem for a Diver*, p. 2-3.
2 Jackie Warner, and Fred Park, *Requiem for a Diver*, p. 6.
3 Jackie Warner, and Fred Park, *Requiem for a Diver*, p. 92.
4 Jackie Warner, and Fred Park, *Requiem for a Diver*, p. 92.
5 Jackie Warner, and Fred Park, *Requiem for a Diver*, p. 93.
6 In the six years the author worked offshore in the North Sea, contact with a Diving Inspector occurred only once.
7 Jackie Warner, and Fred Park, *Requiem for a Diver*, p. 8.
8 The Burgoyne Report, *Offshore Safety*, p. 230.
9 Walker/Guiel FAI, p. 337-338.
10 Walker/Guiel FAI, p. 339.
11 The Burgoyne Report, *Offshore Safety*, p. 292.
12 The Burgoyne Report, *Offshore Safety*, p. 117.
13 The Burgoyne Report, *Offshore Safety*, p. 111.
14 The Burgoyne Report, *Offshore Safety*, p. 232.
15 Howell FAI, p. 6.
16 Howell FAI, p. 66.
17 Howell FAI, p. 180.
18 Howell FAI, p. 171.

19 Howell FAI, p. 73, 181.
20 Howell FAI, p. 74.
21 Howell FAI, p. 182.
22 Howell FAI, p. 102.
23 Howell FAI, p. 184.
24 Howell FAI, p. 184.
25 Howell FAI, p. 206.
26 Howell FAI, p. 194, 199.
27 Howell FAI, p. 184.
28 Howell FAI, p. 79.
29 Howell FAI, p. 80, 165.
30 Howell FAI, p. 83.
31 Howell FAI, p. 86, 108.
32 Howell FAI, p. 165.
33 Howell FAI, p. 86, 166.
34 Howell FAI, p. 83, 167.
35 Howell FAI, p. 187.
36 Howell FAI, p. 186-187.
37 Howell FAI, p. 165, 168.
38 Howell FAI, p. 128.
39 Howell FAI, p. 188.
40 Howell FAI, p. 172.
41 Howell FAI, p. 198.
42 Howell FAI, p. 89-91.
43 Howell FAI, p. 215.
44 Howell FAI, p. 216.
45 Howell FAI, p. 170.
46 Howell FAI, p. 215.
47 Howell FAI, p. 216-217.
48 Howell FAI, p. 216.
49 Howell FAI, p. 216, 219.
50 Howell FAI, p. 195.
51 Howell FAI, p. 225.
52 Howell FAI, p. 164-165.
53 Howell FAI, p. 84, 118-119.
54 Howell FAI, p. 208.
55 Howell FAI, p. 194.
56 Howell FAI, p. 224.
57 Howell FAI, p. 225.
58 Howell FAI, p. 131,146, 149.
59 Howell FAI, p. 137-138.
60 Howell FAI, p. 121,187, 190.
61 Howell FAI, p. 223.
62 Howell FAI, p. 73, 78.
63 Howell FAI, p. 74.
64 Howell FAI, p. 223.

65 Howell FAI, p. 206.
66 Howell FAI, p. 191.
67 Howell FAI, p. 190.
68 Howell FAI, p. 148.
69 Howell FAI, p. 63.
70 Howell FAI, p. 220.
71 Health and Safety records, June 23, 2000; also retold in Jackie Warner, and Fred Park, *Requiem for a Diver*, p. 50; and W. G. Carson, *The Other Price of Britain's Oil*, p. 67.

Chapter 28: Notes

1 Commander "Jackie" Warner, Personal Interview, May 4, 1996.
2 Jackie Warner, and Fred Park, *Requiem for a Diver*, p. 8.
3 W. G. Carson, *The Other Price of Britain's Oil*, p. 6.
4 W. G. Carson, *The Other Price of Britain's Oil*, p. 152-153.
5 The Burgoyne Report, *Offshore Safety*, p. 228.
6 Jackie Warner, and Fred Park, *Requiem for a Diver*, p. 92.
7 Commander "Jackie" Warner, Personal Interview, May 4, 1996.
8 Ellis FAI, p. 36, 216.
9 Merchant Shipping Act 1974, Section 16 (1) (b); Confirmed in an e-mail from the Health and Safety Executive to the author, April 5, 2001.
10 Ellis FAI, p. 297-300.
11 The Burgoyne Report, *Offshore Safety*, p. 35.
12 The Burgoyne Report, *Offshore Safety*, p. 228.
13 The Burgoyne Report, *Offshore Safety*, p. 228.
14 W. G. Carson, *The Other Price of Britain's Oil*, p. 134, 178-179.
15 Commander "Jackie" Warner, Personal Interview, May 4, 1996.
16 Department of Energy, Diving Safety Memorandum 10/1975.
17 Department of Energy, Diving Safety Memorandum 8/1976.
18 Letter to the author from the Health and Safety Executive, March 24, 2000.
19 Commander "Jackie" Warner, Personal Interview, May 4, 1996.
20 Department of Energy, Diving Safety Memorandum 7/1975.
21 Department of Energy, Diving Safety Memorandum 2/1976.
22 See Appendix.
23 Jackie Warner, "Diving Bell Design: Safe deployment, operation and recovery," *Underwater Systems Design*, April/May 1981, p. 11.
24 Clarke FAI, p. 46.
25 Clarke FAI, p. 31, 34, 35, 65, 66, 110, 130.
26 Mineral Workings (Offshore Installations) Act 1971, Section 7 (1).
27 The Offshore Installations (Diving Operations) Regulations 1974, Regulation 6 (1) (b) (i), (iii).
28 The Offshore Installations (Diving Operations) Regulations 1974, Regulation 16 2 (d).
29 The Offshore Installations (Diving Operations) Regulations 1974, Regulation 19 (a).

30 Meehan/Spensley FAI, p. 7, 15, 24.
31 Moore FAI, p. 70-71.
32 The Offshore Installations (Diving Operations) Regulations 1974, Regulation 19 (a).
33 Department of Energy, Diving Safety Memorandum 5/1977.
34 *Press and Journal*, September 14, 1977.
35 *Press and Journal*, September 14, 1977.
36 Jackie Warner, and Fred Park, *Requiem for a Diver*, p. 51.
37 AODC document 065.
38 Jackie Warner, and Fred Park, *Requiem for a Diver*, p. 33.
39 Jackie Warner, "Diving Bell Design: Safe deployment, operation and recovery," *Underwater Systems Design*, April/May 1981, p. 11; Jackie Warner, and Fred Park, *Requiem for a Diver*, p. 33.
40 Jackie Warner, and Fred Park, *Requiem for a Diver*, p. 33.
41 *Press and Journal*, September 14, 1977.
42 *Press and Journal* September 13, 1977.

Chapter 29: Notes

1 Prangley/Ward FAI, p. 475-476, 512.
2 *Brown Book,* 1980, p. 31.
3 Prangley/Ward FAI, p. 158, 180.
4 Prangley/Ward FAI, p. 180.
5 Prangley/Ward FAI, p. 176.
6 Prangley/Ward FAI, p. 123, 143.
7 Prangley/Ward FAI, p. 393-394.
8 Prangley/Ward FAI, p. 459-460.
9 Prangley/Ward FAI, p. 138, 460, 463.
10 Prangley/Ward FAI, p. 142.
11 Prangley/Ward FAI, p. 468.
12 Prangley/Ward FAI, p. 468-469.
13 Prangley/Ward FAI, p. 134.
14 Prangley/Ward FAI, p. 508.
15 Prangley/Ward FAI, p. 508.
16 Prangley/Ward FAI, p. 475, 511.
17 Prangley/Ward FAI, p. 475.
18 Prangley/Ward FAI, p. 483-484.
19 Prangley/Ward FAI, p. 483.
20 Prangley/Ward FAI, p. 374.
21 Prangley/Ward FAI, p. 431.
22 Department of Energy, Diving Safety Memorandum 2/1979.
23 Prangley/Ward FAI, p. 553.
24 Prangley/Ward FAI, p. 547-548.
25 Prangley/Ward FAI, p. 558.
26 Prangley/Ward FAI, p. 559.
27 Prangley/Ward FAI, p. 560.
28 Commander "Jackie" Warner, Personal Interview, May 4, 1996.

29 *The Sunday Times*, November 18, 1979.
30 Lord Robens, *Safety and Health at* Work, p. 64.
31 Jackie Warner and Fred Park, *Requiem for a Diver,* p. 22.

Chapter 30: Notes

1 The title of this chapter is taken from *Discipline and Punish: The Birth of the Prison*, Michel Foucault, p. 78.
2 Susan R. Moody and Jacqueline Tombs, *Prosecution in the public interest*, p. 89.
3 Susan R. Moody and Jacqueline Tombs, *Prosecution in the public interest*, p. 59.
4 W. G. Carson, *The Other Price of Britain's Oil*, p. 231-232.
5 W. G. Carson, *The Other Price of Britain's Oil*, p. 232.
6 W. G. Carson, *The Other Price of Britain's Oil*, p. 7.
7 W. G. Carson, *The Other Price of Britain's Oil*, p. 250.
8 W. G. Carson, *The Other Price of Britain's Oil*, p. 266.
9 W. G. Carson, *The Other Price of Britain's Oil*, p. 266-267.
10 W. G. Carson, *The Other Price of Britain's Oil*, p. 266.
11 W. G. Carson, *The Other Price of Britain's Oil*, p. 250.
12 W. G. Carson, *The Other Price of Britain's Oil*, p. 266.
13 The Burgoyne Report, *Offshore Safety*, p. 61.
14 *The Chronicle* p. 1093.
15 Jackie Warner and Fred Park, *Requiem for a Diver,* p. 33.
16 W. G. Carson, *The Other Price of Britain's Oil*, p. 248-249.
17 *Brown Book,* 1982, p. 60.
18 W. G. Carson, *The Other Price of Britain's Oil*, p. 267.
19 W. G. Carson, *The Other Price of Britain's Oil*, p. 267-268.
20 *Glasgow Herald,* June 24, 1981.
21 Susan R. Moody and Jacqueline Tombs, *Prosecution in the public interest*, p. 32, 82.
22 Gareth Watkins, *The Health and Safety Handbook*, p. 3.
23 Case No. T960640. *Regina v. Kenneth Roberts*, July 2, 1997.
24 *Press and Journal,* May 10, 1985.
25 *Glasgow Herald*, June 24, 1981.
26 The Burgoyne Report, *Offshore Safety*, p. 229.
27 *The Sunday Times*, November 18, 1979.
28 Walker/Guiel FAI p. 1936.
29 Walker/Guiel FAI p. 1871.
30 W. G. Carson, *The Other Price of Britain's Oil*, p. 234.
31 Letter from George Arnoux to the author September 20, 1996.

Chapter 31: Notes

1 Gerald Sterns, Telephone Interview, September 16, 1999.
2 Letter from Sterns's office to Sheila, Roger and Shirley Guiel, September 29, 1981.
3 Peter Anderson, Personal Interview, May 15, 1996.
4 Execution of Summons, July 26, 1982.
5 SI 923, The Submarine Pipe-lines (Diving Operations) Regulations 1976, regulation 3(2).

6 Court of Session, Scotland, Summons, Reference CS 258/4473/1989, July 23, 1982, p. 4.
7 Summons, p. 7.
8 Summons, p. 8.
9 Summons, p. 9.
10 Summons, p. 10.
11 Summons, p. 10.
12 Summons, p. 11.
13 Defences, Infabco Diving Services Ltd, October 22, 1982, p. 2.
14 Defences, IDS, October 22, 1982, p. 3, 6-7.
15 Defences, IDS, October 22, 1982, p. 7.
16 Defences, IDS, October 22, 1982, p. 5.
17 Defences, IDS, October 22, 1982, p. 3.
18 Court of Session, *Answers*, IDS, July 29, 1986, p. 5.
19 Defences, IDS, October 22, 1982, p. 5.
20 Defences, IDS, October 22, 1982, p. 5.
21 Defences, IDS, October 22, 1982, p. 8.
22 Defences, Møllerodden, October 21, 1982, p. 2; Defenses, Wilhelmsen, October 1982, p. 4; Defences, BNOC, October 1982, p. 8.
23 *Press and Journal*, January 26, 1983.
24 *Press and Journal*, February 10, 1983.
25 *Press and Journal*, April 6, 1983.
26 Letter from Peter Anderson to Jeanne Walker, April 5, 1983.
27 Jeanne Walker, Personal Interview, March 2, 1999.
28 Tom Belcher, Telephone Interview, March 15, 1999.
29 Letter from Bill Hemmings to Jeanne Walker, February 29, 1980.
30 Letter from Bill Hemmings to Jeanne Walker, June 4, 1980.
31 Court of Session, Closed Record, November 9, 1983, p 21.
32 Rick and Sheila Borsari, Personal Interview, July 20, 1996.
33 Jeanne Walker, Personal Interview, March 2, 1999.
34 Gerald Sterns, Telephone Interview, September 16, 1999.
35 Gerald Sterns, Telephone Interview, September 16, 1999.
36 Court of Session, *Answers*, IDS, July 29, 1986, p. 6.
37 Letter from Philip J. Stuto (President of Investigative Consultants International, Inc.) to Gerald Sterns, August 18, 1986.
38 Rod Pelzel, Personal Interview, November 12, 1995.
39 Telex from Simpson and Marwick to Don Cox, September 20, 1986.
40 Court of Session, *Minute of Tender*, September 22, 1986, p. 2-3.
41 Court of Session, *Answers*, Møllerodden, 1986, p. 1.
42 Court of Session, *Minute*, October 3, 1986, p. 2.
43 Court of Session, *Minute*, October 7, 1986, p. 2.
44 Telex from Simpson and Marwick to Don Cox, October 14, 1986.
45 Sheila Borsari, Telephone Interview, January 20, 1999.
46 E-mail correspondence from the Health and Safety Executive to the author, May 26, 2000.
47 *New York Times*, November 7, 1991.

48 *New York Times*, November 6, 1991.
49 *Forbes*, March 30, 1992, p. 44.
50 *Panorama* Transcript. *How Safe is your Pension*, June 29, 1992, p. 2.
51 *Panorama* Transcript. *How Safe is your Pension*, June 29, 1992, p. 2-3.
52 *Panorama* Transcript. *How Safe is your Pension*, June 29, 1992, p. 2.
53 Panorama Transcript. How Safe is your Pension, June 29, 1992, p. 5.
54 *Panorama* Transcript. *How Safe is your Pension*, June 29, 1992, p. 5.
55 *Panorama* Transcript. *How Safe is your Pension*, June 29, 1992, p. 6.
56 *Panorama* Transcript. *How Safe is your Pension*, June 29, 1992, p. 3.
57 *Panorama* Transcript. *How Safe is your Pension*, June 29, 1992, p. 3.
58 *Panorama* Transcript. *How Safe is your Pension*, June 29, 1992, p. 4.
59 *Panorama* Transcript. *How Safe is your Pension*, June 29, 1992, p. 4.
60 *Panorama* Transcript. *How Safe is your Pension*, June 29, 1992, p. 4.
61 *Panorama* Transcript. *How Safe is your Pension*, June 29, 1992, p. 3.
62 Ken Duell, Telephone Interview, October 29, 1996.

Appendix: Notes

1 Walker/Guiel FAI, p. 1191, 1197-1198, 1280.
2 Walker/Guiel FAI, p. 1392-1395, 1398, 1465-1467.
3 Walker/Guiel FAI, p. 1580.
4 Walker/Guiel FAI, p. 1177, 1191, 1280.
5 John Brook, Personal Interview, May 3, 1996.
6 Colin J. Sanderson's Trial Notes, p. 19.
7 Tim Newson, Personal Interview, May 28, 1996.
8 Personal Interview. Gary Legere, September 23, 1995, April 4, 1998.
9 Walker/Guiel FAI, p. 809.
10 Walker/Guiel FAI, p. 1280.
11 Walker/Guiel FAI, p. 667.
12 Walker/Guiel FAI, p. 1467.
13 Walker/Guiel FAI, p. 1395.
14 Walker/Guiel FAI, p. 1669-1670.
15 Walker/Guiel FAI, p. 667.

Bibliography and Works Cited

Books

Arnold, Guy. *Britain's Oil.* London: Hamish Hamilton, 1978.

Bathurst, Bella. *The Lighthouse Stevensons.* New York: HarperCollins, 1999.

Benn, Tony. *Conflicts of Interest: Diaries 1977-80.* London: Hutchinson, 1990.

Bond, Captain George F. *Papa Topside.* Annapolis: Naval Institute Press, 1993.

Bove, Alfred A., and Jefferson C. Davis. *Diving Medicine.* Philadelphia: W. B. Saunders Company, 1990.

Carmichael, Ian. *Sudden Deaths and Fatal Accident Inquiries: Scots Law and Practice.* Edinburgh: W. Green/Sweet & Maxwell, 1993.

Carson, W. G. *The Other Price of Britain's Oil.* New Brunswick: Rutgers University Press, 1982.

Chapman, Roger. *No Time On Our Side.* Sidney: Gray's Publishing, 1975.

Foucault, Michel. *Discipline and Punish: The Birth of the Prison.* New York: Pantheon Books, 1977.

Groome, Francis H. *Ordnance Gazetteer of Scotland.* London: The Caxton Publishing Company, 1903.

Hamilton, Adrian. *Oil: The Price of Power.* London: Michael Joseph/Rainbird, 1986.

Hennessy, Peter. *Whitehall.* New York: The Free Press, 1989.

Johnston, Moira. *The Last Nine Minutes: The Story of Flight 981.* New York: William Morrow and Company, Inc., 1976.

Lloyd, Evan L. *Hypothermia and Cold Stress.* Kent: 1986.

M'Millan, William. *Scottish Symbols.* Paisley: Alexander Gardner.

MacKay, D. I., and G. A. Mackay. *The Political Economy of the North Sea.* Boulder: Westview Press, 1975.

McDonnell, Frances. *The Adventures of Peter Williamson.* St. Andrews: McDonnell, 1994.

Moody, Susan R., and Jacqueline Tombs. *Prosecution in the public interest.* Edinburgh: Scottish Academic Press, 1982.

Newberry, W. G. *Handbook for Riggers.* Canada: 1977.

Paterson, John. *Behind the Mask.* Aldershot: Ashgate/Dartmouth, 2000.

Sheehan, A. V. *Criminal Procedure in Scotland and France.* Edinburgh: HMSO, 1975.

Sisman, David. *The Professional Diver's Handbook.* London: Submex, 1982.

Sontag, Sherry, and Christopher Drew. *Blind Man's Bluff.* New York: Public Affairs,1998.

Stair Memorial Encyclopedia, *The Laws of Scotland*, Volume 9.

Stranks, Jeremy. *Health and Safety Law.* London: Financial Times–Prentice Hall, 1998.

Street, Harry. *The Law of Torts.* London: Butterworths, 1976.

The Europa World Year Book, 1996.

US Navy Diving Manual, 1973.

Walker, David M. *The Scottish Legal System.* Edinburgh: W. Green & Son Ltd., 1981.

Warner, Jackie, and Fred Park. *Requiem for a Diver.* Glasgow: Brown, Son & Ferguson, Ltd., 1990.

Watkins, Gareth. *The Health and Safety Handbook.* London: Sweet & Maxwell, 1997.

White, Robin M., and Ian D. Willock. *The Scottish Legal System.* Edinburgh: Butterworths, 1993.

Woolfson, Charles, John Foster, and Matthias Beck. *Paying for the Piper: Captial and Labour in Britain's Offshore Oil Industry.* London: Mansell, 1996.

Wyness, Fenton. *City by the Grey North Sea: Aberdeen.* Aberdeen: Impulse Books, 1972.

Yergin, Daniel. *The Prize.* New York: Simon & Schuster, 1991.

Magazines/periodicals/equipment manuals

Forbes, March 30, 1992.

Hughes, David. *Underwater Systems Design*, April/May 1981.

Karm High Pressure Hydraulic Equipment, *Umbilical Wheel Control Console*, January 18, 1979.

Munro, Sir Leslie. *Hugo Grotius on Freedom of the Seas.* United Nations Review, vol. 4, February 1958.

Ocean Industry, April 1977.

Oil and Gas International Yearbook 1977/78.

Roustabout, April 1978-April 1981, volumes 82, 100, 117.

Time Magazine, *International Law: The Three-Mile Limit*, May 5, 1958.

Van Beest Green Pin Catalogue, *Green Pin: Top Quality Lifting Gear.*

Warner, Jackie. "Diving Bell Design: Safe deployment, operation and recovery," Underwater Systems Design, April/May 1981.

Yearbook of the United Nations 1958, *The Law of the Sea.*

Diaries

Dave Lane's diary.

Don Higham's diary.

John Macadam's diary.

Lacey Lahren's diary.

Richard Walker's diary.

Rod Pelzel's diary.

Sheila Borsari's diary.

Dive logs

Burgess/Wilson BNOC dive log.

Wildrake dive log, BNOC compilation.

Stena Welder dive log.

Dave Lane's dive log.

Eddy Frank's dive log.

John Brook's dive log.

John Joly's dive log.

Roger Palmer's dive log.

Steve Lobban's dive log.

Thistle Alpha Helicopter Movement Report, August 8, 1979.

Thistle Alpha Radio Log.

Wildrake Daily Vessel Report, August 8, 1979.

Notes and personal papers

Alexander, Chuck. *Thistle A, Single Anchor Leg Mooring Project Assessment, March 1977.*

Alexander, Chuck. *Thistle A, Single Anchor Leg Mooring Repair and Reinstallation, November 1979.*

Allen, Mike. Written statement to the Aberdeen Sheriff Court.

Bain, Alan. Sheriff Clerk Depute, Aberdeen. Notes on "Open Doors Day."

Gunter/Almon Police Statement.

Møllerodden design drawings of the *M/S Wildrake* diving system, 1978.

Murray, Stewart: Offshore Installation Manager, Thistle Alpha. OIM notes, August 8, 1979.

Olson, Dave. *Report from Sat Technicians of Their Activities on August 8, 1979.*

Parker, Bob. Trip Notes to Aberdeen, Scotland.

Sanderson, Colin J. Trial Notes.

Transcription of a Tape Recording between BNOC legal advisor Bob McNaught/Frank Surrell and Chris Roper/Randy Gunter, at St. Machar House, Aberdeen, August 10, 1979.

Acts and regulations

Continental Shelf (Jurisdictional) Order 1968.

Continental Shelf Act 1964.

Criminal Procedure (Scotland) Act 1975.

Fatal Accidents and Sudden Deaths Inquiry (Scotland) Act 1906

Fatal Accidents and Sudden Deaths Inquiry (Scotland) Act 1976.

Fatal Accidents Inquiry (Scotland) Act 1895.

Health and Safety at Work etc. Act 1974.

Merchant Shipping Act 1974.

Mineral Workings (Offshore Installations) Act 1971.

Petroleum (Production) Act 1934.

Petroleum and Submarine Pipe-lines Act 1975.

Sheriff Courts (Scotland) Act 1907.

SI 1960/688. Factories: The Diving Operations Special Regulations, 1960.

SI 1972/1542. The Offshore Installations (Logbooks and Registration of Death) Regulations 1972.

SI 1973/1842. The Offshore Installations (Inspectors and Casualties) Regulations 1973.

SI 1974/1229. The Offshore Installations (Diving Operations) Regulations 1974.

SI 1975/116. The Merchant Shipping (Diving Operations) Regulations 1975.

SI 1976/923. The Submarine Pipe-lines (Diving Operations) Regulations 1976.

SI 1977/191 (S. 29) The Fatal Accidents and Sudden Deaths Inquiry Procedure (Scotland) Rules 1977.

SI 923, The Submarine Pipe-lines (Diving Operations) Regulations 1976.

Court documents

Court of Session, *Answers*, IDS, July 29, 1986.

Court of Session, *Answers*, Møllerodden, 1986.

Court of Session, Closed Record, November 9, 1983.

Court of Session, *Minute of Tender*, September 22, 1986.

Court of Session, *Minute*, October 3, 1986.

Court of Session, *Minute*, October 7, 1986.

Court of Session, Summons, Reference CS 258/4473/1989, July 23, 1982.

Court of Session, *Answers by the Fourth Defenders to Minute of Amendment for the Fifth Defenders*, 1986.

Defences, BNOC, October 1982.

Defences, Infabco Diving Services Ltd, October 22, 1982.

Defences, Møllerodden, October 21, 1982.

Defenses, Wilhelmsen, October 1982.

Execution of Summons, July 26, 1982.

H. M. Advocate v. Infabco Diving Services, Limited, December 15, 1980.

H. M. Advocate v. Sub Sea Offshore Limited and Air Products Limited, May 9, 1985

Jeanne Walker et al., vs. Infabco Diving Services et al., Wrongful Death Case No. 8003347, United States District Court Central District of California, July 30, 1980.

Jeanne Walker et al., vs. Infabco Diving Services et al., Wrongful Death Case No. 132961, Superior Court of the State of California in and for the county of Santa Barbara, July 30, 1980.

Ready Mixed Concrete (South East) Ltd. v. Minister of Pensions and National Insurance, 1967.

Regina v. Kenneth Roberts, Case No. T960640. July 2, 1997.

Shirley Guiel et al. vs. Infabco Diving Services et al. Affidavit of Roger Guiel, *Civil Case No. 8003347* R, United States District Court in and for the Central District of California, March 31, 1981.

Shirley Guiel et al. vs. Infabco Diving Services et al. Affidavit of Shirley Guiel, *Civil Case No. 8003347* R, United States District Court in and for the Central District of California, March 31, 1981.

Shirley Guiel et al. vs. Infabco Diving Services et al. Affidavit of Sheila Borsari, *Civil Case No. 8003347* R, United States District Court in and for the Central District of California, March 31, 1981.

Shirley Guiel et al., vs. Infabco Diving Services et al., Wrongful Death Case No. 8003347, United States District Court Central District of California, July 30, 1980.

Shirley Guiel et al., vs. Infabco Diving Services et al., Wrongful Death Case No. 132961, Superior Court of the State of California in and for the county of Santa Barbara, July 30, 1980.

Documentaries

Panorama. *How Safe is your Pension*, June 29, 1992.

Thames Television, TV EYE, *The Last Dive*, December 17, 1981.

Government reports and studies

Burgoyne Report. *Offshore Safety*. Command paper 7841, London: HMSO, 1980.

Department of Energy, *Development of the Oil and Gas Resources of the United Kingdom (Brown Book)*, London: HMSO 1973, 1975, 1977, 1978, 1980.

Department of Energy, *Diving Safety Memorandum* 1974-1979.

Department of Energy, *United Kingdom Offshore Oil and Gas Policy*. Command Paper 5696. London: HMSO, 1974.

Department of Energy. *North Sea Costs Escalation Study*, Energy Paper Number 7. London: HMSO, 1976.

Department of Trade and Industry, *North Sea Oil and Gas: A Report to Parliament.* London: HMSO, 1973.

Golden, Surgeon Lt. Commander Frank. *Hypothermia: a Problem for North Sea Industries.* Journal of Occupational Medicine, 1976.

Golden, Surgeon Lt. Commander Frank. *Recognition and Treatment of Immersion Hypothermia.* Proceedings of the Royal Society of Medicine, Vol. 66, October 1973.

Health and Safety Executive diving records, 1971-1996.

Lord Robens. *Safety and Health at Work*. Command paper 5034, London: HMSO, 1972.

Marcus, P., D. Robertson, and R. Langford. *Metallised Plastic Sheeting for Use in Survival.* Aviation, Space, and Environmental Medicine, 1977.

Norwegian Underwater Institute Report No. 2-80: *Project Polar Bear I: Testing of diver thermal protection in a simulated "lost bell."*

Norwegian Underwater Institute, *Deep-Ex '80, Project II, Thermal Studies in Diving*, December 16, 1980.

Pellow, Don. A report by Pellow Engineering Services, *Failure of the Wildrake Bell Connection*, October 28, 1998.

Scottish Development Department, *North Sea Oil Discussion Paper*, April 1973.

UK Offshore Operators Association Limited, *Exploration and Development of UK Continental Shelf Oil*, Energy Commission Paper Number 17, October 1978.

Fatal Accident Inquiries

Baldwin, Roger, and Peter Henry Michael Holmes. *Transcript of Evidence,* March 30, 1977.

Clarke, John. *Transcript of Evidence,* June 23, 1975.

Dimmer, John. *Transcript of Evidence,* December 2, 1974

Ellis, Clay Don. *Transcript of Evidence,* June 29, 1977

Hoffman, Graig. *Transcript of Evidence,* June 26, 1978.

House, Timothy. *Determinations*, April 29, 1974.

Howell, John. *Transcript of Evidence,* June 6, 1977

Meehan, Charles, and Howard Spensley. *Transcript of Evidence,* January 30, 1978.

Minn, Htun. *Determinations*, March 28, 1972.

Moore, Michael. *Transcript of Evidence,* September 12, 1977.

Prangley, Gerard Anthony, and Lothar Michael Ward. *Transcript of Evidence,* November 14, 1979.

Walker, Richard Arthur, and Victor Francis Guiel Jr. *Transcript of Evidence,* May 11, 1981.

North Sea Fatalities
Feb 1971-Feb 1984
All Sectors

1	Feb-01-71	Michael Lally
2	Mar-01-71	Michael Brushneed
3	Nov-01-71	Htun Min
4	May-01-72	Robert Taylor
5	Aug-28-73	Paul Havlena
6	Dec-01-73	Timothy House
7	Jan-17-74	Pier Skipness
8	Jan-17-74	Robert Smyth
9	Mar-01-74	Mark Bartholemy
10	Mar-30-74	William Norris
11	Jul-05-74	John Dimmer
12	Aug-27-74	P. Kelly
13	Oct-14-74	John Clarke
14	Oct-15-74	Gary Shields
15	Dec-02-74	David Keane
16	Dec-17-74	J. Phillips
17	Feb-06-75	John Martin
18	Mar-01-75	Kevin Wilson
19	Mar-23-75	A. Alvestad
20	Jun-14-75	George Turner
21	Jul-16-75	Peter Walsh
22	Jul-16-75	Peter Carson
23	Sep-09-75	Roger Baldwin
24	Sep-09-75	Peter Holmes
25	1975	Unknown
26	Jan-12-76	John Howell
27	Jan-17-76	Clay Ellis
28	May-03-76	Anthony Dobson
29	May-12-76	Nicholas Hubert
30	May-13-76	C. Dymott
31	Jul-14-76	R. Dupuy
32	Nov-04-76	Howard Spensley
33	Nov-04-76	Charles Meehan
34	Dec-24-76	Michael Moore
35	May-10-77	G. Hoffman
36	Aug-20-77	D. Sansalone
37	Oct-17-77	P. Azzopardi
38	Oct-03-77	R. Murphy
39	Oct-14-77	Claud Cailleux
40	Jan-78	D. Hoover
41	Nov-26-78	Michael Ward
42	Nov-26-78	Tony Prangley
43	May-05-79	Barry Eke
44	Aug-08-79	Richard Walker
45	Aug-08-79	Victor Guiel
46	1980	0 Fatalities
47	1981	0 Fatalities
48	Oct-29-82	D Phillips
49	Mar-16-83	Jarle Teersen
50	Jun-02-83	RM Wallace
51	Nov-83	EA Coward
52	Nov-83	RP Lucas
53	Nov-83	T Hellevick
54	Nov-83	B Bergersen
55	Nov-83	W Crammond
56	Aug-15-84	Mark Dawson
57	Feb-21-84	David Bomar
58	Feb-21-84	Thomas Mackey

Note: The names of the deceased were collected from multiple sources, primarily government records and books.

Index

G

H

M

N

Y